Radio Access Network Slicing and Virtualization for 5G Vertical Industries

Radio Access Network Slicing and Virtualization for 5G Vertical Industries

Edited by

Lei Zhang
University of Glasgow, Glasgow, UK

Arman Farhang
Maynooth University, Ireland

Gang Feng
University of Electronic Science and Technology of China, China

Oluwakayode Onireti
University of Glasgow, Glasgow, UK

WILEY

IEEE PRESS

This edition first published 2021

Registered Offices
John Wiley & Sons, Inc., 111 River Street, Hoboken, NJ 07030, USA
John Wiley & Sons Ltd, The Atrium, Southern Gate, Chichester, West Sussex, PO19 8SQ, UK

Editorial Office
The Atrium, Southern Gate, Chichester, West Sussex, PO19 8SQ, UK

For details of our global editorial offices, customer services, and more information about Wiley products visit us at www.wiley.com.

Wiley also publishes its books in a variety of electronic formats and by print-on-demand. Some content that appears in standard print versions of this book may not be available in other formats.

Library of Congress Cataloging-in-Publication Data

Names: Zhang, Lei (Engineering teacher) editor. | Farhang, Arman, editor. | Feng, Gang (Engineering teacher) editor. | Onireti, Oluwakayode, editor.
Title: Radio access network slicing and virtualization for 5G vertical industries / Lei Zhang, Arman Farhang, Gang Feng, Oluwakayode Onireti.
Description: Hoboken, NJ, USA : Wiley, 2021. | Includes bibliographical references and index.
Identifiers: LCCN 2020024172 (print) | LCCN 2020024173 (ebook) | ISBN 9781119652380 (hardback) | ISBN 9781119652458 (adobe pdf) | ISBN 9781119652472 (epub)
Subjects: LCSH: 5G mobile communication systems. | Multiple access protocols (Computer network protocols)
Classification: LCC TK5103.25 .Z53 2021 (print) | LCC TK5103.25 (ebook) | DDC 621.3845/6–dc23
LC record available at https://lccn.loc.gov/2020024172
LC ebook record available at https://lccn.loc.gov/2020024173

Cover Design: Wiley
Cover Image: © alexsl/Getty Images

Set in 9.5/12.5pt STIXTwoText by SPi Global, Chennai, India

10 9 8 7 6 5 4 3 2 1

Contents

About the Editors

Dr. Lei Zhang is a Senior Lecturer (associate professor) at the University of Glasgow, UK. He received his PhD from the University of Sheffield, UK. His research interests include wireless communication systems and networks, blockchain technology, radio access network slicing (RAN slicing), Internet of Things (IoT), multi-antenna signal processing, MIMO systems, etc. He has 19 patents granted/filed in more than 30 countries/regions including US/UK/EU/China/Japan etc. Dr. Zhang has published 2 books and 100+ peer-reviewed papers. He received IEEE Communication Society TAOS Best Paper Award 2019. He is a Technical Committee Chair of 5th International conference on UK-China Emerging Technologies (UCET) 2020. He was the Publication and Registration Chair of IEEE Sensor Array and Multichannel (SAM) 2018, Co-chair of Cyber-C Blockchain workshop 2019. He is an associate editor of IEEE Internet of Things (IoT) Journal, IEEE Wireless Communications Letters and Digital Communications and Networks. Dr. Zhang is a senior member of IEEE.

Dr. Arman Farhang received the PhD degree from the Trinity College Dublin (TCD), Dublin, Ireland, in 2016. He was a research fellow with the Irish National Telecommunications Research Centre (CONNECT), Trinity College Dublin, Dublin, Ireland, from 2016 to 2018. He was a Lecturer in the School of Electrical and Electronic Engineering at University College Dublin (UCD), Ireland, for a short period in 2018. He joined the Department of Electronic Engineering at Maynooth University in September 2018 as a Lecturer. Dr. Farhang also holds an adjunct professor position at TCD. He is an associate investigator in the CONNECT Centre, where he leads the research around the topic of waveforms for 5G and beyond. He has published over 50 peer-reviewed publications on the topic of waveforms. He serves as an associate editor of "EURASIP Journal on Wireless Communications and Networking." Dr. Farhang is a member of the organization committee of the IEEE Communication Society flagship conference ICC 2020. He has also served as a TPC member of many well-reputed IEEE conferences and workshops. His research interests include wireless communications, digital signal processing for communications, multiuser communications, and multi-antenna and multicarrier systems.

Gang Feng received the BEng and MEng degrees in electronic engineering from the University of Electronic Science and Technology of China (UESTC) in 1986 and 1989, respectively, and the PhD degree in information engineering from The Chinese University of Hong Kong in 1998. In December 2000, he joined the School of Electric and Electronic Engineering, Nanyang Technological University, Singapore, as an Assistant Professor and was promoted as an Associate Professor in October 2005. He is currently a Professor with the National

Key Laboratory of Science and Technology on Communications, UESTC. He has extensive research experience and has published widely in wireless networking research. A number of his articles have been highly cited. His research interests include next generation mobile networks, mobile cloud computing, and AI-enabled wireless networking. He has also received a number of paper awards, including the recent IEEE ComSoc TAOS Best Paper Award and the ICC Best Paper Award in 2019.

Oluwakayode Onireti received the MSc degree (distinction) in mobile and satellite communications in 2009, and the PhD degree in electronics engineering in 2012, from the Institute for Communication Systems (ICS, formally known as CCSR) of the University of Surrey. He is currently a Lecturer at the James Watt School of Engineering, University of Glasgow. He was actively involved in European Commission (EC)-funded projects such as ROCKET and in the award-winning EARTH project. His other project involvements include the EPSRC-funded DARE (distributed autonomous and resilient emergency management systems) project, the QNRF-funded QSON project, and industry-funded projects such as the Energy Proportional EnodeB for LTE-Advanced and Beyond project. His main research interests include self-organizing cellular networks, energy-efficient networks, multiple-input multiple-output systems, wireless blockchain networks, network slicing, and cooperative communications. He has published more than 60 technical papers in scholarly journals and international conferences. He has served as technical program committee member of several IEEE conferences and as reviewer for several IEEE and other top journals. He received an Exemplary Reviewer Award from the IEEE Transaction on Wireless Communications in 2017.

Preface

The move toward an always-connected society, where people and machines concurrently interact with multiple devices, requires future networks to be highly flexible, combining multiple standards and architectures, while simultaneously providing services to multiple users of various types and traffic requirements (massive machine type communications (mMTC), vehicle-to-vehicle (V2V) communications, etc.). Moreover, future networks are expected to provide orders of magnitude improvement to such heterogeneous networks in key technical requirements such as throughput, number of connected devices, latency, and reliability.

It is cumbersome to design a unified all-in-one radio that meets the extreme requirements for all types of services. For instance, an MTC service might require an multicarrier system with smaller subcarrier spacing (thus longer symbol duration in the time domain) than the current standards to support massive delay-tolerant devices, while V2V communication necessitates significantly larger frequency subcarrier spacing (thus shorter symbol duration) to satisfy the stringent delay requirements and provide robustness against Doppler spread. Moreover, some other configurations, such as waveform, should be dynamically optimized/selected to adapt to the traffic type, wireless channel, and user mobility. On the contrary, designing separate service systems that run on separate infrastructures make the operation and management of the system highly complex, expensive, and inefficient. In addition, many studies have shown that fixed spectrum allocation is wasteful since the license holders do not utilize the full spectrum continuously. Thus, to guarantee the required performance for each individual use case, the physical layer (PHY) configurations should be delicately optimized and medium access control (MAC) layer radio resources should be allocated on demand.

Radio access network (RAN) slicing is proposed to efficiently support all the aforementioned scenarios. RAN slicing enables design, deployment, customization, and optimization of isolated virtual sub-networks, or slices on a common physical network infrastructure, as such, to accommodate all fifth generation (5G) vertical industries like massive Internet of Things (IoT) and vehicle-to-everything (V2X) communications. From the radio perspective, one viable RAN solution to support diverse requirements in 5G-and-beyond systems is to multiplex multiple types of services in one baseband system in orthogonal time and/or frequency resources, with either physical (e.g. using guard interval or guard band) or algorithmic (e.g. filtering or precoding the data) isolation to avoid/alleviate the interference between them. Frequency division multiplexing (FDM) is preferred in Third Generation

Partnership Project (3GPP) for multiplexing different services due to several advantages such as forward compatibility and ease of supporting services with different latency requirements.

With such desired degree of freedom (DoF) for RAN slicing, it is expected to reap the diversity gain by optimizing air interface of individual slices. However, cohabitation of the individually optimized services, each with a different numerology Each numerology is defined by a set of parameters for the multicarrier waveform such as subcarrier-spacing, symbol-length, and the length of cyclic prefix., in one system leads to several technical challenges to both physical and multiple access control layers of the overall system. A fundamental challenge is how the mismatch among services with different numerologies affects signal detection, channel estimation, and equalization. In addition, the mismatch will also result in PHY configurations disorder and misalignment among services/slices. This, from the slice level viewpoint, will bring intricate interference. From the resource allocation perspective, such a high degree of heterogeneity among services fosters a complex radio system composed of three resource layers. To address the abovementioned issues, waveforms with low out-of-band emissions are important in multi-service systems, as isolation between the signals corresponding to different services is required. Therefore, there is a common consensus on the need to introduce alternative technologies that complement orthogonal frequency division multiplexing (OFDM) by a more flexible and effective air interface that better serves the challenging requirements of future networks than OFDM. To this end, in the recent years, several candidate waveforms have emerged, e.g., filtered OFDM (F-OFDM), universal filtered multicarrier (UFMC), generalized frequency division multiplexing (GFDM), filter bank multicarrier (FBMC) and orthogonal time frequency space (OTFS) modulation.

As one of the pillars that underpins multi-service communications in future networks, the first part of this book presents the recent advances in waveform design and provides an in-depth insight to the reader about different waveforms and their relationship with OFDM. Furthermore, the first part of the book provides a new perspective that facilitates straightforward understanding of channel equalization and the application of the new waveforms in provision of services with different requirements. It also facilitates derivation of efficient structures for synthesis and analysis of these waveforms. From the mixed numerologies coexistence perspective, the first part of the book will discuss signal detection, channel estimation, equalization, and resource allocation. These include inter-numerology-interference (INI) analysis of OFDM and its variants, e.g. windowed orthogonal frequency division multiplexing (W-OFDM), F-OFDM, new signal detection methods, effective resource allocation, and intelligent handoff algorithms while considering INI issues. The second part of the book will cover RAN slicing and virtualization and their applications to V2X communications, ultra-reliable low-latency communications (URLLC), etc. Additionally, the book will discuss resource sharing and optimization. Flexible function split and the design of access control and handoff policies for RAN slicing will also be discussed, followed by RAN function split, spectrum sharing, and optimizing resource allocation.

The book is suitable for telecom engineers and industry actors to aid them identify realistic and cost-effective concepts that are uniquely tailored to support specific vertical industries. Moreover, researchers, professors, doctorate and postgraduate students would also benefit from this book, as it enables them to identify open issues and classify their research based on the existing literature. While beginners can learn about the novel techniques to characterize different modulation schemes, experienced researchers, scientists, and experts from industry can understand the extensive theoretical design fundamentals, research trends, and in particular, various perspectives on the air interface design for the future networks. Additionally, the book aims at providing in-depth knowledge to 5G stakeholders, regulators, institutional actors, and research agencies on how 5G networks can seamlessly support vertical industries and aid them in making the right choice of the techniques/architectures while designing the next generation of wireless communication systems.

The book is split into two parts: Part I is focused on the PHY layer aspects including modulation and signaling methods, channel estimation, and mixed numerology, and Part II is focused on layers higher than PHY.

Part I

The mismatch between different services in 5G-and-beyond networks leads to inter-carrier interference (ICI) and thus Chapter 1 focuses on ICI. The authors first provide an overview of OFDM, which has been deemed as the signal waveform for both uplink and downlink transmission for the 5G wireless network. The authors then discuss the impact of inter-ICI on the carrier-to-interference power ratio (CIR) performance of OFDM system. The chapter then introduces four effective ICI cancellation schemes without explicitly estimating ICI coefficients based on mirror-mapping rules and their resulting CIR performance. Finally, the chapter concludes by demonstrating the benefits of the four ICI cancellation schemes for OFDM underwater acoustic (UWA) communications that suffers from strong ICI.

Chapter 2 describes single subband matrix form for F-OFDM where all the linear convolution operations are converted into matrix multiplications to derive a well-channelized signal. The chapter presents the analytical derivations of the in-band interference that is generated due to the mismatch between different services, including ICI, forward inter-symbol interference (ISI), and backward ISI. Furthermore, the chapter presents a low-complexity block-wise parallel interference cancellation (BwPIC) algorithm to tackle the issue caused by the filtering operations. Finally, numerical results are included to show the effectiveness of the BwPIC algorithm.

In Chapter 3, the authors describe the INI model of W-OFDM system in the context of mixed numerologies. The chapter establishes a theoretical model of the INI, and derives the analytical expression of its power as a function of the channel frequency response of interfering subcarriers, the spectral distance separating the aggressor and the victim subcarriers, and the overlapping windows generated by the interferer's transmitter window and the victim's receiver window. Furthermore, the chapter presents numerical results to show the signal-to-interference ratio (SIR) gain that can be achieved with W-OFDM when compared with the normal OFDM.

Chapter 4 describes GFDM, which was one of the candidate waveforms for 5G, proposed as a generalized form of OFDM. The chapter highlights the role of GFDM in the implementation of multicarrier systems. Further, the chapter presents a unified time–frequency representation of GFDM, which allows a better understanding of GFDM structure and enables a comprehensive analysis. The chapter then introduces GFDM modem implementation in a practical architecture, which is more flexible than OFDM. This structure provides additional degrees of freedom for the processing of conventional and new waveforms. Finally, for practical exploitation of GFDM, the chapter introduces integration of GFDM modem as a precoding technique on top of OFDM system.

Chapter 5 presents a short review of FBMC techniques. The fundamental theory behind construction of FBMC waveforms with maximum compactness in time and frequency are presented. Moreover, a number of effective methods for designing the prototype filters for this class of multicarrier systems are reviewed. The topics of synchronization and tracking are treated in detail, and equalization and computational complexity are discussed briefly. In terms of applications, this chapter brings up a number of appealing features of FBMC waveforms that make them an ideal choice in the emerging area of massive multiple-input multiple-output (MIMO) networks.

Chapter 6 covers the basics of OTFS modulation and presents its discrete-time formulation. An in-depth analysis of the channel impact on transmit data symbols is also provided. This brings deeper insights into OTFS. The derived formulation in this chapter reveals that the precoding and post-processing units can be combined with OFDM modulator and demodulator, respectively, which leads to a simplified modem structure. Through computational complexity analysis, this chapter shows that in realistic scenarios, OTFS modem becomes simpler than OFDM to implement.

Chapter 7 provides insights into waveform parameter assignment considering the current waveform design in 5G new radio (NR). This chapter first reviews the literature on waveform parameter assignment in the context of multi-numerology RAN slicing. Then the waveform parameter options for the 5G NR are introduced. Finally, the chapter discusses numerology assignment, waveform processing, and the related joint optimization issues.

Part II

The authors in Chapter 8 highlight the advantages and the issues of network slicing when it is implemented in a dynamic spectrum-sharing manner. The authors start by discussing the motivations behind spectrum sharing in future networks. This is then followed by a historical overview of the approaches to spectrum sharing. Then the concept of network slicing, different types of network slicing, and network slicing in RANs are introduced. Finally, the chapter defines the concept of isolation and presents results on isolation using connection admission control (CAC).

Chapter 9 discusses access control and handoff policy design for RAN slicing where a unified framework for access control is presented. Additionally, user access control policies to select admissible users for optimizing the quality of service (QoS) and the number of admissible users is designed. This is followed by the investigation of the handoff issue for mobile users in RAN slicing. Then a multi-agent reinforcement learning based smart handoff policy is presented. This will reduce handoff cost while maintaining user QoS requirements in RAN slicing. Finally, the performance of the designed policy in reducing the handoff cost is numerically evaluated and compared with traditional policies without learning.

Chapter 10 presents a mechanism to address slice recovery and reconfiguration in the unified framework. This chapter first formulates a recovery problem to address the failures of substrate networks and the uncertainties of traffic demands in a RAN slice. Based on that, two robust RAN slicing algorithms are proposed for slice recovery and reconfiguration under stochastic demands. This is followed by an in-depth assessment of the failure rate, average load of substrate attainable with the presented algorithm. Finally, a tradeoff between the robustness of slices and the average load of the links is established, which can be efficiently managed and controlled.

In Chapter 11, a flexible function split over ethernet for enabling RAN slicing in future networks is described. The chapter starts by introducing Cloud-RAN as one of the enablers of 5G. It then introduces functional split to Cloud-RAN with a more flexible placement of baseband functionality. This is then followed by a description of the fronthaul reliability and slicing when multipath is deployed at the fronthaul. Then the flexible functional split and the multipath packet-based fronthaul for RAN slicing are evaluated via both experimental and numerical results.

Chapter 12 discusses service-oriented RAN support of network slicing to meet different key performance indicators (KPIs) for different vertical services. The chapter starts by presenting the general concept and principles of service-oriented RAN slicing. The chapter then describes the RAN system deployment scenarios, namely deployments on a single frequency band, deployment over multifrequency bands, and hybrid deployment. This chapter also introduces key technologies enabling service-oriented RAN slicing.

In Chapter 13, 5G network slicing principles and enabling technologies for the V2X ecosystem are discussed. The chapter introduces the current and future V2X applications and the related requirements, together with the available V2X communication technologies, with a focus on the latest progress in the 3GPP and concerning radio interfaces and architecture design. The emerging segment routing technology is also investigated as a further facilitator of network slicing for V2X communications. Furthermore, the chapter summarizes the main lessons learnt from research and development activities, while also highlighting challenges and future perspectives in the promising area of network slicing for V2X communications.

Chapter 14 presents an optimal resource allocation scheme that maximizes the spectral efficiency (SE) by communication and control co-design, where both URLLC and control convergence rate requirements are considered. The scheme enables the use of the optimal resource to support URLLC while guaranteeing the required control performance level. The chapter presents the analysis of the relationship between the control convergence rate and communication requirements and the impact of both on the URLLC quality. Finally, an iterative method to obtain the optimal resource allocation that maximizes the SE is developed. Results are included to show the effectiveness of this method.

Lei Zhang, Arman Farhang, Gang Feng, Oluwakayode Onireti
22 June 2020

This work was supported in part by the U.K. Engineering and Physical Sciences Research Council (EPSRC) under Grant Number EP/S02476X/1, and Science Foundation Ireland (SFI) and is co-funded under the European Regional Development Fund under Grant Number 13/RC/2077.

List of Contributors

Huseyin Arslan
Electrical-Electronics Engineering
Istanbul Medipol University
Turkey

and

Electrical Engineering
University of South Florida
USA

Claudia Campolo
Dipartimento DIIES
Universitá Mediterranea di Reggio
Calabria
Italy

Marwa Chafii
ETIS UMR 8051
Université Paris Seine, Université de
Cergy-Pontoise
France

Bo Chang
National Key Laboratory of Science and
Technology on Communications
University of Electronic Science and
Technology of China (UESTC)
China

Xiang Cheng
School of Electronics Engineering and
Computer Science
Peking University
China

Xilin Cheng
NXP Semiconductor
USA

Laurie Cuthbert
School of Applied Sciences
Macao Polytechnic Institute, Macao SAR
China

Gerhard Fettweis
Vodafone Chair Mobile Communication
Systems
Technische Universität Dresden
Germany

Behrouz Farhang-Boroujeny
Electrical and Computer Engineering
Department
University of Utah
USA

Arman Farhang
Department of Electronic Engineering
Maynooth University, Co.
Ireland

Gang Feng
University of Electronic Science and Technology of China
China

Feng Han
Huawei Technologies Co., LTD.

Muhammad Ali Imran
James Watt School of Engineering
University of Glasgow
UK

Yinghao Jin
Huawei Technologies Co., LTD.

Jun Li
School of Electronics and Communication Engineering
Guangzhou University
China

Liying Li
Department of Mathematics, Physics and Electrical Engineering
Northumbria University
UK

Zhongju Li
Vodafone Chair Mobile Communication Systems
Technische Universität Dresden
Germany

Yue Liu
School of Applied Sciences
Macao Polytechnic Institute, Macao SAR
China

Toktam Mahmoodi
Centre for Telecommunications Research
Department of Engineering
King's College London
UK

Juquan Mao
Institute for Communication Systems (ICS)
Home of 5G Innovation Centre
University of Surrey
UK

Antonella Molinaro
Dipartimento DIIES
Universitá Mediterranea di Reggio Calabria
Feo di Vito
Italy

Ghizlane Mountaser
Centre for Telecommunications Research
Department of Engineering
King's College London
UK

Ahmad Nimr
Vodafone Chair Mobile Communication Systems
Technische Universität Dresden
Germany

Vincenzo Sciancalepore
EC Laboratories Europe GmbH
Heidelberg
Germany

Yao Sun
James Watt School of Engineering
University of Glasgow
UK

Wei Tan
Huawei Technologies Co., LTD.

Miaowen Wen
School of Electronic and Information Engineering
South China University of Technology
China

Ruihan Wen
Southwest Minzu University
China

and

University of Electronic Science and Technology of China
China

Pei Xiao
Institute for Communication Systems (ICS)
Home of 5G Innovation Centre
University of Surrey
UK

Bowen Yang
James Watt School of Engineering
University of Glasgow
UK

Chenchen Yang
Huawei Technologies Co., LTD.

Xu Yang
School of Applied Sciences
Macao Polytechnic Institute, Macao SAR
China

Ahmet Yazar
Electrical-Electronics Engineering
Istanbul Medipol University
Turkey

Lei Zhang
James Watt School of Engineering
University of Glasgow
UK

Xiaoying Zhang
Institute of Electronic Science
National University of Defense Technology
China

Guodong Zhao
James Watt School of Engineering
University of Glasgow
UK

List of Abbreviations

2D	Two-dimensional
3GPP	Third Generation Partnership Project
4G	Fourth generation
5G	Fifth generation
5GAA	5G Automotive Association
5GC	5G core network
A/D	Analog-to-digital
ACI	Adjacent channel interference
ACI	Adjacent carrier interference
ACSR	Adjacent conjugate symbol repetition
AF	Application function
AMF	Access and mobility management function
AN	Access network
API	Application programming interface
AR	Augmented Reality
ARQ	Automatic repeat request
AS	Application server
ASR	Adjacent symbol repetition
ATM	Asynchronous Transfer Mode
AUSF	Authentication Server Function
AWGN	Additive white Gaussian noise
BBU	Baseband unit
BEM	Basis expansion model
BER	Bit error rate
BFGC	Block-fading Gaussian channel
b-ISI	Backward inter-symbol interference
BS	Base station
BSS	Business support system
BWP	Bandwidth part
BwPIC	Block-wise parallel interference cancellation
CA/DC	Carrier aggregation/dual connectivity
CAC	Connection admission control
CDMA	Code division multiple access

CFO	Carrier frequency offset
CFR	Channel frequency response
CIR	Carrier-to-interference power ratio
CMT/FBMC	Cosine modulated multitone based filter bank multicarrier
CN	Core network
CP	Cyclic prefix
CP-OQAM	Cyclic prefix-based offset quadrature amplitude modulation
CPRI	Common public radio interface
CR	Cognitive radio
C-RAN	Cloud-radio access network
CS	Cyclic suffix
CSI	Channel state information
CSMA/CA	Carrier sense multiple access/collision avoidance
CU	Central unit
C-V2X	Cellular-vehicle-to-everything
D/A	Digital-to-analog
DAB	Digital audio broadcasting
DC	Data center
DFT	Discrete Fourier transform
DL	Downlink
DMT	Discrete multitone
DN	Data network
DSA	Dynamic spectrum access
DSB	Double side-band
DSCP	Differentiated Services Code Point
E2E	End-to-end
eMBB	Enhanced mobile broadband
eMBMS	Evolved multimedia broadcast multicast service
eNB	E-UTRAN NodeB
EPS	Evolved Packet System
ETSI	European Telecommunications Standards Institute
FBMC	Filter bank multicarrier
FD	Frequency domain
FDE	Frequency-domain equalization
FDCP	Frequency-domain cyclic prefix
FDM	Frequency division multiplexing
FFT	Fast Fourier transform
f-ISI	Forward inter-symbol interference
FMC	Fixed mobile convergence
FMT	Filtered multitone
FMT/FBMC	Filtered multitone based filter bank multicarrier
F-OFDM	Filtered orthogonal frequency division multiplexing
FRP	Failure recovery problem
FTD	Fractional time delay
GB	Guard band

GFDM	Generalized frequency division multiplexing
GR	Group report
GTP	GPRS Tunneling Protocol
GTP-U	GPRS Tunneling Protocol-user
HD	High definition
IBI	Inter-block interference
IC	Interference cancellation
ICI	Intercarrier interference
IDFT	Inverse discrete Fourier transform
IFFT	Inverse fast Fourier transform
INI	Inter-numerology interference
IoT	Internet of Things
IOTA	Isotropic orthogonal transform algorithm
IP	Internet protocol
ISG	Industry Specification Group
ISI	Inter-symbol interference
ITS	Intelligent Transportation System
ITU	International Telecommunication Union
ITU-R	ITU Radiocommunication Sector
ITU-T	ITU Telecommunication Standardization Sector
KPI	Key performance indicator
LCM	Least common multiplier
LEO	Low Earth Orbit
LMMSE	Linear minimum mean squared error
LTE	Long-Term Evolution
MAC	Medium access control
MANO	Management and orchestration
MCJT	Mirror conjugate transmission
MCM	Multicarrier modulation
MCMC	Markov chain Monte Carlo
MCSR	Mirror conjugate symbol repetition
MCVT	Mirror conversion transmission
MEC	Mobile edge computing
MIMO	Multiple-input multiple-output
mIoT	Massive Internet of Things
MIP	Mixed integer problem
ML	Machine learning
MMSE	Minimum mean squared error
mMTC	Massive machine type communications
MNO	Mobile network operator
MPLS	Multi-Protocol Label Switching
MSR	Mirror symbol repetition
MT	Mobile terminal
MUI	Multiuser interference
MVNO	Mobile virtual network operator

NaaS	Networks as a service
NAS	Non-access stratum
NEF	Network Exposure Function
NF	Network function
NFV	Network function virtualization
NG-RAN	Next generation radio access network
NOA	Network operator A
NOB	Network operator B
NOMA	Non-orthogonal multiple access
NR	New radio
NRF	Network Repository Function
NSI	Network slice instance
NSSAI	Network Slice Selection Assistance Information
NSSF	Network Slice Selection Function
O&M	Operation and maintenance
OEM	Original equipment manufacturer
Ofcom	Office of Communications
OFDM	Orthogonal frequency division multiplexing
OFDMA	Orthogonal frequency division multiple access
OOB	Out-of-band
OoBE	Out of band emissions
OP	Orthogonal precoding
OQAM	Offset quadrature amplitude modulation
OQAM/FBMC	Offset quadrature amplitude modulation based filter bank multicarrier
OSA	Opportunistic spectrum access
OSS	Operation support system
OTFS	Orthogonal time frequency space modulation
PA	Power amplifier
PAPR	Peak-to-average power ratio
PAM	Pulse amplitude modulated
PCF	Policy Control Function
PDCP	Packet data convergence protocol
PDP	Packet data protocol
PDU	Packet data unit
PHY	Physical layer
PLMN	Public land mobile network
PN	Pseudo-random noise
PrU	Protected user
PSD	Power spectral density
PSK	Phase shift keying
PU	Primary user
QAM	Quadrature amplitude modulation
QoE	Quality of experience
QoS	Quality of service

QPSK	Quadrature phase shift keying
QU	Qualified user
QUR	Qualified user ratio
RAE	Resource allocation entity
RAN	Radio access network
RAT	Radio access technology
RB	Resource block
RC	Raised cosine
RF	Radio frequency
RLC	Radio link control
ROBUST	Robust failure recovery problem
RRC	Radio resource control
RRH	Remote radio head
RRM	Radio resource management
RSU	Road-side unit
S/P	Serial-to-parallel
SC	Single carrier
SC-FDMA	Single carrier-frequency division multiple access
SCS	Subcarrier spacing
SD	Slice differentiator
SDN	Software-defined networking
SDP	Semidefinite programming problem
SE	Spectral efficiency
SFFT	Symplectic finite Fourier transform
SGW	Serving gateway
SGW-LBO	Serving gateway-local breakout
SI	System information
SIC	Successive interference cancellation
SID	Segment identifier
SINR	Signal-to-interference-plus-noise ratio
SIR	Signal to interference ratio
SISO	Single-input single-output
SL	Segment list
SLA	Service level agreement
SMF	Session Management Function
SMT	Staggered multitone
SN	Substrate network
SNR	Signal-to-noise ratio
S-NSSAI	Single Network Slice Selection Assistance Information
SQAM	Staggered quadrature amplitude modulation
SRH	Segment routing header
SRRC	Square-root raised-cosine
SRv6	Segment Routing Ipv6
SST	Slice service type
SU	Secondary user

TCP	Transmission control protocol
TD	Time domain
TDD	Time division duplexing
TDM	Time division multiplexing
TE	Traffic engineering
TEID	Tunnel endpoint identifier
TN	Transport network
T-R	Transmitter–receiver
TTI	Transmission time interval
TV	Time varying
UDM	Unified data management
UDP	User datagram protocol
UE	User equipment
UFMC	Universal filtered multicarrier
UL	Uplink
UP	User plane
UPF	User plane function
URLLC	Ultra-reliable low-latency communications
UWA	Underwater acoustic
V2C	Vehicle-to-cloud
V2I	Vehicle-to-infrastructure
V2N	Vehicle-to-network
V2P	Vehicle-to-pedestrian
V2V	Vehicle-to-vehicle
V2X	Vehicle-to-everything
VDSL	Very high speed digital subscriber line
VNF	Virtual network function
VNS	Variable neighborhood search
VPN	Virtual private network
VR	Virtual reality
VRU	Vulnerable road user
VSB	Vestigial side-band
VUE	Vehicular user equipment
WLAN	Wireless local area network
W-OFDM	Windowed orthogonal frequency division multiplexing
ZF	Zero-forcing

Part I

Waveforms and Mixed-Numerology

1

ICI Cancellation Techniques Based on Data Repetition for OFDM Systems

Miaowen Wen[1,], Jun Li[2], Xilin Cheng[3] and Xiang Cheng[4]*

[1] *School of Electronic and Information Engineering, South China University of Technology, China*
[2] *Research Center of Intelligent Communication Engineering, School of Electronics and Communication Engineering, Guangzhou University, China*
[3] *NXP Semiconductor, San Jose, CA, USA*
[4] *School of Electronics Engineering and Computer Science, Peking University, China*

1.1 OFDM History

Orthogonal frequency division multiplexing (OFDM), as a special case of multicarrier transmission, is an elegant solution to the problem of high date rate transmission. OFDM was first developed to transmit data streams without inter-symbol interference (ISI) and intercarrier interference (ICI) in the 1960s [1]. A big breakthrough for efficiently implementing the OFDM system was made in the 1970s, when discrete Fourier transform (DFT) was applied to perform baseband modulation and demodulation in OFDM [2]. Another breakthrough in OFDM is the emergence of cyclic prefix (CP) in the 1980s [3], which maintains orthogonality of the transmitted signals over multipath fading channels. In the 1990s, OFDM was exploited for wideband data communications. The first application in the commercial use of OFDM was digital audio broadcasting (DAB) in the 1980s and 1990s, where OFDM guarantees highly reliable data transmission over a high-velocity and complex environment. At the beginning of the 2000s, wireless local area network (WLAN) applied the OFDM technique to the physical layers [4]. In recent years, OFDM has been widely used for fourth generation (4G) and fifth generation (5G) wireless systems to increase the utilization of spectrum resources as well as to combat frequency-selective fading.

The advantages of OFDM are summarized as follows:

- Resistance to frequency-selective fading
- Elimination of ISI and ICI
- Efficient use of the available spectrum
- Recovery of symbol lost by adequate channel coding and interleaving
- Enabling one-tap channel equalization

However, every coin has two sides. The disadvantages of OFDM lie in the following:

- Large dynamic range of transmitted signal, or peak to average power ratio (PAPR)
- Sensitivity to carrier frequency offset (CFO) and Doppler

* Corresponding Author: Miaowen Wen; eemwwen@scut.edu.cn

Radio Access Network Slicing and Virtualization for 5G Vertical Industries, First Edition.
Edited by Lei Zhang, Arman Farhang, Gang Feng, and Oluwakayode Onireti.

1.2 OFDM Principle

Consider an OFDM system of bandwidth B. The entire frequency band is divided into N subbands, each of bandwidth $B_N = B/N$. Given the channel coherence bandwidth B_c satisfying $B_N \ll B_c < B$, which usually holds for a sufficiently large value of N, the frequency-selective fading channel is converted into multiple-frequency flat fading subchannels (or say subcarriers). This can be also explained in the time domain as follows. The symbol duration of the modulated signal on each subcarrier is given by $T_N \approx 1/B_N$. When $B_N \ll B_c$, the symbol duration T_N is much larger than the channel delay spread $T_{\max} \approx 1/B_c$, which indicates that each subcarrier experiences flat fading. Therefore, as long as all subcarriers are orthogonal to each other, high data-rate interference-free parallel transmission can be achieved. How to ensure subcarrier orthogonality and perform the discrete implementation of OFDM will be discussed in detail in this section.

1.2.1 Subcarrier Orthogonality

In an OFDM system, the substream on the k-th subcarrier X_k is linearly modulated relative to the subcarrier frequency f_k, and the modulated signals associated with all the subcarriers are summed together to form the transmitted signal as

$$x(t) = \sum_{k=0}^{N-1} X_k e^{j2\pi f_k t}, \tag{1.1}$$

where $0 \le t \le T_N, f_k = f_c + k\Delta f$ with Δf representing the subcarrier interval, and f_c denoting the basic carrier frequency.

To ensure orthogonality among all subcarriers, the frequency tones $\{f_0, \dots, f_{N-1}\}$ must satisfy the following condition: given any two subcarriers k and p,

$$\frac{1}{T_N}\int_0^{T_N} e^{j2\pi f_k t} \cdot e^{-j2\pi f_p t} dt = \begin{cases} 0, & k \neq p, \\ 1, & k = p, \end{cases} \tag{1.2}$$

where $0 \le k, p \le N-1$. The solution to Eq. (1.2) turns out to be $\Delta f = n/T_N$, where n is a nonzero integer. The choice of $n = 1$ is preferred as it leads to a most compact spectrum. A snapshot of the OFDM spectrum with $N = 8$ is given in Figure 1.1. Figure 1.1 shows that the spectrum of one subcarrier signal overlaps that of the others. It is expected that as N increases to a large value, the overall spectrum tends to be confined by a rectangle with a range of $N/T_N = B$.

At the receiver, without taking into account the effects of the channel and noise, that is, the received signal $y(t)$ is equal to the transmitted signal $x(t)$, the output signal on the k-th subcarrier Y_k is given by

$$\begin{aligned} Y_k &= \frac{1}{T_N}\int_0^{T_N} y(t)e^{-j2\pi f_k t} dt \\ &= \frac{1}{T_N}\sum_{p=0}^{N-1} X_p \int_0^{T_N} e^{j2\pi(p-k)\Delta f t} dt. \end{aligned} \tag{1.3}$$

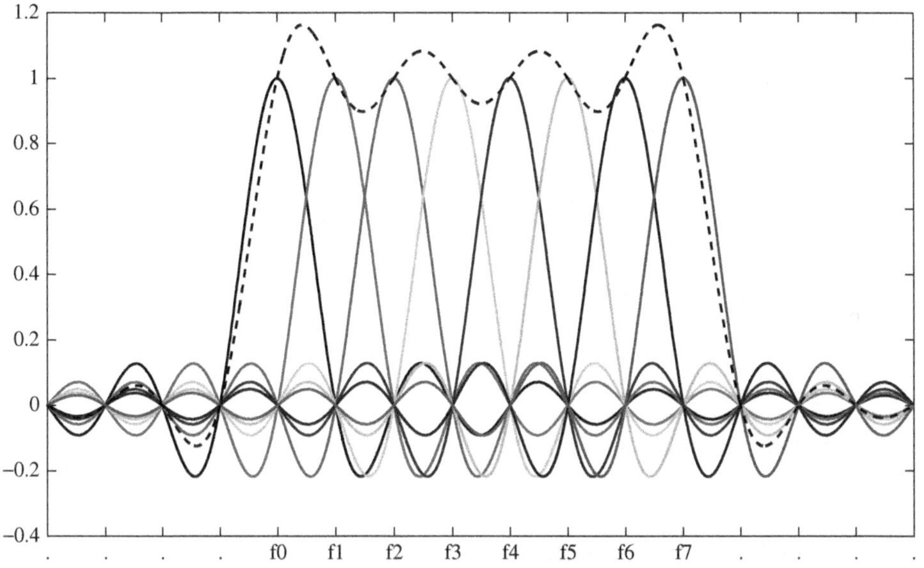

Figure 1.1 Spectrum of OFDM signals with $N = 8$, depicted by the dashed line.

Because of the orthogonality in Eq. (1.2), Eq. (1.3) can be simplified as

$$Y_k = X_k. \tag{1.4}$$

From the above, it is clear that the subcarrier orthogonality ensures interference-free parallel data transmission.

Subcarrier orthogonality can be also interpreted in the frequency domain. In the frequency domain, all subcarriers should be orthogonal so as to separate out the corresponding input signal without interference. According to the Nyquist criterion, no ISI occurs when the condition that the coherence bandwidth is no less than $1/T_N$. This implies that the minimum frequency separation required for subcarriers to remain maintain orthogonality over the symbol interval $[0, T_N]$ is $1/T_N$. Obviously, the minimum frequency interval should be set to be $1/T_N$ to save the frequency resource in OFDM.

1.2.2 Discrete Implementation

According to the Nyquist criterion, to recover a signal of bandwidth B without causing any distortion, the sampling frequency must be no less than B. Consider the sampling time interval to be $\Delta t = 1/B$. The time-domain discrete transmitted signal can be obtained from Eq. (1.1) as

$$x_n \triangleq x(n\Delta t) = \frac{1}{\sqrt{N}} \sum_{k=0}^{N-1} X_k e^{j\frac{2\pi nk}{N}}, \tag{1.5}$$

where $n = 0, 1, \ldots, N-1$. Obviously, the output of Eq. (1.5) is the inverse discrete Fourier transform (IDFT) of $\{X_0, X_1, \ldots, X_{N-1}\}$.

Denote $y_n \triangleq y(n\Delta t)$ as the sampled time-domain received signal. Similarly, without taking into account the channel and noise, the discrete frequency-domain signal can be obtained from Eq. (1.3) by

$$Y_k = \frac{1}{\sqrt{N}} \sum_{n=0}^{N-1} y_n e^{-j\frac{2\pi nk}{N}} = \frac{1}{N} \sum_{k=0}^{N-1} X_k \sum_{n=0}^{N} e^{j\frac{2\pi}{N}pn} e^{-j\frac{2\pi}{N}kn}, \tag{1.6}$$

which is in fact the DFT of $\{y_0, y_1, \ldots, y_{N-1}\}$. The subcarrier orthogonality in Eq. (1.2) also applies to the discrete-time representation:

$$\frac{1}{N} \sum_{n=0}^{N} e^{j\frac{2\pi}{N}pn} e^{-j\frac{2\pi}{N}kn} = \begin{cases} 1, & k = p, \\ 0, & k \neq p, \end{cases} \tag{1.7}$$

such that $Y_k = X_k$.

In summary, the OFDM modulation and demodulation can be realized by IDFT and DFT, respectively, which is very simple by virtue of the computationally efficient inverse fast Fourier transform (IFFT) and fast Fourier transform (FFT) algorithms.

1.2.3 OFDM in Multipath Channel

When taking into account the multipath channel $\mathbf{h} = [h_0, h_1, \ldots, h_{L-1}]$, where L is the number of channel taps, the orthogonality of subcarriers will be unfortunately broken, as shown in Figure 1.2. A CP can be used to avoid ISI when an OFDM signal is transmitted over the multipath channel, where the CP is essentially an identical copy of the last portion of the OFDM symbol. This CP preserves the orthogonality of the subcarriers and prevents ISI between successive OFDM symbols. Specifically, the CP of length $\mu \geq L$, which is composed of the last μ samples of the OFDM symbol, namely $\{x_{N-\mu}, \ldots, x_{N-1}\}$, is appended to the beginning of the OFDM symbol. After adding the CP, the new time-domain OFDM symbol yields

$$\tilde{\mathbf{x}} = [\tilde{x}_{-\mu}, \ldots, \tilde{x}_{-1}, \tilde{x}_0, \ldots, \tilde{x}_{N-1}]^T = [x_{N-\mu}, \ldots, x_{N-1}, x_0, \ldots, x_{N-1}]^T. \tag{1.8}$$

The effect of CP is illustrated in Figure 1.2, which shows that subcarrier orthogonality is maintained provided that the length of CP is no less than the channel delay spread.

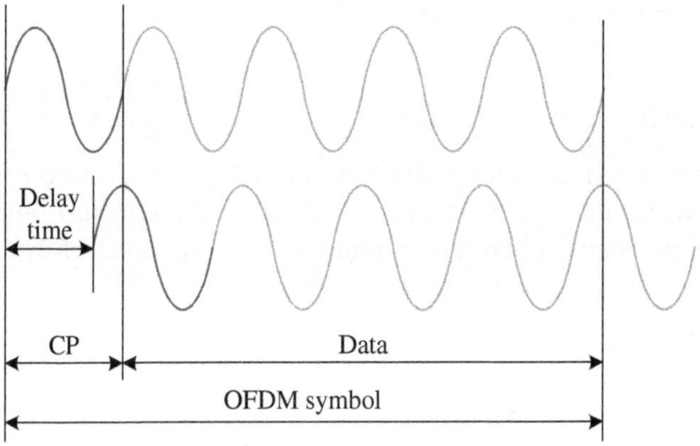

Figure 1.2 The multipath and CP effects.

After passing through the channel, the received signal without noise is

$$y_n = \tilde{x}_n * h_n = \sum_{k=0}^{\mu-1} h_k \tilde{x}_{n-k} = \sum_{k=0}^{\mu-1} h_k x_{<n-k>_N} = x_n \otimes h_n, \tag{1.9}$$

where $< n - k>_N$ denotes $n - k$ modulo N. This means that by appending a CP to the channel input, the linear convolution associated with the channel impulse response becomes a cyclic convolution.

Let us take another look at the relation in Eq. (1.9) via matrix representation. The channel output including the CP can be expressed in matrix form:

$$\begin{bmatrix} y_{-\mu} \\ \vdots \\ y_{-1} \\ y_0 \\ \vdots \\ y_{N-2} \\ y_{N-1} \end{bmatrix} = \begin{bmatrix} h_0 & 0 & \cdots & \cdots & 0 & \cdots & 0 \\ \vdots & h_0 & 0 & \cdots & 0 & \cdots & 0 \\ h_{L-1} & \vdots & \ddots & \ddots & \vdots & \ddots & \vdots \\ 0 & h_{L-1} & \ddots & h_0 & 0 & \ddots & 0 \\ 0 & 0 & \ddots & \vdots & h_0 & \ddots & \vdots \\ \vdots & 0 & \ddots & h_{L-1} & \vdots & h_0 & 0 \\ 0 & \cdots & 0 & 0 & h_{L-1} & \cdots & h_0 \end{bmatrix} \begin{bmatrix} x_{N-\mu} \\ \vdots \\ x_{N-1} \\ x_0 \\ \vdots \\ x_{N-2} \\ {}_{N-1} \end{bmatrix}. \tag{1.10}$$

The channel matrix of dimensions $(N + \mu) \times (N + \mu)$ in Eq. (1.10) is a Toeplitz matrix. Removing the CP leads to

$$\begin{bmatrix} y_0 \\ y_1 \\ \vdots \\ y_{L-1} \\ y_L \\ \vdots \\ y_{N-1} \end{bmatrix} = \begin{bmatrix} h_0 & 0 & \cdots & 0 & h_{L-1} & \cdots & h_1 \\ h_1 & h_0 & 0 & \cdots & 0 & \ddots & \vdots \\ \vdots & \ddots & \ddots & \ddots & \ddots & \ddots & h_{L-1} \\ h_{L-1} & \cdots & h_1 & h_0 & 0 & \cdots & 0 \\ 0 & h_{L-1} & \cdots & h_1 & h_0 & \cdots & 0 \\ \vdots & \ddots & \ddots & \vdots & \vdots & \ddots & \vdots \\ 0 & \cdots & 0 & h_{L-1} & h_{L-2} & \cdots & h_0 \end{bmatrix} \begin{bmatrix} x_0 \\ x_1 \\ \vdots \\ x_{L-1} \\ x_L \\ \vdots \\ x_{N-1} \end{bmatrix}. \tag{1.11}$$

The channel matrix of dimensions $N \times N$ in Eq. (1.11) now becomes a circular matrix, which can be decomposed into the multiplication of an IFFT matrix, a diagonal matrix, and an FFT matrix.

Benefiting from the cyclic convolution, the subcarrier signal Y_k can be directly obtained by X_k and the corresponding frequency response H_k from

$$Y_k = \mathrm{DFT}\{x_n \otimes h_n\} = X_k H_k, \tag{1.12}$$

where $k = 0, 1, \ldots, N - 1$ and

$$H_k = \sum_{l=0}^{L-1} h_l e^{-j\frac{2\pi kl}{N}}. \tag{1.13}$$

The overall block diagram of an OFDM transceiver is shown in Figure 1.3. At the transmitter, the incoming bit stream is divided into N substreams via a serial-to-parallel (S/P) converter. Then, all bit substreams are independently mapped to the corresponding constellation points by amplitude phase modulation, such as phase shift keying (PSK) and quadrature amplitude modulation (QAM), yielding $X_0, X_1, \ldots, X_{N-1}$. An N-point IFFT operation follows, converting the signal from frequency domain to time domain. After that, a CP

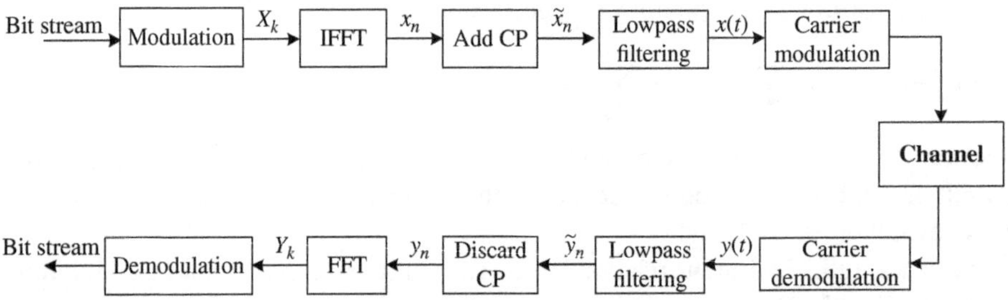

Figure 1.3 Block diagram of OFDM systems.

of length μ is appended to the time-domain signal, which is further passed through a low-pass filter to generate the baseband transmitted signal. Finally, the baseband transmitted signal is converted up to the passband by a carrier frequency f_c, and sent over the channel. At the receiver, the received signal is first converted down to the baseband by using a local carrier signal of carrier frequency f_c'. Then, the receiver follows the reverse baseband processing procedure at the transmitter to finally obtain the estimated bit stream.

1.3 Carrier Frequency Offset Effect

Ideally, f_c is identical to f_c'. However, due to hardware imperfection and/or Doppler, this may not hold, resulting in a CFO $\Delta f_c = f_c - f_c'$. Denote ϵ as the normalized CFO, which is given by

$$\epsilon = \frac{\Delta f_c}{\Delta f}. \tag{1.14}$$

To investigate the effect of CFO, we assume a fixed ϵ. From Eq. (1.9), the time-domain received signal can be written as

$$\begin{aligned}
y_n &= \frac{1}{\sqrt{N}}\left(\sum_{l=0}^{L-1} h_l \tilde{x}_{n-l}\right) e^{j\frac{2\pi\epsilon n}{N}} + w_n \\
&= \frac{1}{\sqrt{N}}\left[\sum_{l=0}^{L-1} h_l \sum_{k=0}^{N-1} X_k e^{j\frac{2\pi k(n-l)}{N}}\right] e^{j\frac{2\pi\epsilon n}{N}} + w_n \\
&= \frac{1}{\sqrt{N}}\sum_{k=0}^{N-1} X_k \left(\sum_{l=0}^{L-1} h_l e^{-j\frac{2\pi kl}{N}}\right) e^{j\frac{2\pi(k+\epsilon)}{N}} + w_n \\
&= \frac{1}{\sqrt{N}}\sum_{k=0}^{N-1} X_k H_k e^{j\frac{2\pi(k+\epsilon)}{N}} + w_n,
\end{aligned} \tag{1.15}$$

where w_n is the additive white Gaussian noise (AWGN) and $n = 0, 1, \dots, N-1$.

After DFT, the received signals in the frequency domain can be expressed as

$$Y_m = S_0 H_m X_m + \sum_{k=0,k\neq m}^{N-1} S_{k-m} H_k X_k + W_m, \quad m = 0, 1, \dots, N-1, \tag{1.16}$$

where W_m is the noise in the frequency domain, and

$$S_k = \frac{\sin(\pi\varepsilon)}{N\sin(\pi(k+\varepsilon)/N)} e^{j\left(\pi\varepsilon\left(1-\frac{1}{N}\right)-\frac{\pi k}{N}\right)} \tag{1.17}$$

is the ICI coefficient [5, 6].

1.3.1 Properties of ICI Coefficients

The ICI coefficient S_k is a periodic function with period N. The amplitude of S_k is plotted in Figure 1.4 with $\varepsilon = 0.1$ and $N = 1000$. It can be observed that for large values of $|k|$, $|S_k|$ goes to 0. This means that ICI mainly comes from neighboring subcarriers. The phase of S_k is $\angle S_k = \pi\left[\varepsilon\left(1-\frac{1}{N}\right)-\frac{k\%N}{N}\right]$ for $0 < \varepsilon < 1$, which is plotted in Figure 1.5 with $N = 1000$. From its expression or Figure 1.5, it is obvious that for small values of $|k|$, $\angle S_k \approx \pi\varepsilon$ and $\angle S_{-k} \approx \pi\varepsilon - \pi$. This implies $S_k + S_{-k} \approx 0$. In addition, for small values of ε, $\angle S_k \approx 0$ and $\angle S_{-k} \approx -\pi$. This means $S_k + S^*_{-k} \approx 0$. It is worth noting that $S_k + S_{-k} \approx 0$ is a more accurate approximation than $S_k + S^*_{-k} \approx 0$. As we will see in Section 1.4, the mirror-mapping-based schemes are designed on the basis of the aforementioned properties.

1.3.2 Carrier-to-Interference Power Ratio

The carrier-to-interference power ratio (CIR) is a widely used metric for evaluating the system ICI power level without considering the noise power. Suppose that the transmitted data symbols are mutually independent. According to Eq. (1.16), the instantaneous CIR of the OFDM system can be derived as

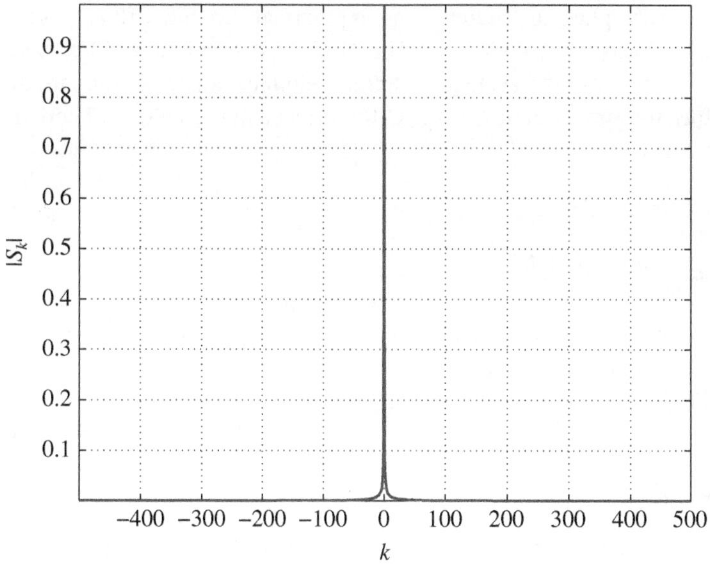

Figure 1.4 The amplitude of S_k.

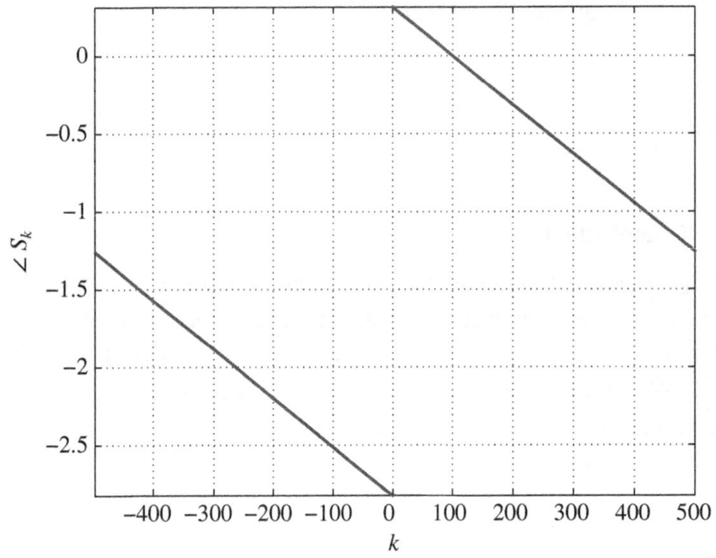

Figure 1.5 The phase of S_k.

$$\mathrm{CIR}_{\mathrm{OFDM,\,Inst}} = \frac{|S_0 H_m|^2}{\sum_{k=0,k\neq m}^{N-1} |S_{k-m} H_k|^2}. \tag{1.18}$$

The average CIR can be calculated by averaging the above instantaneous CIR expression over the distribution of the channel gains. However, the calculation of the average CIR using multiple integral is overly complicated. Therefore, an approximate average CIR expression can be derived by taking the average of the numerator and the denominator of Eq. (1.18) separately. It is a good predictor of the average CIR and is simple enough to compare different schemes [7]. Thus, the approximate average CIR is utilized in this chapter. Then, the CIR of the OFDM system is

$$\mathrm{CIR}_{\mathrm{OFDM}} = \frac{|S_0|^2 \mathbb{E}\{|H_m|^2\}}{\sum_{k=0,k\neq m}^{N-1} |S_{k-m}|^2 \mathbb{E}\{|H_k|^2\}}. \tag{1.19}$$

Owing to the periodic property of S_k and the identities $\mathbb{E}\{|H_k|^2\} = \mathbb{E}\{|H_m|^2\} = \eta$ and $\sum_{k=0}^{N-1} |S_k|^2 = 1$, Eq. (1.19) can be simplified as

$$\mathrm{CIR}_{\mathrm{OFDM}} = \frac{|S_0|^2}{\sum_{k=1}^{N-1} |S_k|^2} = \frac{|S_0|^2}{1 - |S_0|^2}. \tag{1.20}$$

From Eqs. (1.17) and (1.20), an accurate approximation of the CIR can be obtained for a sufficiently large value of N as $\mathrm{CIR}_{\mathrm{OFDM}} \approx \mathrm{sinc}^2(\varepsilon)/(1 - \mathrm{sinc}^2(\varepsilon))$. It is clear that the CIR of the OFDM system is very sensitive to the CFO.

1.4 ICI Cancellation Techniques

ICI can seriously degrade the performance of OFDM systems. To tackle this problem, several ICI countermeasures have been proposed over the past few years. These techniques can be generally classified into three categories. In the first category, the basis expansion model (BEM) is considered for ICI mitigation [8–10]. In the second category, the ICI is treated explicitly by first estimating the ICI coefficients followed by various ICI cancellation algorithms (see e.g. [11–14]). In the third category, a low-complexity strategy, is taken to treat ICI implicitly. The adjacent-mapping-based ICI cancellation method proposed by Zhao and Haggman [6] is a representative where each data symbol is transmitted over two adjacent subcarriers. Depending on the conversion or the conjugate relation between adjacent subcarrier pairs carrying the same information, there are the adjacent symbol repetition (ASR) scheme and the adjacent conjugate symbol repetition (ACSR) scheme. These schemes can implicitly cancel ICI by combining received signals on adjacent subcarrier pairs.

This section introduces four ICI cancellation schemes belonging to the third category, called mirror symbol repetition (MSR), mirror conjugate symbol repetition (MCSR), mirror conversion transmission (MCVT), and mirror conjugate transmission (MCJT). Based on the interesting relationship between the interference from a mirror subcarrier pair, they apply data repetition within OFDM symbols according to some carefully designed subcarrier mapping rules and operations such that after combining the mirror subcarrier pairs carrying the same information, the CIR can be significantly improved. The MSR and MCSR schemes repeat the data within one OFDM symbol, while the MCVT and MCJT schemes repeat the data across two consecutive OFDM symbols. All these four ICI cancellation schemes feature the simplicity of implementation and the effectiveness of ICI mitigation.

1.4.1 One-Path Cancellation with Mirror Mapping

Figure 1.6 depicts the system architecture of the ICI one-path cancellation schemes with mirror mapping. Compared with the plain OFDM, the ICI one-path cancellation schemes have two additional modules, i.e. the ICI self-canceling modulation before the IFFT operation at the transmitter and the ICI self-canceling demodulation after the FFT operation at the receiver.

For the ICI self-canceling modulation, the input modulated data symbols are first grouped into transmit blocks. Each block consists of $(N/2-1)$ modulated data symbols $\{X\}_{k=1}^{N/2-1}$, which are then mapped onto N subcarriers using the one-to-two mirror-mapping rule as follows:

$$\tilde{X}_k = \begin{cases} 0, & k = 0, N/2 \\ X_k, & k = 1, 2, \ldots, N/2-1 \\ \mathcal{O}(X_{N-k}), & k = N/2+1, \ldots, N-1, \end{cases} \tag{1.21}$$

where $\{\tilde{X}_k\}_{k=0}^{N-1}$ are the actual transmitted data symbols on the OFDM subcarriers. $\mathcal{O}(x)$ is defined as the mapping operation that reflects the relationship between the two modulated data symbols with the same information. Note that $\tilde{X}_{N-k} = \mathcal{O}(\tilde{X}_k)$, $k = 1, 2, \ldots, N/2-1$ represents mirror mapping.

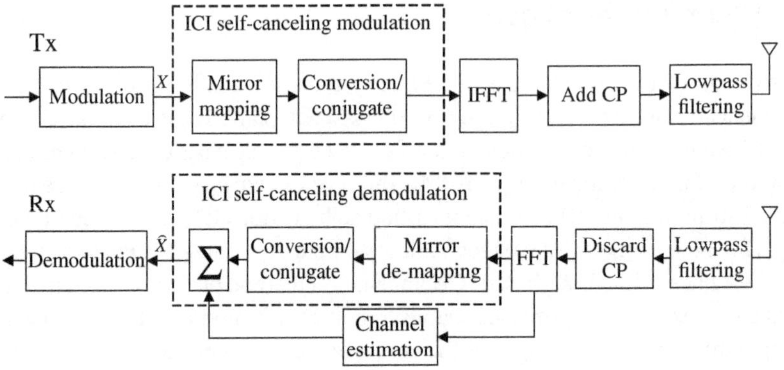

Figure 1.6 Block diagram of an OFDM transceiver with the ICI self-cancellation modules using mirror mapping in the baseband.

After the ICI self-canceling demodulation, the received signals on subcarrier m, $m \in \{1, 2, \ldots, N/2-1\}$ and its corresponding mapped subcarrier pair $(m' = N - m)$ will carry the same data information. This signal redundancy renders it possible to improve the ICI mitigation performance through a coherent combining technique:

$$\hat{X}_m = \frac{H_m^* Y_m + \mathcal{O}(H_{N-m}^* Y_{N-m})}{|H_m|^2 + |H_{N-m}|^2}, \tag{1.22}$$

where H_m, H_{N-m}, Y_m, and Y_{N-m} are the channel frequency response (CFR) and the received signals at subcarrier m and its corresponding subcarrier pair $(m' = N - m)$, respectively. Note that Eq. (1.22) is essentially maximum ratio combining (MRC).

1.4.1.1 MSR Scheme

For MSR scheme, the conversion operation is applied, which can be represented as $\mathcal{O}(x) = -x$. The 0-th and $N/2$-th subcarriers are vacant in order to meet the opposite polarity condition. The transmitted data symbols in the frequency-domain (FD) after the ICI self-canceling modulation are $\tilde{X}_1 = -\tilde{X}_{N-1} = X_1$, $\tilde{X}_2 = -\tilde{X}_{N-2} = X_2$, $\ldots$, $\tilde{X}_{N/2-1} = -\tilde{X}_{N/2+1} = X_{N/2-1}$, and $\tilde{X}_0 = \tilde{X}_{N/2} = 0$. According to Eq. (1.22), the decision variable at the m-th subcarrier ($m \in \{1, \ldots, N/2-1\}$) becomes

$$\begin{aligned}
\hat{X}_m &= H_m^* Y_m - H_{N-m}^* Y_{N-m} \\
&= (S_0|H_m|^2 + S_0|H_{N-m}|^2 - S_{-2m}H_m^* H_{N-m} - S_{2m}H_{N-m}^* H_m)X_m \\
&\quad + \sum_{k=1,k\neq m}^{N/2-1} (S_{k-m}H_m^* H_k + S_{m-k}H_{N-m}^* H_{N-k} - S_{-k-m}H_m^* H_{N-k} - S_{m+k}H_{N-m}^* H_k)X_k \\
&\quad + H_m^* W_m - H_{N-m}^* W_{N-m}.
\end{aligned}$$

The factor $1/(|H_m|^2 + |H_{N-m}|^2)$ is removed from the decision variable expression, as it does not affect the CIR value. The CIR at the m-th subcarrier can be expressed as

$$\mathrm{CIR}_{\mathrm{MSR}}(m) = \frac{\mathbb{E}\{|S_0(|H_m|^2 + |H_{N-m}|^2) - S_{-2m}H_m^*H_{N-m} - S_{2m}H_{N-m}^*H_m|^2\}}{\sum_{k=1,k\neq m}^{N/2-1} \mathbb{E}\{\ |S_{k-m}H_m^*H_k + S_{m-k}H_{N-m}^*H_{N-k} - S_{-k-m}H_m^*H_{N-k} - S_{m+k}H_{N-m}^*H_k|^2\}}.$$

For the flat fading channel, i.e. $L = 1$, the CIR becomes

$$\mathrm{CIR}_{\mathrm{MSR}}\ (\ m\ ) = \frac{|2S_0 - S_{2m} - S_{-2m}|^2}{\sum_{k=1,k\neq m}^{N/2-1} |S_{k-m} + S_{-(k-m)} - S_{k+m} - S_{-(k+m)}|^2}. \tag{1.23}$$

See Ref. [15] for proof. As $S_k + S_{-k} \approx 0$, the CIR of the MSR scheme can be markedly improved. The CIR under the multipath Rayleigh fading channel and the detailed derivation are available in [15].

The average CIR of the MSR scheme is then given by

$$\mathrm{CIR}_{\mathrm{MSR}} = \frac{2}{N-2} \sum_{m=1}^{N/2-1} \mathrm{CIR}_{\mathrm{MSR}}(m). \tag{1.24}$$

1.4.1.2 MCSR Scheme

For the MCSR scheme, conjugate operation is adopted, which can be represented as $\mathcal{O}(x) = x^*$. Accordingly, we have $\tilde{X}_1 = \tilde{X}_{N-}1^* = X_1$, $\tilde{X}_2 = \tilde{X}_{N-2}{}^* = X_2$, …, $\tilde{X}_{N/2-1} = \tilde{X}_{N/2+1}{}^* = X_{N/2-1}$, and $\tilde{X}_0 = \tilde{X}_{N/2} = 0$. From Eq. (1.22), the decision variable at the m-th subcarrier is given by

$$\begin{aligned}
\hat{X}_m &= H_m^*Y_m + H_{N-m}Y_{N-m}^* \\
&= (S_0|H_m|^2 + S_0^*|H_{N-m}|^2)X_m + \sum_{k=1,k\neq m}^{N/2-1} (S_{k-m}H_m^*H_k + S_{m-k}^*H_{N-m}H_{N-k}^*)X_k \\
&\quad + \sum_{k=N/2+1}^{N-1} (S_{k-m}H_m^*H_k + S_{m-k}^*H_{N-m}H_{N-k}^*)X_{N-k}^* + H_m^*W_m + H_{N-m}W_{N-m}^*.
\end{aligned}$$

Then, the CIR of the MCSR scheme can be expressed as

$$\mathrm{CIR}_{\mathrm{MCSR}}(m) = \frac{\mathbb{E}\{|S_0|H_m|^2 + S_0^*|H_{N-m}|^2|^2\}}{\sum_{k=1,k\notin\{m,N/2\}}^{N-1} \mathbb{E}\{|S_{k-m}H_m^*H_k + S_{m-k}^*H_{N-m}H_{N-k}^*|^2\}}. \tag{1.25}$$

For the flat fading channels, it can be readily verified that the CIR of the MCSR scheme is

$$\mathrm{CIR}_{\mathrm{MCSR}}(m) = \frac{4\Re\{S_0\}^2}{\sum_{k=1,k\notin\{N/2-m,N-m\}}^{N-1} |S_k + S_{-k}^*|^2}. \tag{1.26}$$

As $S_k + S_{-k}^* \approx 0$, the CIR of the MCSR scheme is improved compared with that of the plain OFDM. The CIR under the multipath Rayleigh fading channel is the same as that under flat

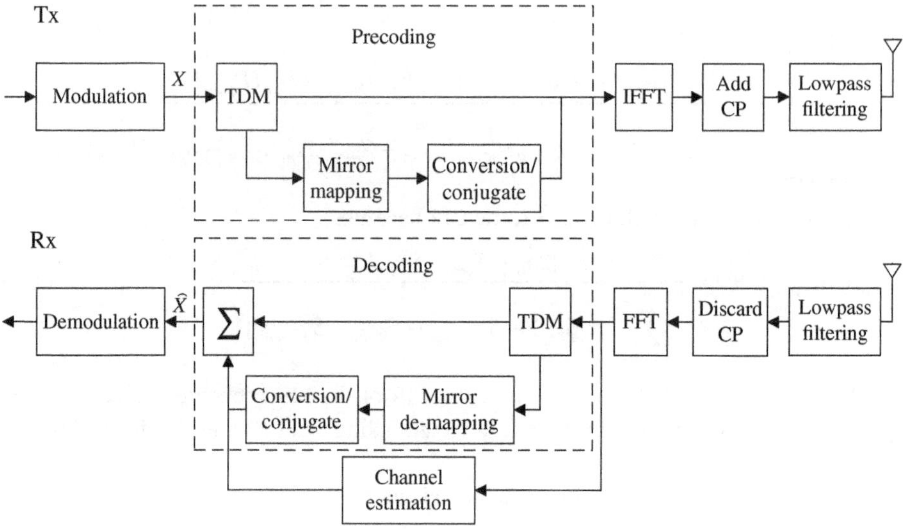

Figure 1.7 Block diagram of an OFDM transceiver with the ICI two-path cancellation modules using mirror mapping in the baseband.

fading channels, indicating that the CIR of the MCSR scheme is not affected by the channel length L. See Ref. [15] for proof.

Accordingly, the average CIR of the MCSR scheme is given by

$$\text{CIR}_{\text{MCSR}} = \frac{2}{N-2} \sum_{m=1}^{N/2-1} \text{CIR}_{\text{MCSR}}(m). \tag{1.27}$$

1.4.2 Two-Path Cancellation with Mirror Mapping

Figure 1.7 depicts the system architecture of the ICI two-path cancellation schemes with mirror mapping. The two-path cancellation schemes transmit data symbols in two consecutive OFDM symbols, which are usually referred to as two independent paths separated by time division multiplexing (TDM). Evidently, the main additional operations due to the introduction of two-path cancellation schemes are integrated inside the precoding and decoding modules.

In general, for the two-path cancellation schemes with mirror mapping, at the precoding module, one OFDM symbol input $\{X_k\}_{k=0}^{N-1}$ will become two OFDM symbol outputs $\{\bar{X}_k^{(1)}\}_{k=0}^{N-1}$ and $\{\bar{X}_k^{(2)}\}_{k=0}^{N-1}$, where the first OFDM symbol $\{\bar{X}_k^{(1)}\}_{k=0}^{N-1}$ is identical to the input OFDM symbol, i.e. $\bar{X}_k^{(1)} = X_k$, and the second OFDM symbol $\{\bar{X}_k^{(2)}\}_{k=0}^{N-1}$ obeys the subcarrier mirror-mapping rule and can be obtained as $\bar{X}_k^{(2)} = \mathcal{O}(X_{N-k})$, $k = \{0, \dots, N-1\}$. For ease of exposition, if the conversion operation is utilized for the mapping operation, i.e. $\mathcal{O}(x) = -x$, we refer to it as the MCVT scheme, and if the conjugate operation is used, i.e. $\mathcal{O}(x) = x^*$, we call it the MCJT scheme.

After the deliberate design of the transmitted signal in the precoding module, the received signals at the m-th subcarrier and the $(N-m)$-th subcarrier of the first OFDM symbol and the second OFDM symbol, respectively, will carry the same data information. Therefore, at the decoding module, it is reasonable to use MRC for decoding, yielding

$$\hat{X}_m = \frac{H_m^* \bar{Y}_m^{(1)} + \mathcal{O}\{H_{N-m}^* \bar{Y}_{N-m}^{(2)}\}}{|H_m|^2 + |H_{N-m}|^2}, \tag{1.28}$$

where H_m is the CFR at subcarrier m, and $\bar{Y}_m^{(i)}$ is the received signal in the FD corresponding to the i-th transmitted OFDM symbol ($i \in \{1, 2\}$).

From the above, it can be found that compared with the one-path cancellation approach, the two-path cancellation approach is conveniently compatible with the traditional OFDM transceiver design without any modifications.

1.4.2.1 MCVT Scheme

For the MCVT scheme, conversion operation is adopted. The two consecutive transmitted OFDM symbols are of the form $\bar{X}^{(1)} = [X_0, X_1, \dots, X_{N-1}]$ and $\bar{X}^{(2)} = [-X_0, -X_{N-1}, \dots, -X_1]$. From Eq. (1.28), the decision variable at the m-th subcarrier ($m = 0, \dots, N-1$) is given by

$$\begin{aligned}\hat{X}_m &= H_m^* \bar{Y}_m^{(1)} - H_{N-m}^* \bar{Y}_{N-m}^{(2)} \\ &= [S_0(\varepsilon)|H_m|^2 + S_0(\varepsilon + \Delta\varepsilon)|H_{N-m}|^2]X_m \\ &\quad + \sum_{k=0,k\neq m}^{N-1} [S_{k-m}(\varepsilon)H_m^* H_k + S_{m-k}(\varepsilon + \Delta\varepsilon)H_{N-m}^* H_{N-k}]X_k \\ &\quad + H_m^* W_m^{(1)} - H_{N-m}^* W_{N-m}^{(2)},\end{aligned} \tag{1.29}$$

where the CFO of the first symbol is ε, the CFO of the second symbol is $\varepsilon + \Delta\varepsilon$, and $W_m^{(p)}$ ($p = 1, 2$) is the noise at the m-th subcarrier of the p-th path.

According to Eq. (1.29), the CIR of the MCVT scheme can be expressed as

$$\mathrm{CIR}_{\mathrm{MCVT}} = \frac{\mathbb{E}\{|S_0(\varepsilon)|H_m|^2 + S_0(\varepsilon + \Delta\varepsilon)|H_{N-m}|^2|^2\}}{\sum_{k=0,k\neq m}^{N-1} \mathbb{E}\{|S_{k-m}(\varepsilon)H_m^* H_k + S_{m-k}(\varepsilon + \Delta\varepsilon)H_{N-m}^* H_{N-k}|^2\}}. \tag{1.30}$$

For the flat fading channel, the CIR of the MCVT scheme becomes

$$\mathrm{CIR}_{\mathrm{MCVT}} = \frac{|S_0(\varepsilon) + S_0(\varepsilon + \Delta\varepsilon)|^2}{\sum_{k=1}^{N-1} |S_k(\varepsilon) + S_{-k}(\varepsilon + \Delta\varepsilon)|^2}. \tag{1.31}$$

1.4.2.2 MCJT Scheme

For the MCJT scheme, conjugate operation is adopted. In this case, the two consecutive transmitted OFDM symbols are given by $\bar{X}^{(1)} = [X_0, X_1, \dots, X_{N-1}]$ and $\bar{X}^{(2)} = [X_0^*, X_{N-1}^*, \dots, X_1^*]$. According to Eq. (1.28), the decision variable at the m-th ($m \in \{0, 1, \dots, N-1\}$) subcarrier is given by

$$\begin{aligned}\hat{X}_m &= H_m^* \bar{Y}_m^{(1)} + H_{N-m} \bar{Y}_{N-m}^{(2)*} \\ &= [S_0(\varepsilon)|H_m|^2 + S_0^*(\varepsilon + \Delta\varepsilon)|H_{N-m}|^2]X_m \\ &+ \sum_{k=0,k\neq m}^{N-1} [S_{k-m}(\varepsilon)H_m^* H_k + S_{m-k}^*(\varepsilon + \Delta\varepsilon)H_{N-m}H_{N-k}^*]X_k \\ &+ H_m^* W_m^{(1)} + H_{N-m} W_{N-m}^{(2)*}.\end{aligned} \tag{1.32}$$

According to Eq. (1.32), the CIR of the MCVT scheme can be expressed as

$$\text{CIR}_{\text{MCJT}} = \frac{\mathbb{E}\{|S_0(\varepsilon)|H_m|^2 + S_0^*(\varepsilon + \Delta\varepsilon)|H_{N-m}|^2|^2\}}{\sum_{k=0,k\neq m}^{N-1} \mathbb{E}\{|S_{k-m}(\varepsilon)H_m^* H_k + S_{m-k}^*(\varepsilon + \Delta\varepsilon)H_{N-m}H_{N-k}^*|^2\}}. \tag{1.33}$$

For the flat fading channel, the CIR of the MCJT scheme is given by

$$\text{CIR}_{\text{MCJT}} = \frac{|S_0(\varepsilon) + S_0^*(\varepsilon + \Delta\varepsilon)|^2}{\sum_{k=1}^{N-1} |S_k(\varepsilon) + S_{-k}^*(\varepsilon + \Delta\varepsilon)|^2}. \tag{1.34}$$

Similarly to MCSR, we see here that the denominator in the CIR expression is the summation over $S_k(\varepsilon) + S_{-k}^*(\varepsilon + \Delta\varepsilon)$, which is approximately zero for $k \neq 0$ and small $\Delta\varepsilon$. In addition, by comparing with Eq. (1.26), it can be found $\text{CIR}_{\text{MCJT}} \approx \text{CIR}_{\text{MCSR}}$ when $\Delta\varepsilon = 0$.

Finally, it is worth noting that in flat fading channels, the CFRs at all subcarriers are same, and the CFR coefficients are inherently canceled in the CIR expression. Thus, the CIR of each scheme, namely Eqs. (1.20), (1.23), (1.26), (1.31), and (1.34), has exactly the same expression as that in AWGN channels.

1.4.3 CIR Comparison

An OFDM system with $N = 1024$ subcarriers is considered for comparing the CIR performance among plain OFDM, adjacent-mapping-based schemes, and mirror-mapping-based schemes.

Figure 1.8 presents the CIR results of flat fading channels, i.e. $L = 1$. For the ICI two-path cancellation schemes, CIR results with $\Delta\varepsilon = 0$ and $\Delta\varepsilon = 0.03$ are presented. From Figure 1.8, the following facts can be observed:

- All mirror-mapping-based schemes express much better CIR performance than plain OFDM.
- MSR and MCSR outperform ASR and ACSR, respectively. This shows that the mirror-mapping rule has better ICI suppression capability than the adjacent-mapping rule. The main reason is that for the mirror-mapping-based schemes, the interference from neighbor subcarriers is sufficiently suppressed, while there is still some residual interference from neighbor subcarriers for the adjacent-mapping-based schemes.
- With $\Delta\varepsilon = 0$, the CIR curves of the MCSR scheme and the MCJT scheme coincide, and the CIR of the MSR scheme is close to that of the MCVT scheme. This is because the same operation is adopted for both schemes.
- With $\Delta\varepsilon = 0$, the CIR performance of the conversion-based schemes, namely MSR and MCVT, is better than that of the conjugate-based schemes, namely MCSR and MCJT.

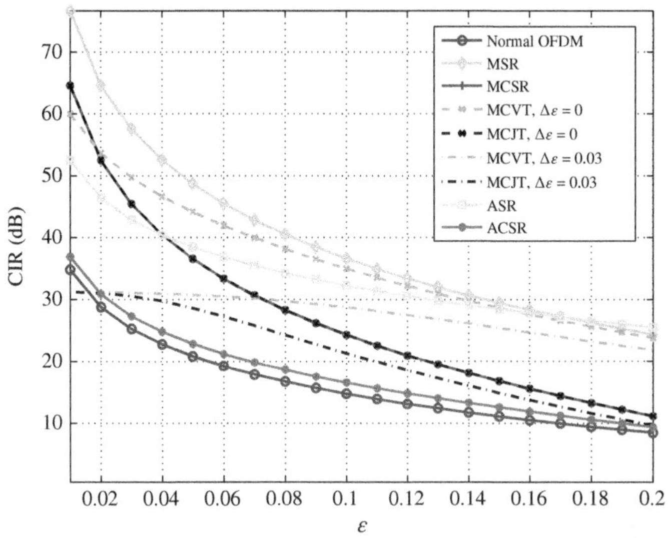

Figure 1.8 CIR comparison among the plain OFDM, adjacent-mapping-based schemes, and mirror-mapping-based schemes for different values of ε over flat fading channels.

This is because in their CIR expressions $S_k + S^*_{-k} \approx 0$ is a rougher approximation than $S_k + S_{-k} \approx 0$.

- With $\Delta\varepsilon = 0.03$, the CIR performance of the MCVT scheme and the MCJT scheme is degraded due to the CFO deviation between the first OFDM symbol and the second OFDM symbol.

Figures 1.9 and 1.10 present the CIR results of multipath Rayleigh fading channels. Assume that each channel tap has the same power, i.e. $\sigma_0^2 = \sigma_1^2 = \cdots = \sigma_{L-1}^2$. The effect of the channel length L is investigated. From Figures 1.9 and 1.10, it can be observed that for the multipath Rayleigh fading channels, mirror-mapping-based schemes have better CIR performance compared with the plain OFDM. The CIR performance of the plain OFDM and the conjugate-based schemes, namely MCSR and MCJT, is not affected by the channel length L. For conversion-based schemes, namely MSR and MCVT, the CIR performance is degraded with increase in channel length L. For OFDM-based communications, to maintain the spectral efficiency, N is usually much larger than the channel length L. Thus, conversion-based schemes can effectively reduce ICI for the long OFDM symbol design.

In summary, based on the CIR results of flat fading channels and multipath Rayleigh fading channels, we conclude that mirror-mapping-based schemes can effectively mitigate ICI.

1.5 Experiment on Sea

Fast time varying, phase variation, and Doppler shift constitute the features of the underwater acoustic (UWA) communication channel. As a consequence, the effect of ICI becomes significantly critical for OFDM-based UWA communication systems. The applicability of mirror-mapping-based schemes in UWA communications is verified in this section by sea experiments.

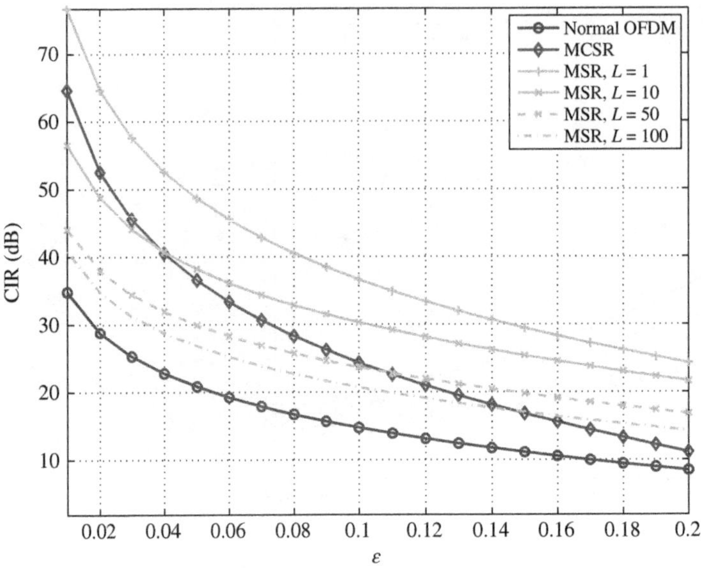

Figure 1.9 CIR comparison among the plain OFDM, the MSR scheme, and the MCSR scheme for different values of ε over multipath Rayleigh fading channels.

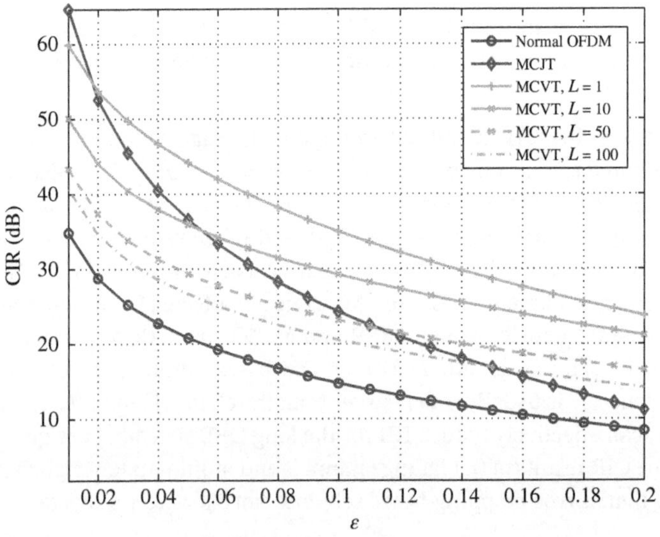

Figure 1.10 CIR comparison among the plain OFDM, the MCVT scheme, and the MCJT scheme for different values of ε over multipath Rayleigh fading channels.

1.5.1 Experiment Settings

The experiment was conducted in a sea area about 3 km east of Gushan, Taiwan, on 21–22 May 2013, which is illustrated in Figure 1.11. Three nodes were deployed, i.e. node 4, node 5, and node 9, each of which consisted of one transducer and four hydrophones.

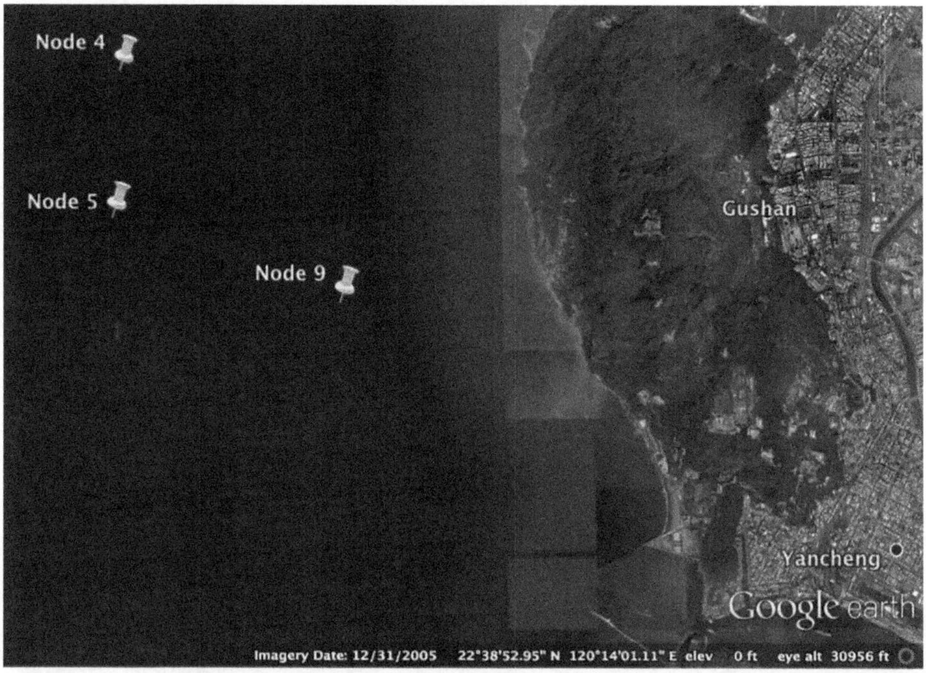

Figure 1.11 Geographical locations of transceiver nodes. The GPS coordinates of nodes 4, 5, and 9 are (N22.66038, E120.21450), (N22.64844, E120.21405), and (N22.64169, E120.23450), respectively. Their relative distances are 1549.58 m (from node 4 to node 5), 2187.60 m (from node 5 to node 9), and 3377.72 m (from node 9 to node 4). Source: Map data: google, © Digital Globe, © Kingway Ltd, © GeoForce Technologies, © TerraMetrics.

The sea depth is around 20 m and the node depth is around 10 m. The transducer and the hydrophones may drift due to waves. During the sea test, the nodes transmitted with each other and the received data packets were recorded.

The basic system parameters are provided in Table 1.1. The bandwidth of the system is 5.36 kHz. The total number of subcarriers is 1600, within which there are 1278 data subcarriers, 214 pilot subcarriers, and 108 null subcarriers. The guard interval has a length of 50 ms, which is much longer than the maximum channel delay spread. The pilot subcarriers are used for channel estimation. For the ICI two-path cancellation schemes, the pilots are uniformly inserted among the data subcarriers. However, for the ICI self-cancellation schemes, to avoid loss of spectral efficiency, the mirror-mapped pilot structure has to be adopted. Thus, the ICI self-cancellation schemes do not facilitate OFDM system standardization. In addition, quadrature phase shift keying (QPSK) modulation is adopted for the mirror-mapping-based schemes.

To demonstrate the performance of the mirror-mapping-based schemes, there are two benchmark candidates, both of which have the same spectral efficiency as the mirror-mapping-based schemes. The first one is the plain OFDM with BPSK modulation (OFDM-B). The second one is the OFDM scheme with half of the total subcarriers occupied by QPSK data symbols and each of the data symbols surrounded by two null subcarriers (OFDM-QH). For OFDM-QH, the ICI from direct neighbors is removed. Figure 1.12

Table 1.1 System parameters.

Sampling rate at the transmitter	48 kHz
Sampling rate at the receiver	48 kHz
Signal bandwidth	5.36 kHz
Carrier frequency	17 kHz
Number of total subcarriers	1600
Number of data subcarriers	1278
Number of pilot subcarriers	214
Number of null subcarriers	108
Subcarrier spacing	3.35 Hz
OFDM symbol duration	299 ms
Guard interval	50 ms
Number of hydrophones	4

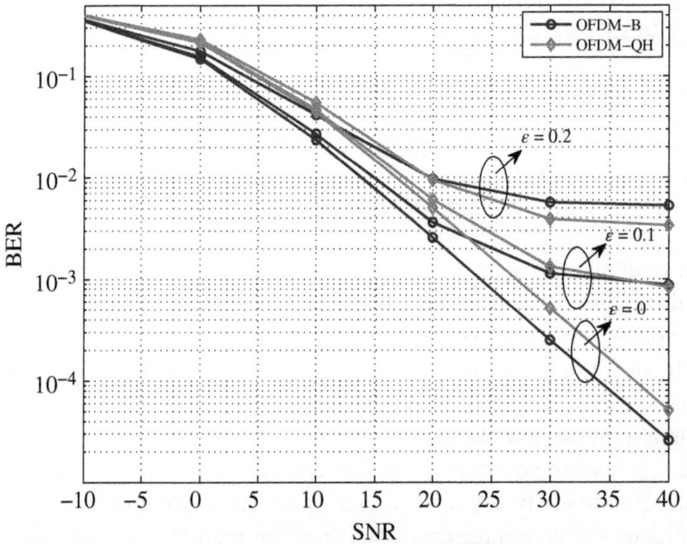

Figure 1.12 BER performance comparison between OFDM-QH and OFDM-B for CFOs $\varepsilon = 0, 0.1, 0.2$. The total subcarrier number is $N = 1024$, the bandwidth is 6 kHz, and the channel spread length is 9 ms. For OFDM-QH, Gray code is adopted for QPSK modulation.

compares the bit error rate (BER) performance of OFDM-B and OFDM-QH under different CFOs. It is found that OFDM-B has lower BER than OFDM-QH in low-medium signal-to-noise-ratio (SNR) range (SNR < 20, BER > 10–3). This is because of the higher symbol detection error for QPSK modulation. Considering the high signal attenuation of UWA channels and relatively low SNR at the receive hydrophones, OFDM-B, the one with better BER, is chosen as the benchmark in the sea experiment.

Preamble	Plain OFDM (BPSK) 5 symbols	ICI Cancellation (QPSK) 10 symbols	Postamble

Figure 1.13 Frame structure.

The frame structure for transmission is given in Figure 1.13. The transmitted frame consists of a preamble, transmitted OFDM data symbols, and a postamble. We allocate 5 OFDM symbols with BPSK modulation for the plain OFDM and 10 OFDM symbols with QPSK modulation for either the ICI self-cancellation schemes or the two-path cancellation schemes. Therefore, 17 OFDM symbols are involved in each frame in the experiment. In addition, Gray code is adopted for QPSK modulation. Depending on the ICI cancellation structure at the last 10 OFDM symbols, there are four different frames, implementing the MSR scheme, the MCSR scheme, the MCVT scheme, and the MCJT scheme, respectively. In the experiment, ignoring preamble and CP overhead, the data rate of mirror-mapping-based schemes and OFDM-B is 4.27 kbps.

For time synchronization, the received signal is correlated with the preamble and the postamble to obtain the starting and ending time of the received data frame. Then, the resampling factor is calculated by comparing the received signal length and the transmitted signal length. Finally, to remove the major Doppler effect, the received signal is resampled according to the resampling factor [16].

The CFR of each subcarrier can be estimated via the received signals on the pilot subcarriers. The CFRs on the pilot subcarriers are estimated first. Then the CFRs on the data subcarriers can be obtained through the piecewise cubic spline interpolation. To combat the time-varying feature of UWA channels, channel estimation is done for each OFDM symbol.

1.5.2 Experiment Results

To obtain the BER results and enable a fair comparison of different schemes, the packages with indices "M0000043.DAT," "M0000044.DAT," "M0000046.DAT," "M0000047.DAT," "M0000049.DAT," "M0000050.DAT," "M0000052.DAT," and "M0000053.DAT" are utilized with their information provided in Table 1.2. They are transmitted consecutively with similar receive SNR levels and through the same transmitter–receiver (T-R) pair. Figure 1.14 illustrates the BER performance of OFDM-B, the mirror-mapping-based ICI cancellation schemes, and the plain OFDM with QPSK modulation (OFDM-Q). The BER of OFDM-Q is calculated by directly decoding the mirror-mapping-based schemes without combining data subcarrier pairs, and thus ICI is not suppressed. As expected, all mirror-mapping-based schemes have lower BER than OFDM-B and OFDM-Q for all hydrophones. This confirms that mirror-mapping-based schemes can achieve superior ICI mitigation in OFDM UWA communications. Unlike with the CIR results, it is observed that all four schemes have similar BER performance. This is not surprising since the CIR performance is not directly related to the BER performance [17]. The theoretical analyses of the proposed mirror-mapping-based schemes over frequency-selective UWA channels in terms of BER will be considered as our future work.

Table 1.2 Packet information.

Packet index	Scheme	Effective SNR	Transmit time	T-R pair
M0000043.DAT	MCJT	12.47	18:16:00 05/21/13	N5-N9
M0000044.DAT	MCJT	12.55	18:16:25 05/21/13	N5-N9
M0000046.DAT	MCSR	12.27	18:16:55 05/21/13	N5-N9
M0000047.DAT	MCSR	12.15	18:17:21 05/21/13	N5-N9
M0000049.DAT	MSR	12.21	18:17:51 05/21/13	N5-N9
M0000050.DAT	MSR	12.29	18:18:16 05/21/13	N5-N9
M0000052.DAT	MCVT	12.31	18:18:46 05/21/13	N5-N9
M0000053.DAT	MCVT	12.36	18:19:11 05/21/13	N5-N9

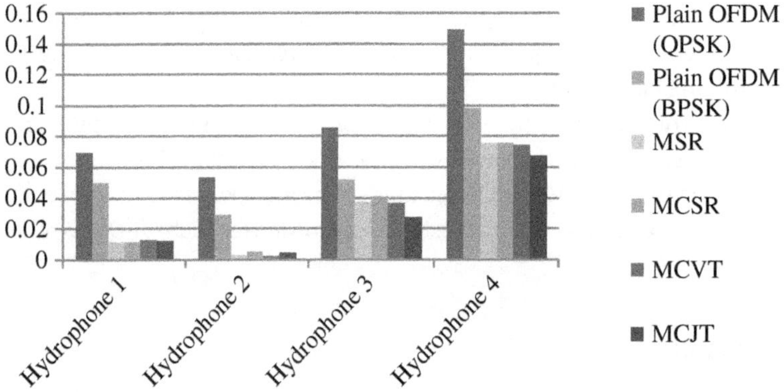

Figure 1.14 BER performance of the plain OFDM and the mirror-mapping-based schemes.

1.6 Summary

In this chapter, four effective low-complexity mirror-mapping-based ICI cancellation schemes without explicitly estimating ICI coefficients or CFO were introduced. ICI self-cancellation schemes, namely MSR and MCSR, are robust against the CFO deviation between the first OFDM symbol and the second OFDM symbol. However, they require the mirror-mapped pilot structure, which does not facilitate OFDM system standardization. In addition, compared with conjugate-based schemes, conversion-based schemes, namely MSR and MCVT, have better CIR performance. However, their CIR performance is degraded for multipath Rayleigh fading channels with large channel length. Thus, scheme selection depends on the actual system requirement and channel conditions. Finally, all mirror-mapping-based schemes were tested in a recent sea experiment conducted in Taiwan in May 2013. Decoding results showed that the proposed mirror-mapping-based schemes provide much lower BER than plain OFDM. This confirms that mirror-mapping-based schemes are very effective for ICI mitigation.

References

1 Chang, R.W. (1966). Synthesis of band-limited orthogonal signals for multichannel data transmission. *Bell System Technical Journal* 45 (10): 1775–1796.

2 Weinstein, S. and Ebert, P. (1971). Data transmission by frequency-division multiplexing using the discrete Fourier transform. *IEEE transactions on Communication Technology* 19 (5): 628–634.

3 Peled, A. and Ruiz, A. (1980). Frequency domain data transmission using reduced computational complexity algorithms. ICASSP'80. IEEE International Conference on Acoustics, Speech, and Signal Processing, Volume 5. IEEE, pp. 964–967.

4 Hwang, T., Yang, C., Wu, G. et al. (2008). OFDM and its wireless applications: a survey. *IEEE transactions on Vehicular Technology* 58 (4): 1673–1694.

5 Moose, P.H. (1994). A technique for orthogonal frequency division multiplexing frequency offset correction. *IEEE Transactions on Communications* 42 (10): 2908–2914.

6 Zhao, Y. and Haggman, S.G. (2001). Intercarrier interference self-cancellation scheme for OFDM mobile communication systems. *IEEE Transactions on Communications* 49 (7): 1185–1191.

7 Lee, J., Lou, H., Toumpakaris, D., and Cioffi, J.M. (2006). SNR analysis of OFDM systems in the presence of carrier frequency offset for fading channels. *IEEE Transactions on Wireless Communications* 5 (12): 3360–3364.

8 Qu, F. and Yang, L. (2008). Basis expansion model for underwater acoustic channels? OCEANS 2008, Quebec City, QC, Canada, September 2008, pp. 1–7.

9 Leus, G. and van Walree, P.A. (2008). Multiband OFDM for covert acoustic communications. *IEEE Journal on Selected Areas in Communications* 26 (9): 1662–1673.

10 Mason, S.F., Berger, C.R., Zhou, S. et al. (2009). Receiver comparisons on an OFDM design for Doppler spread channels. OCEANS 2009-EUROPE, Bremen, Germany, May 2009, pp. 1–7.

11 Tu, K., Fertonani, D., Duman, T.M. et al. (2011). Mitigation of intercarrier interference for OFDM over time-varying underwater acoustic channels. *IEEE Journal of Oceanic Engineering* 36 (2): 156–171.

12 Huang, X. and Wu, H. (2007). Robust and efficient intercarrier interference mitigation for OFDM systems in time-varying fading channels. *IEEE Transactions on Vehicular Technology* 56 (5): 2517–2528.

13 Tang, Z., Cannizzaro, R.C., Leus, G., and Banelli, P. (2007). Pilot-assisted time-varying channel estimation for OFDM systems. *IEEE Transactions on Signal Processing* 55 (5): 2226–2238.

14 Huang, J., Zhou, S., Huang, J. et al. (2011). Progressive inter-carrier interference equalization for OFDM transmission over time-varying underwater acoustic channels. *IEEE Journal of Selected Topics in Signal Processing* 5 (8): 1524–1536.

15 Cheng, X., Wen, M., Cheng, X., and Yang, L. (2015). Effective mirror-mapping-based intercarrier interference cancellation for OFDM underwater acoustic communications. *Ad Hoc Networks (Elsevier)* 34: 5–16.

16 Li, B., Zhou, S., Stojanovic, M. et al. (2008). Multicarrier communication over underwater acoustic channels with nonuniform Doppler shifts. *IEEE Journal of Oceanic Engineering* 33 (2): 198–209.

17 Beaulieu, N.C. and Tan, P. (2007). On the use of correlative coding for OFDM. 2007 IEEE International Conference on Communications, Glasgow, UK, June 2007, pp. 756–761.

2

Filtered OFDM: An Insight into Intrinsic In-Band Interference

Juquan Mao[1], Lei Zhang[2] and Pei Xiao[1]

[1] *Institute for Communication Systems (ICS), Home of 5G Innovation Centre, University of Surrey, UK*
[2] *James Watt School of Engineering, University of Glasgow, UK*

2.1 Introduction

Orthogonal frequency division multiplexing (OFDM) has strengths such as robustness against multipath fading, a simple implementation based on fast Fourier transform (FFT) algorithms, and perfect compatibility with multiple-input multiple-output (MIMO) technique. With these advantages, OFDM is extensively adopted in modern communication systems. However, due to limited spectrum localization, OFDM has significant limitations in challenging new spectrum use scenarios, such as asynchronous multiple access, as well as mixed numerology cases aiming to use adjustable numerologies, such as subcarrier spacing (SCS), symbol length, and cyclic prefix (CP) length, in support of diverse service requirements [1, 2].

To address the mentioned limitations of OFDM, several candidate waveforms and their variants, such as filter bank multicarrier (FBMC) [3, 4], generalized frequency division multiplexing (GFDM) [5, 6], and universal filtered multicarrier (UFMC) [7] are being studied and promoted. The comparisons among these schemes with respects to different criteria can be found in [2, 8, 9]. Regarding OFDM based advanced waveform candidates, filtered OFDM (f-OFDM) schemes are receiving considerable attention due to their ability to address the mentioned issues while maintaining a high level of commonality with legacy OFDM systems [10, 11]. Specifically, the system bandwidth is divided into arbitrary numbers of subbands, each containing a legacy OFDM signal. Consequently, f-OFDM is capable of retaining the advantages of OFDM while avoiding its limitations. First, filtering is applied to every single subband to suppress out-of-band (OOB) emissions; thus the guard band can be reduced with a better localized spectrum. Second, numerology can be optimized independently for a certain type of service within each subband; thus services with different technical requirements are flexibly supported. Third, thanks to the filtering, the synchronization requirement is also relaxed due to the reduced sidelobe, making interference from asynchronous transmissions more tolerable.

Many aspects of f-OFDM, such as general framework and methodology, design and implementation, and field trials, have been reported in literature [10–18]. Much of the

Radio Access Network Slicing and Virtualization for 5G Vertical Industries, First Edition.
Edited by Lei Zhang, Arman Farhang, Gang Feng, and Oluwakayode Onireti.

research has studied system performance based on simulations with specific parameters [10, 11, 14]. Zhang et al. progressed further by deriving a system model in [19] to quantitatively analyze interference in f-OFDM systems, in which the channel matrix is divided into three parts to decompose total inferences into inter-symbol interference (ISI) and adjacent carrier interference (ACI). To better suppress OOB emissions, the filters employed in f-OFDM systems are usually very long (up to half of FFT size [11, 13]), which inevitably leaves the systems prone to in-band interference. The existing works in the literature indicate that it has a trivial influence on system performance for medium to wide subbands [17, 18], and few studies have investigated the performance degradation in narrow subband systems.

In this chapter, we will

- present an analytical model for 3GPP-compatible single band f-OFDM systems, in which the receiver has no knowledge of the transmitter filter [20]. Based on the model, the in-band interference signal is decomposed into ICI, forward ISI (f-ISI), and backward ISI (b-ISI), such that the impact of each interference component can be studied individually;
- derive an analytical metric to quantify the interference level as a function of several system parameters. This work leads to the development of the condition for in-band interference-free systems, and provides a practical approach for the selection of cyclic redundancy (CR) length to balance system efficiency and receiver complexity. A low-complexity block-based parallel interference cancellation algorithm is then proposed for the in-band interference mitigation.
- show simulation results for evaluating the performance of f-OFDM with different settings in subband width where the performance of narrow subbands is studied in comparison with wide subbands.

The rest of the chapter is organized as follows: Section 2.2 presents the single subband f-OFDM transceiver structure and describes the system model. Section 2.3 analyzes the in-band interference and introduces approaches to suppress the interference. Section 2.4 presents numerical results. Finally, Section 2.5 concludes the chapter.

2.1.1 Notations

$\mathcal{E}\{\cdot\}$ denotes the expectation operator. Vectors and matrices are denoted by boldface lowercase and uppercase characters, receptively. The operators $(\cdot)^*$, $(\cdot)^T$, and $(\cdot)^H$ represent complex conjugate, transpose, and conjugate transpose, respectively. The $M \times M$ identity matrix is denoted by $\mathbf{I}_M$. diag($\mathbf{x}$) is a function returning a square diagonal matrix with the elements of vector $\mathbf{x}$ on the main diagonal, while diag($\mathbf{A}$) returns a column vector of the main diagonal elements of matrix $\mathbf{A}$. $|\cdot|$ denotes the magnitude of a complex number, while $||\cdot||$ is the Frobenius norm of a matrix. $\circledast$ denotes the circular convolution, and $\mathbb{R}$ and $\mathbb{C}$ represent real and complex space, respectively.

2.2 System Model for f-OFDM SISO System

In this section, we describe a system model for single band f-OFDM systems in which the interference (including ICI, f-ISI, and b-ISI) induced by filtering operations are investigated.

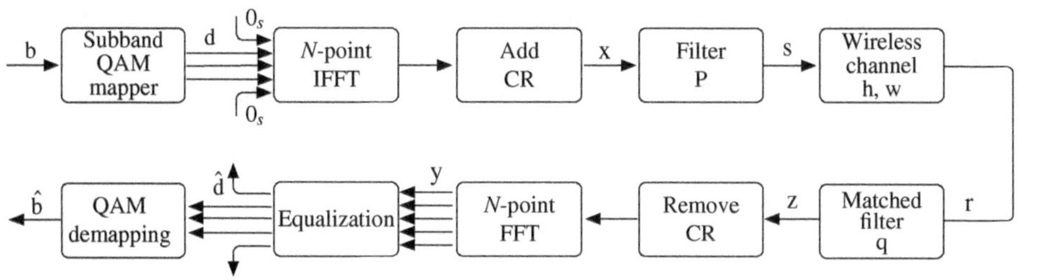

Figure 2.1 Block diagram of the f-OFDM transceiver.

These interference components are referred as in-band interference. The interference from the other subbands in the system (inter-subband interference) is out of the scope of our work.[1] Assume M consecutive subcarriers in the range of $\mathcal{M} = \{M_0, M_0 + 1, \ldots, M_0 + M - 1\}$ are assigned to the subband with a corresponding waveform shaping filter denoted as vector $\mathbf{p}$ in the time domain and $\ddot{\mathbf{p}}$ in the frequency domain.

Considering the block diagram depicted in Figure 2.1, a sequence of information bits $\mathbf{b}$ is fed into a quadrature amplitude modulation (QAM) mapper to obtain symbols from a 2^μ-valued complex constellation, where μ is the modulation order. Representing the complex symbols in K consecutive OFDM symbols in a sub-frame as a length MK vector gives

$$\mathbf{d} = [\mathbf{d}_0^T, \mathbf{d}_1^T, \ldots, \mathbf{d}_{K-1}^T]^T, \tag{2.1}$$

where $\mathbf{d}_k = [d_{k,0}, d_{k,1}, \ldots, d_{k,M-1}]^T \in \mathbb{C}^{M\times 1}$ with the individual element $d_{k,m}$ corresponding to the data symbol transmitted on the m-th subcarrier in the k-th OFDM symbol. The data symbols contained in $\mathbf{d}_k$ are assumed to be independent and identically distributed (i.i.d.) with $\mathcal{E}\{\mathbf{d}_k\mathbf{d}_k^H\} = \sigma_s^2\mathbf{I}$, where σ_s^2 is the average transmission power of QAM symbols.

The transmitter and receiver procedures are described as follows:

(1) *Inverse fast Fourier transform* (IFFT) and CR appending. An N-point ($N > M$) IFFT operation is performed on per OFDM symbol basis, followed by the addition of a CR of length N_{cr} for eliminating/mitigating ISIs. The CR is made up of two parts, cyclic prefix (CP) of length N_{cp} and cyclic suffix (CS) of length N_{cs} ($N_{\text{cr}} = N_{\text{cp}} + N_{\text{cs}}$). To be compatible with CP-OFDM, CR can be implemented as an extended CP that incorporates CP and CS at the transmitter, and the FFT window at the receiver is moved forward by the length of CS. The CP is adopted for combating the ISI induced by the dispersive nature of channels and the ISI introduced by filter forward spreading, while the CS focuses on the alleviation of the ISI induced by the filter backward spreading. The k-th OFDM symbol can be expressed in the form of matrix multiplication as

$$\mathbf{x}_k = \rho_{\text{cr}}\mathbf{T}_{\text{cr}}\mathbf{F}\mathbf{d}_k \in \mathbb{C}^{L\times 1}, \tag{2.2}$$

where $\mathbf{x}_k$ is a vector of dimension $L = N + N_{\text{cr}}$. $\mathbf{F}$ is an $N \times M$ submatrix of N-point IFFT matrix defined by its element on the n-th row and m-th column as $F_{n,m} = \sqrt{1/N}\exp\{\frac{j2\pi n(m+M_0)}{N}\}$, and it is unitary as $\mathbf{F}^H\mathbf{F} = \mathbf{I}_M$. $\mathbf{T}_{\text{cr}} = [\mathbf{I}_{\text{cp}}^T, \mathbf{I}_N, \mathbf{I}_{\text{cs}}^T]^T$ is the $L \times N$ matrix inserting the CP and CS, with $\mathbf{I}_{\text{cp}}$ and $\mathbf{I}_{\text{cs}}$ containing the last and the first

1 Interested users are referred to [21], which gives a thorough analysis on inter-subband interference.

N_{cp} rows of the identity matrix $\mathbf{I}_N$, respectively. ρ_{cr} is the power normalization factor defined as $\rho_{cr} = \sqrt{N/L}$.

(2) *Transmitter filtering*. By applying a spectrum shaping filter, the actual transmitted f-OFDM signal is obtained as

$$\mathbf{s} = \mathbf{x} \circledast \mathbf{p}, \tag{2.3}$$

where $\mathbf{x} = [\mathbf{x}_0, \mathbf{x}_1, \dots, \mathbf{x}_{K-1}]^T$, and

$$\mathbf{p} = [p_0, p_1, \dots, p_{N_p}]^T \tag{2.4}$$

is a length $N_p + 1$ vector describing the impulse response of the transmit filter. The filter is designed to be centered in the assigned subband with a width equivalent to the bandwidth of the subband. N_p is typically chosen as

(1) an even number so that the time domain symmetry of the filters is ensured,

(2) less than or equal to half of the FFT size, i.e. $N_p \leq N/2$ to keep the signal dispersed due to the transmitter and receiver filtering under one f-OFDM symbol duration, such that the f-OFDM symbol of interest only suffers ISI from its direct neighbors.

As illustrated in Figure 2.2, the signal spreads bidirectionally and inflicts interference between adjacent OFDM symbols. The forward/backward ISI are generated by the forward/backward spreading of the previous/next OFDM symbols to the current symbol window, while the ICI comes from the energy loss of the current OFDM symbol due to its bidirectional spreading. To facilitate the interference analysis caused by filtering, we derive the equivalent matrix form of a linear filtering process. The L received samples relative to the k-th OFDM symbol are grouped in the vector $\mathbf{s}_k$, thus obtaining

$$\mathbf{s}_k = \mathbf{P}^u \mathbf{x}_{k-1} + \mathbf{P}^m \mathbf{x}_k + \mathbf{P}^l \mathbf{x}_{k+1}, \tag{2.5}$$

where the $L \times L$ matrix $\mathbf{P}^u$ spreads the $(k-1)$-th OFDM symbol into the core window of the k-th OFDM symbol and causes the f-ISI. It is a strictly upper triangular matrix defined by its (i,j)-th element as

$$P^u{}_{i,j} = \begin{cases} p_{\frac{N_p}{2}+L+i-j}, & j \geq i + L - \frac{N_p}{2} \\ 0, & j < i + L - \frac{N_p}{2} \end{cases}. \tag{2.6}$$

The matrix $\mathbf{P}^m$ is a Toeplitz matrix specified by its first column $\left[p_{\frac{N_p}{2}}, \dots, p_{N_p}, \mathbf{0}_{1\times(L-\frac{N_p}{2}-1)}\right]^T$ and first row $\left[p_{\frac{N_p}{2}}, \dots, p_0, \mathbf{0}_{1\times(L-\frac{N_p}{2}-1)}\right]^T$. It is a matrix in which all nonzero entries are on the main diagonal and the first $\frac{N_p}{2}$ diagonals above and below.

The matrix $\mathbf{P}^l$ extends the $(k+1)$-th OFDM symbol into the core window of the k-th OFDM symbol and leads to the b-ISI. It is a strictly lower triangular matrix defined by its (i,j)-th element

$$P^l{}_{i,j} = \begin{cases} p_{\frac{N_p}{2}-1-L+i-j}, & i \geq j + L - \frac{N_p}{2} + 1 \\ 0, & i < j + L - \frac{N_p}{2} + 1 \end{cases}. \tag{2.7}$$

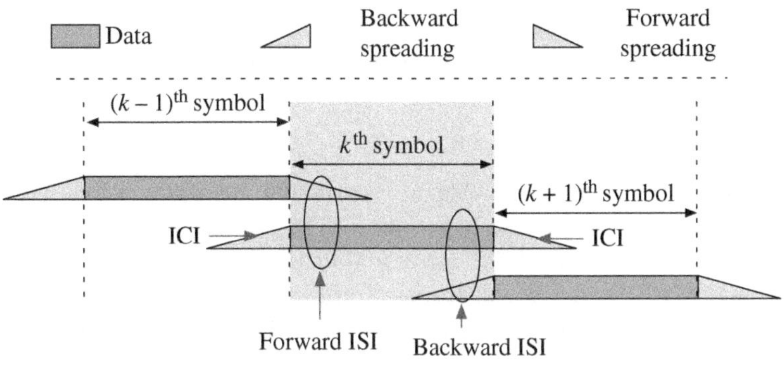

Figure 2.2 f-OFDM filtering illustration.

(3) *Passing the channel.* An $(N_{\text{ch}}+1)$-tap $(N_{\text{ch}} \leq N_{\text{cp}})$ channel is assumed to have an impulse response

$$\mathbf{h} = [h_0, h_1, \ldots, h_{N_{\text{ch}}}]^T. \tag{2.8}$$

After passing the above channel followed by an addition of additive white Gaussian noise (AWGN), the L received samples corresponding to the k-th OFDM symbol can be written as

$$\mathbf{r}_k = \mathbf{H}^u \mathbf{s}_{k-1} + \mathbf{H}^m \mathbf{s}_k + \mathbf{w}_k, \tag{2.9}$$

where $\mathbf{H}^m$ is a Toeplitz matrix with $[\mathbf{h}^T, \mathbf{0}_{1\times(L-N_{\text{ch}}-1)}]^T \in \mathbb{C}^{L\times 1}$ as its first column and $[h_0, \mathbf{0}_{1\times(L-1)}]^T \in \mathbb{C}^{L\times 1}$ as its first row.

$\mathbf{H}^u$ is the matrix relative to the channel spreading, and it is a strictly upper triangular matrix defined by its (i,j)-th element as

$$H^u{}_{i,j} = \begin{cases} h_{L+i-j}, & j \geq i + L - N_{\text{ch}} \\ 0, & \text{otherwise} \end{cases}. \tag{2.10}$$

$\mathbf{w}_k$ is the Gaussian noise vector with each element having zero mean and variance σ_n^2.

(4) *Filtering at the receiver side.* 3GPP suggests that spectral confinement techniques, such as filtering or windowing for a waveform at the transmitter, should be transparent to the receiver [20]. Therefore, we assume that the receiver has no knowledge of the transmitter filtering and define a length $N_q + 1$ receiver filter as

$$\mathbf{q} = [q_0, q_1, \cdots, p_{N_q}]^T. \tag{2.11}$$

The L samples corresponding to the k-th OFDM symbol passing through the receiver filter are grouped into the vector $\mathbf{z}_k$, obtaining

$$\mathbf{z}_k = \mathbf{Q}^u \mathbf{r}_{k-1} + \mathbf{Q}^m \mathbf{r}_k + \mathbf{Q}^l \mathbf{r}_{k+1}, \tag{2.12}$$

where $\mathbf{Q}^u, \mathbf{Q}^m, \mathbf{Q}^l$ are all $L \times L$ matrices defined in a similar approach as $\mathbf{P}^u, \mathbf{P}^m, \mathbf{P}^l$, respectively. Their definitions are omitted to conserve space.

After some basic algebraic manipulations, $\mathbf{z}_k$ can be expressed as

$$\mathbf{z}_k = \mathbf{\Theta}_{\text{pre}} \mathbf{x}_{k-1} + \mathbf{\Theta} \mathbf{x}_k + \mathbf{\Theta}_{\text{next}} \mathbf{x}_{k+1} + \tilde{\mathbf{w}}_k, \tag{2.13}$$

where the definitions of $\mathbf{\Theta}_{\text{pre}}$, $\mathbf{\Theta}$, $\mathbf{\Theta}_{\text{next}}$, and $\tilde{\mathbf{w}}_k$ can be found in Appendix 2.A.

(5) *CP removal and DFT*. By applying matrix $\mathbf{R}_{\rm cr} = [\mathbf{0}_{N\times N_{\rm cp}}, \mathbf{I}_N, \mathbf{0}_{N\times N_{\rm cs}}] \in \mathbb{B}^{N\times L}$ to $\mathbf{z}_k$ in (2.A.1), the CR is removed. After FFT at the receiver, we obtain $\mathbf{y}_k = \mathbf{F}^H\tilde{\mathbf{y}}_k = \mathbf{F}^H\mathbf{R}_{\rm cr}\mathbf{z}_k$, which can be rearranged based on (2.13) and (2.2) as

$$\mathbf{y}_k = \boldsymbol{\Psi}_{\rm pre}\mathbf{d}_{k-1} + \boldsymbol{\Psi}\mathbf{d}_k + \boldsymbol{\Psi}_{\rm next}\mathbf{d}_{k+1} + \hat{\boldsymbol{w}}_k, \tag{2.14}$$

where $\boldsymbol{\Psi}_{\rm pre} = \rho_{\rm cr}\mathbf{F}^H\mathbf{R}_{\rm cr}\boldsymbol{\Theta}_{\rm pre}\mathbf{T}_{\rm cr}\mathbf{F}$ is the $M\times M$ matrix that produces f-ISI, while $\boldsymbol{\Psi}_{\rm next} = \rho_{\rm cr}\mathbf{F}^H\mathbf{R}_{\rm cr}\boldsymbol{\Theta}_{\rm next}\mathbf{T}_{\rm cr}\mathbf{F}$ produces b-ISI, and $\hat{\boldsymbol{w}}_k = \mathbf{F}^H\mathbf{R}_{\rm cr}\tilde{\mathbf{w}}_k$ is the Fourier transformed AWGN vector. $\boldsymbol{\Psi} = \rho_{\rm cr}\mathbf{F}^H\mathbf{R}_{\rm cr}\boldsymbol{\Theta}\mathbf{T}_{\rm cr}\mathbf{F}$ is a $M\times M$ matrix transforming the desired signal. It is generally not a strict diagonal matrix unless $N_{\rm cr} > N_p + N_q + N_{\rm ch}$ (will be discussed in detail in Section 2.3.1). We decompose $\boldsymbol{\Psi}$ into a diagonal matrix $\boldsymbol{\Psi}_\Lambda = {\rm diag}({\rm diag}(\boldsymbol{\Psi}))$ and another matrix $\boldsymbol{\Psi}_e$ as $\boldsymbol{\Psi} = \boldsymbol{\Psi}_\Lambda + \boldsymbol{\Psi}_e$. The $M\times M$ matrix $\boldsymbol{\Psi}_e$ with all its diagonal elements being zero, and its non-diagonal elements, copied from the same positions of the matrix $\boldsymbol{\Psi}$, produces the ICI. By substituting $\boldsymbol{\Psi}$ with $\boldsymbol{\Psi}_\Lambda + \boldsymbol{\Psi}_e$, (2.14) can be rearranged as

$$\mathbf{y}_k = \underbrace{\boldsymbol{\Psi}_\Lambda\mathbf{d}_k}_{\text{desired signal}} + \underbrace{\boldsymbol{\Psi}_e\mathbf{d}_k}_{\text{ICI}} + \underbrace{\boldsymbol{\Psi}_{\rm pre}\mathbf{d}_{k-1}}_{\text{forward ISI}} + \underbrace{\boldsymbol{\Psi}_{\rm next}\mathbf{d}_{k+1}}_{\text{backward ISI}} + \underbrace{\hat{\boldsymbol{w}}_k}_{\text{noise}}. \tag{2.15}$$

(6) *Equalization and detection*. The signal expressed in (2.15) can be equalized using the classic equalization methods with a trade-off between complexity and performance. The simplest method is one-tap equalization using diagonal matrix $\boldsymbol{\Psi}_\Lambda$, obtained as $\hat{\mathbf{d}}_k = \boldsymbol{\Psi}_\Lambda^{-1}\mathbf{y}_k$, which can be conducted independently on each subcarrier. However, one-tap equalization can only achieve sub-optimal performance due to the ignored intrinsic interference. More advanced equalization methods will be discussed in Section 2.3.4.

2.3 In-Band Interference Analysis and Discussion

2.3.1 Channel Diagonalization and In-Band Interference-Free Systems

In interference-free systems, the channel can be fully diagonalized so that an elegant one-tap equalization can be utilized to achieve optimal performance. In the sequel, we will derive the condition to achieve interference-free f-OFDM systems. The in-band interference signal to the k-th f-OFDM symbol can be decomposed into three components based on (2.15) as

$$\mathbf{I}_k = \underbrace{\boldsymbol{\Psi}_e\mathbf{d}_k}_{\text{ICI}} + \underbrace{\boldsymbol{\Psi}_{\rm pre}\mathbf{d}_{k-1}}_{\text{forward ISI}} + \underbrace{\boldsymbol{\Psi}_{\rm post}\mathbf{d}_{k+1}}_{\text{backward ISI}}. \tag{2.16}$$

The matrix $\boldsymbol{\Psi}_{\rm pre}$ produces the filtering/channel f-ISI. It is a strictly upper triangular matrix, of which only the top $N_p + N_{\rm ch}$ rows have nonzero elements. The proof can be found in Appendix 2.B. when $\mathbf{R}_{\rm cr}$ is applied to remove the nonzero rows of $\boldsymbol{\Theta}_{\rm pre}$, the length of CP can be chosen as

$$N_{\rm cp} \geq N_p + N_{\rm ch} \tag{2.17}$$

to force a zero-valued $\boldsymbol{\Psi}_{\rm pre}$ such that zero f-ISI is ensured.

The matrix $\boldsymbol{\Psi}_{\text{next}}$ generates the filtering b-ISI. It is a strict lower triangular matrix of which only the bottom N_q rows have nonzero elements. The proof is omitted to conserve space since it can be easily done in a similar manner to $\boldsymbol{\Theta}_{\text{pre}}$. Similarly, the length of CS can be chosen as

$$N_{\text{cs}} \geq N_q \tag{2.18}$$

to secure a zero b-ISI.

When (2.17) and (2.18) are met, the adding CR converts the linear convolution $\mathbf{q} * \mathbf{h} * \mathbf{p} * \mathbf{x}$ into a circular convolution $\mathbf{q} \circledast \mathbf{h} \circledast \mathbf{p} \circledast \mathbf{x}$. From the definition of convolution theorem, circular convolution in the time domain leads to multiplication in the frequency domain. Denote $\ddot{\mathbf{p}}, \ddot{\mathbf{h}}, \ddot{\mathbf{q}}$ as DFT of $\mathbf{p}, \mathbf{h}, \mathbf{q}$ in the subband; the received data symbol vector $\mathbf{y}_k$ in (2.14) can be reformed as

$$\mathbf{y}_k = \Lambda \mathbf{d}_k + \hat{\boldsymbol{w}}_k, \tag{2.19}$$

where the $M \times M$ diagonal matrix $\Lambda = \mathbf{QHP}$ with $\mathbf{Q} = \text{diag}(\ddot{\mathbf{q}}), \mathbf{H} = \text{diag}(\ddot{\mathbf{h}}), \mathbf{P} = \text{diag}(\ddot{\mathbf{p}})$.

The Eq. (2.19) implies that $\boldsymbol{\Psi} = \boldsymbol{\Psi}_\Lambda = \Lambda$ and the ICI generating matrix $\boldsymbol{\Psi}_e = 0$. Thus, zero in-band ICI is guaranteed.

With the discussion above, we form the following Proposition:

Proposition 2.1 *Consider a single-band f-OFDM system depicted in Figure 2.1 with the transmitter, the channel, and the receiver filter being defined in (2.4),(2.8),(2.11). It is an in-band interference-free system under the condition of perfect synchronization at the receiver if*

$$N_{cp} \geq N_p + N_{ch} \quad \text{and} \quad N_{cs} \geq N_q, \tag{2.20}$$

and the received data symbol can be expressed as in (2.19).

2.3.2 In-Band Interference Power

When the interference-free condition in (2.20) is violated, i.e. $N_{\text{cp}} < N_p + N_{\text{ch}}$ or $N_{\text{cs}} < N_q$, the system will not be strictly orthogonal. In the time domain, the signal from adjacent OFDM symbols spreads into the core window of the OFDM symbol of interest, which produces the f- and b-ISI. In the frequency domain, the extended part of the interested OFDM symbol falls out of the range of CP/CS and leads to the energy loss. As a result, the matrix $\boldsymbol{\Psi}$ is no longer diagonal, causing ICI. The average power of the desired signal and interference signal including f-ISI, b-ISI, and ICI on all subcarriers in one transmission block can be grouped as an $N \times 1$ vector, obtained as

$$\begin{aligned} \eta_x &= \mathcal{E}\{\text{diag}(\boldsymbol{\Psi}_x \boldsymbol{\Psi}_x^H)\}\sigma_s^2 \\ &(\eta_x, \boldsymbol{\Psi}_x) \in \{(\eta_s, \boldsymbol{\Psi}_\Lambda), (\eta_{\text{f-ISI}}, \boldsymbol{\Psi}_{\text{pre}}), (\eta_{\text{b-ISI}}, \boldsymbol{\Psi}_{\text{next}}), (\eta_{\text{ICI}}, \boldsymbol{\Psi}_e)\}. \end{aligned} \tag{2.21}$$

Forcing the channel matrix to be an identity channel ($N_{\text{ch}} = 1, \mathbf{H}^m = 1, \mathbf{H}^u = 0$), so that the interference induced by the channel can be avoided, and the interference signal from filtering itself can be studied. The average power of the aforementioned signal can be rearranged as

$$\begin{aligned} \eta_x &= \text{diag}\ (\boldsymbol{\Psi}_x \boldsymbol{\Psi}_x^H)\sigma_s^2, \\ &(\eta_x, \boldsymbol{\Psi}_x) \in \{(\eta_s, \boldsymbol{\Psi}_\Lambda), (\eta_{\text{f-ISI}}, \boldsymbol{\Psi}_{\text{pre}}), (\eta_{\text{b-ISI}}, \boldsymbol{\Psi}_{\text{next}}), (\eta_{\text{ICI}}, \boldsymbol{\Psi}_e)\}. \end{aligned} \tag{2.22}$$

The system signal to interference plus noise ratio (SINR) can then be calculated as

$$\gamma = \frac{\eta_s}{\sum_{x\in\{\text{f-ISI,b-ISI,ICI}\}} \eta_x + \text{diag}(\hat{\mathbf{w}}_k \hat{\mathbf{w}}_k^H)}. \quad (2.23)$$

With the provided SINR, some existing works, such as [22, 23], are available for calculating system bit error rate (BER) under various channels and different modulation/coding schemes.

2.3.3 In-Band Interference Mitigation: A Practical Approach for Choosing CR Length

Proposition 2.1 implies that in-band interference-free systems can be achieved by adding a sufficient number of redundant samples. The implementation of CR, although elegant and simple, is not entirely free. It comes with a bandwidth and power penalty. Since N_{cr} redundant samples are transmitted, the actual bandwidth for f-OFDM increases from B to $\frac{N_{\text{cr}}+N}{N}B$. Similarly, an additional N_{cr} samples must be counted against the transmit power budget resulting in a power loss of $10\log_{10}\frac{N_{\text{cr}}+N}{N}$ dB. For an f-OFDM system with stringent frequency localization requirements, the filter length can be chosen up to half of the symbol duration, making the satisfaction of interference-free condition in (2.20) unaffordable with respect to the power and bandwidth loss. On the one hand, interference-free systems are preferred due to the benefit to the computational complexity reduction and the SINR improvement. On the other hand, a highly efficient system requires shorter overhead (CP/CS). Therefore, choosing the size of CR, as a trade-off between the two contradicting parties, forms an optimization problem. However, it is very hard to find an optimal solution due to the multi-objective characteristic of the problem.

A suboptimal attempt in the literature [10] is suggested by setting the length of overhead to the width of the main lobe, due to the fact that the main lobe of a sinc filter carries most of its energy. However, it neglects the fact that filters of different bandwidth vary in the amount of captured by the main lobes. For instance, a wider subband filter has less energy enclosed in the main lobe. We take this into consideration and propose a subband width-dependent approach for choosing the CR length.

For a length $K+1$ filter, defined as $\mathbf{f} = [f_0, f_1 \dots, f_K]^T$, occupying a subband of width equivalent to M subcarriers in a channel of N subcarriers, the energy rate consisting in the k middle samples is defined as

$$\zeta(\mathbf{f}, k) = \frac{\sum_{m=\frac{K}{2}-\frac{k}{2}}^{\frac{K}{2}+\frac{k}{2}} f_m f_m^*}{\sum_{m=0}^{K} f_m f_m^*}. \quad (2.24)$$

Then, the number of overhead can be chosen to ensure that the minimum amount of energy captured by the CR is greater than a predefined value. We define the following proposition:

Proposition 2.2 *Proposition 2.2: Consider a single-band f-OFDM system depicted in Figure 2.1 with the transmitter filter, the channel, and the receiver filter being defined in (2.4),*

(2.8), and (2.11), respectively. It is considered as a nearly in-band interference-free system under the condition of perfect synchronization at the receiver if

$$N_{cp} \geq K_p + N_{ch} \quad \text{and} \quad N_{cs} \geq K_q, \tag{2.25}$$

where K_p and K_q are selected to satisfy

$$\arg\min_{K_p} \zeta(\mathbf{p}, K_p) \geq \alpha \quad \text{and} \quad \arg\min_{K_q} \zeta(\mathbf{q}, K_q) \geq \alpha \tag{2.26}$$

with α being a predefined value, e.g. $\alpha = 0.99$. The received data symbol can then be approximated by (2.19) with trivial interference small enough to be ignored, and the effective channel is nearly diagonal.[2] □

When the condition in (2.25) is met, it can be seen from the numerical results in Figure 2.5 that the power of effective interference, i.e. the maximum total power of ICI, f-ISI, and b-ISI, reduces to the level very close to −30 dB. Equation (2.25) is named as a nearly in-band interference-free condition of f-OFDM systems.

The solutions to the optimization problems in (2.26) can be obtained using a linear search in a sorted list $(1, 2, \dots, K)$. In particular, the linear search sequentially checks each element of the list and evaluates ζ in (2.24) until it finds the first CP length that satisfies the specified condition in (2.26). The denominator of ζ is a constant value (only calculated once), and the numerator is an accumulated term. Therefore, the calculation of ζ at each iteration comprises an addition, a multiplication, and a division, i.e. the computation complexity of each iteration is constant. In the worst cases, it makes K comparisons. However, the optimal CP length can be found close to the width of the first main lobe ($\lfloor N/M \rfloor$) due to the energy distribution nature of the filter. Therefore, the complexity of the search algorithm is $\mathcal{O}(N/M)$.

2.3.4 An Alternative for In-Band Interference Mitigation: Frequency Domain Equalization (FDE)

2.3.4.1 Linear Equalizers

Two representative equalizers, i.e. zero-forcing (ZF) and minimum mean squared error (MMSE), apply an equalization matrix to the current symbol to reverse the effective channel effect. Considering that the received signal of the k-th f-OFDM symbol can be expressed as

$$\mathbf{y}_k = \mathbf{\Psi}\mathbf{d}_k + \text{interference terms} + \hat{\boldsymbol{w}}_k, \tag{2.27}$$

we define ZF and MMSE equalizers in f-OFDM systems as

$$\begin{aligned} \mathbf{E}_{\text{eq}}^{\text{zf}} &= (\mathbf{\Psi}\mathbf{\Psi}^H)^{-1}\mathbf{\Psi}^H, \\ \mathbf{E}_{\text{eq}}^{\text{mmse}} &= \mathbf{\Psi}^H(\mathbf{\Psi}\mathbf{\Psi}^H + \sigma_n^2\mathbf{I})^{-1}. \end{aligned} \tag{2.28}$$

---

2 How close the effective channel to a diagonal matrix can be measured quantitatively by Frobenius norm of the matrix $\mathbf{\Psi}_e$. The smaller the $||\mathbf{\Psi}_e||_F$ is, the closer the effective channel is equivalent to a diagonal matrix. $||\mathbf{\Psi}_e||_F = 0$ indicates a perfect diagonal matrix.

2.3.4.2 Nonlinear Equalizers

A nonlinear equalizer has improved performance relative to linear receivers by employing interference cancellation (IC) techniques. Conventional successive interference cancellation (SIC) algorithms come with significantly high computational complexity due to their high cancellation granularity. We propose a novel interference cancellation algorithm customized for f-OFDM systems, namely, block-wise parallel interference cancellation (BwPIC), which only performs cancellation once for all f-OFDM symbols in a data block per iteration. The algorithm comes with lower complexity than SIC because the cancellation is only carried out on a block basis.

The algorithm cancels the in-band interference of all f-OFDM symbols in one block in parallel. The detail is shown in Algorithm 2.1. It involves a sequence of interference cancellation/equalization/slicing operations. At each iteration of the outer loop, a vector $\tilde{\mathbf{d}} = [\tilde{\mathbf{d}}_0, \ldots, \tilde{\mathbf{d}}_{K-1}]^T$ is updated, and $\tilde{\mathbf{d}}_{\mathbf{pre}}/\tilde{\mathbf{d}}_{\mathbf{next}}$ are derived accordingly. Then the interference corresponding to a whole block of f-OFDM symbols is canceled simultaneously according to (2.14) as

$$\hat{y} = \mathbf{y} - \boldsymbol{\Psi}_e \tilde{\mathbf{d}} - \boldsymbol{\Psi}_{\text{pre}} \tilde{\mathbf{d}}_{\mathbf{pre}} - \boldsymbol{\Psi}_{\mathbf{next}} \tilde{\mathbf{d}}_{\mathbf{next}}; \tag{2.29}$$

however, the equalization and slicing are performed on a single f-OFDM symbol basis at each iteration of the inner loop. One-tap equalization is adopted in the algorithm for reducing computational complexity. The slicing operation approximates an equalized symbol to its nearest QAM point in the constellation.

Algorithm 1 Block-wise Parallel Interference Cancellation (BwPIC).

Inputs: $\mathbf{y}, \boldsymbol{\Psi}_e, \boldsymbol{\Psi}_{\text{pre}}, \boldsymbol{\Psi}_{\text{next}}, I$
output: $\tilde{\mathbf{d}}$
Initialization: $\tilde{\mathbf{d}} = \mathbf{0}_{\mathbf{KM}\times\mathbf{1}}$
for $i = 1; i <= I; i{+}{+}$ **do**
    $\tilde{\mathbf{d}}_{\text{pre}} = [\mathbf{0}_{M\times 1}, \tilde{\mathbf{d}}_{\mathbf{1}}^T, \ldots, \tilde{\mathbf{d}}_{\mathbf{K-1}}^T]^T$,
    $\tilde{\mathbf{d}}_{\text{next}} = [\tilde{\mathbf{d}}_{\mathbf{2}}^T, \ldots, \tilde{\mathbf{d}}_{\mathbf{K}}^T, \mathbf{0}_{M\times 1},]^T$
    $\hat{\mathbf{y}} = \mathbf{y} - \boldsymbol{\Psi}_e \tilde{\mathbf{d}} - \boldsymbol{\Psi}_{\text{pre}} \tilde{\mathbf{d}}_{\text{pre}} - \boldsymbol{\Psi}_{\text{next}} \tilde{\mathbf{d}}_{\text{next}}$ (interference canceling)
    for $k = 1; k <= K; k{+}{+}$ **do**
        $\hat{\mathbf{d}}_k = \mathbf{E}\hat{\mathbf{y}}$ (one-tap equalization)
        $\tilde{\mathbf{d}}_k = \mathbb{Q}(\hat{\mathbf{d}}_k)$ (Slicing)
    end for
end for
return $\tilde{\mathbf{d}}_k$

2.4 Numerical Results

In this section, we consider the evaluation of the following:

(1) The derived system model and interference power induced by filters with different settings
(2) The BER performance of f-OFDM single antenna systems under AWGN channels,[3] and the different performance enhancement techniques represented in Section 2.3.4

3 The reason that AWGN channels are chosen for verifying BER performance is to rule out the impact of the multipath fading channel and focus on the interference produced by filtering.

The following parameters, unless otherwise specified, are adopted for simulations. The f-OFDM system occupies $N = 1024$ subcarriers. The considered multipath fading channel of length $N_f = 8$ is a block-fading Gaussian channel (BFGC), and the duration of a transmitted data block is smaller than the coherence time of the channel. Therefore, the fading envelope is assumed to be constant during the transmission of a block and independent from block to block. The length of a block (a frame) lasts over a duration of 14 OFDM symbols. It is assumed that the channel state information (CSI) is perfectly known at the receiver. A soft-truncated sinc filter defined in [14] is employed at the transmitter and receiver side, with filter length of $N_p = N_q = N/2$ and slope controlling parameters $\alpha_p = 0.6$, $\alpha_q = 0.65$.

2.4.1 Numerical Results for In-Band Interference

The in-band interference in Section 2.3.1–2.3.3 is numerically evaluated and plotted for different values of subband width and CR length.

Figure 2.3 shows an example of power distributions for the desired signal and interference signal (ICI, f-ISI, and b-ISI) on a subcarrier level, evaluated through (2.22) with FFT size of 1024, subband width of 36 subcarriers in Figure 2.3a and 240 subcarriers in Figure 2.3b. In addition, no CP/CS was added for interference alleviation. It is shown in Figure 2.3 that the theoretical value of in-band interference power matches the simulation result. It is clearly visible that the interference in the wider subband has lower power as a whole than that in the narrower band. Figure 2.3a indicates that uneven power is distributed for both the desired signal and interference signal among subcarriers. Compared to other subcarriers,

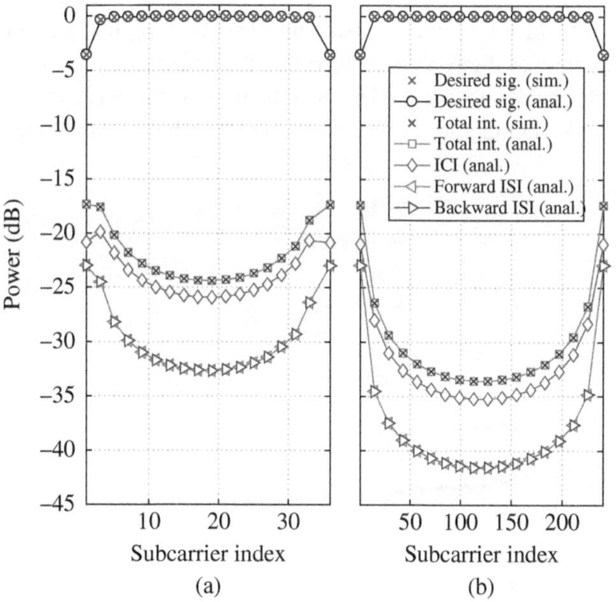

Figure 2.3 Power of desired signal and interference $N = 1024, N_{cr} = 0, N_p = N_q = 512$, $\alpha_p = 0.6, \alpha_q = 0.65$. The three contributions (ICI, forward ISI, and backward ISI) to the total interference are evaluated individually from our analytical expressions, which cannot be fulfilled through simulation; thus only their analytical results are plotted. (a) A narrow subband of 3 RBs (36 subcarriers) and (b) a wider subband of 20 RBs (240 subcarriers).

those near the edge (edge subcarriers) have lower desired signal power while experiencing higher interference. The two overlapping curves corresponding to the interference power from the previous f-OFDM (f-ISI) and next f-OFDM symbol (b-ISI) indicate the same power distribution due to the symmetry of the filters. The trends are also captured in Figure 2.3b for the wider subband.

The maximum, minimum, and average normalized power of ICI/ISI with respect to subcarriers of the same subband are shown in Figure 2.4, where the ISI represents f-ISI or b-ISI as both have the same power distribution indicated in Figure 2.3. These results provide a direct comparison among subbands with varying width from 1 to 50 RBs. It can be seen that the maximum normalized power remains constant as the width of subband grows. However, this is not the case with the minimum and average normalized power, where both decrease as the subband width increases. This implies that narrower subbands are more prone to in-band interference compared to wider subbands. These features apply to both the ICI and ISIs.

Figure 2.5 presents the average effective interference power of six subbands of variable width versus the number of CRs, where $N_{cp} = N_{cs} = N_{cr}/2$. The effective interference here refers to the sum of the ICI, f- and b-ISI. The effect of CR length for alleviating in-band interference is observed from all these curves. The power of the average effective interference decreases as the length of CR increases, and it drops under 25 dB for all six subbands when the number of CRs equals the length of the corresponding filter main lobe due to the fact that most of the filter energy is contained in the main lobe.

The BER performance of f-OFDM under the AWGN channel is evaluated and plotted in Figure 2.6. The results are presented in two cases, $N_{cr} = 0$ in Figure 2.6a and $N_{cr} = 72$ in Figure 2.6b, each having six curves corresponding to a different subband width and a curve representing the BER of legacy OFDM for a benchmark comparison. Taking into consideration the computational complexity at the receiver side, one-tap equalization method is adopted. When the in-band interference is not handled by introducing CR, the performance of f-OFDM, as shown in Figure 2.6a, degrades dramatically compared with OFDM systems.

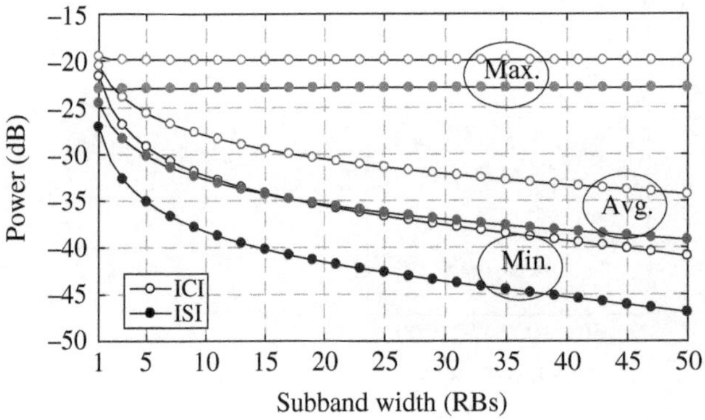

Figure 2.4 Max, min, and average normalized power of ICI/ISI with respect to subcarriers against subbands of different width with $N = 1024, N_{cr} = 0, N_p = N_q = 512, \alpha_p = 0.6, \alpha_q = 0.65$; ISI represents either f-ISI or b-ISI. All curves of the figure are generated from the analytical results.

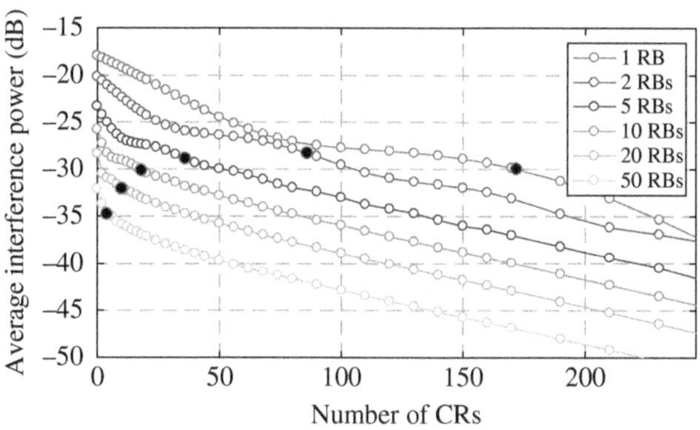

Figure 2.5 Average effective interference power with respect to subcarriers vs. number of CRs with $N = 1024, N_{cr} = 0, N_p = N_q = 512, \alpha_p = 0.6, \alpha_q = 0.65$; the solid black dots indicate the number of CRs equal to the width of the filter main lobe of the corresponding subband. All curves of the figure are generated from the analytical results.

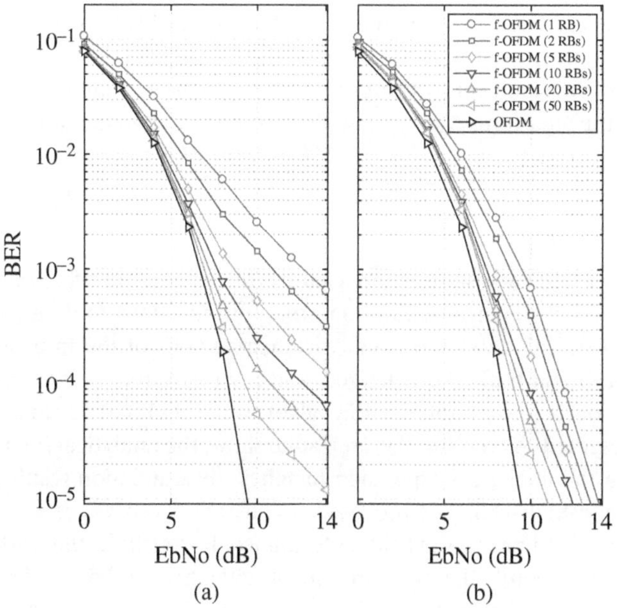

Figure 2.6 Error performance for f-OFDM systems under AWGN channel with QPSK modulation. (a) $N_{cp} = N_{cs} = 0$. (b)$N_{cp} = N_{cs} = 36$.

Moreover, error floors also tend to develop for all subbands. Another interesting observation is that the performance degradation in narrower subbands is higher than that in the wider subbands, again suggesting that narrower subbands suffer more in-band interference. In Figure 2.6b, the BER performance is significantly improved due to the use of the CR. There is still a gap of approximately 2–5 dB, subject to how wide the subband of interest is, which implies that there is still space to improve, especially for narrow subbands.

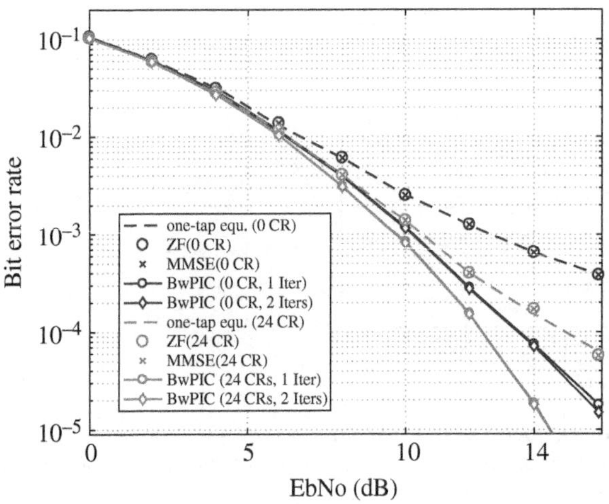

Figure 2.7 Error performance comparison with and without implementation of BwPIC for f-OFDM systems under AWGN channels with QPSK modulation.

The effect of different approaches to interference suppression is shown in terms of BER performance enhancement in Figure 2.7 with two different CR length setting, $N_{cr} = 0$ and $N_{cr} = 24$ in a subband of 12 subcarriers. It can be seen that ZF and MMSE have almost the same BER performance in both CR settings. However, the BER performance with the implementation of BwPIC improves significantly compared to the systems without BwPIC, and the results also show that the algorithm converges with no more than two iterations.

2.5 Conclusion

A single subband matrix form f-OFDM system is modeled in this chapter, in which all the linear convolution operations were converted into matrix multiplications to derive a well-channelized signal. Based on this model, the analytical expressions of the in-band interference, including ICI, f-ISI, and b-ISI, were derived, and the interference-free condition was developed. In addition, a low-complexity FEQ algorithm – BwPIC – to cancel the in-band interference is presented. As the simulation results show, the analytical interference power calculated from the analytical expression matches the simulation results, validating the analytical model established. The proposed BwPIC effectively cancels the interference signal and significantly improves the BER performance. To conclude, the work presented in this chapter provides a useful reference and valuable guidance for the practical deployment of the waveform in wireless systems.

2.A Appendix

2.A.1 Derivation of $\mathbf{z}_k$

Substituting $\mathbf{r}_k$ in Eq. (2.12) with its expression in (2.9), followed by replacing $\mathbf{s}_k$ with its expression in (2.5), we obtain

$$\begin{aligned}
\mathbf{z}_k &= (\mathbf{Q}^u\mathbf{H}^m + \mathbf{Q}^m\mathbf{H}^u)(\mathbf{P}^u\mathbf{x}_{k-2} + \mathbf{P}^m\mathbf{x}_{k-1} + \mathbf{P}^l\mathbf{x}_k) \\
&+ (\mathbf{Q}^m\mathbf{H}^m + \mathbf{Q}^l\mathbf{H}^u)(\mathbf{P}^u\mathbf{x}_{k-1} + \mathbf{P}^m\mathbf{x}_k + \mathbf{P}^l\mathbf{x}_{k+1}) \\
&+ \mathbf{Q}^l\mathbf{H}^m(\mathbf{P}^u\mathbf{x}_k + \mathbf{P}^m\mathbf{x}_{k+1} + \mathbf{P}^l\mathbf{x}_{k+2}) \\
&+ \tilde{\mathbf{w}}_k,
\end{aligned} \tag{2.A.1}$$

where $\tilde{\mathbf{w}}_k = \mathbf{Q}^u\mathbf{w}_{k-1} + \mathbf{Q}^m\mathbf{w}_k + \mathbf{Q}^l\mathbf{w}_{k+1}$ is the filtered AWGN signal.

To simplify (2.A.1) and eliminate redundant terms, we introduce the following properties:

2.A.1 The product of the two strictly upper triangular matrices is zero, and the product of the two strictly lower triangular matrices is also zero.

2.A.2 The product of $\mathbf{P}^u/\mathbf{Q}^u$ and $\mathbf{H}^m$ is a strictly triangular matrix $\mathbf{A}$ with $\mathbf{A}_{i,j} = 0$, when $i > j - [L - \frac{N_p}{2} - (N_{\text{ch}} - 1)]$. The proof can be found in Appendix 2.3.1.

We have $\mathbf{Q}^u\mathbf{H}^u = 0$ under the Property 2.A.1, $\mathbf{Q}^u\mathbf{H}^m\mathbf{P}^u = 0$ and $\mathbf{Q}^l\mathbf{H}^m\mathbf{P}^l = 0$ under Properties 2.A.1 and 2.A.2, thus obtaining a simplified version of $\mathbf{z}_k$

$$\mathbf{z}_k = \boldsymbol{\Theta}_{\text{pre}}\mathbf{x}_{k-1} + \boldsymbol{\Theta}\mathbf{x}_k + \boldsymbol{\Theta}_{\text{next}}\mathbf{x}_{k+1} + \tilde{\mathbf{w}}_k, \tag{2.A.2}$$

where $\boldsymbol{\Theta}_{\text{pre}} = \mathbf{Q}^u\mathbf{H}^m\mathbf{P}^m + \mathbf{Q}^m\mathbf{H}^u\mathbf{P}^m + \mathbf{Q}^m\mathbf{H}^m\mathbf{P}^u$, $\boldsymbol{\Theta} = \mathbf{Q}^u\mathbf{H}^m\mathbf{P}^l + \mathbf{Q}^m\mathbf{H}^u\mathbf{P}^l + \mathbf{Q}^m\mathbf{H}^m\mathbf{P}^m + \mathbf{Q}^l\mathbf{H}^u\mathbf{P}^m + \mathbf{Q}^l\mathbf{H}^m\mathbf{P}^u$, and $\boldsymbol{\Theta}_{\text{next}} = \mathbf{Q}^m\mathbf{H}^m\mathbf{P}^l + \mathbf{Q}^l\mathbf{H}^u\mathbf{P}^l + \mathbf{Q}^l\mathbf{H}^m\mathbf{P}^m$.

2.B Appendix

2.B.1 Proof of $\boldsymbol{\Theta}_{\text{pre}}$ Being a Strict Upper Triangle

As $\boldsymbol{\Theta}_{\text{pre}} = (\mathbf{Q}^u\mathbf{H}^m\mathbf{P}^m + \mathbf{Q}^m\mathbf{H}^m\mathbf{P}^u + \mathbf{Q}^l\mathbf{H}^u\mathbf{P}^u)$, following the same approach as in Appendix 2.3.1, we can prove the following: (i) the product of $\mathbf{Q}^u\mathbf{H}^m\mathbf{P}^m$ is a strictly upper triangular matrix of which only the top $\frac{N_p}{2}$ rows have nonzero elements, (ii) the product of $\mathbf{Q}^m\mathbf{H}^u\mathbf{P}^m$ is a strictly upper triangular matrix of which only the top $\frac{N_p}{2} + N_{\text{ch}}$ rows have nonzero elements, (iii) the product of $\mathbf{Q}^m\mathbf{H}^m\mathbf{P}^u$ is a strictly upper triangular matrix of which only the top $N_p + N_{\text{ch}}$ rows have nonzero elements. Therefore, the sum of three strictly upper triangular matrices results in a strictly upper triangular matrix $\boldsymbol{\Theta}_{\text{pre}}$, and only the top $N_p + N_{\text{ch}}$ rows of it have nonzero elements.

2.C Appendix

2.C.1 Proof of Property 2.A.2

The product of $\mathbf{P}^u/\mathbf{Q}^u$ and $\mathbf{H}^m$ is a strictly triangular matrix $\mathbf{A}$, with $A_{i,j} = 0$, when $i > j - [L - \frac{N_p}{2} - (N_{\text{ch}} - 1)]$. As $\mathbf{P}^u$ and $\mathbf{Q}^u$ are matrices with the same structure, we will only give the detailed steps for proving one of them. The other one can be done in a similar manner.

$$A_{i,j} = \sum_{k=1}^{L} P^{u}{}_{i,k} H^{m}{}_{k,j}$$
$$= \sum_{k=1}^{j+N_{\rm ch}-1} P^{u}{}_{i,k} H^{m}{}_{k,j} + \sum_{k=j+N_{\rm ch}}^{L} P^{u}{}_{i,k} H^{m}{}_{k,j} \tag{2.C.1}$$

Because the condition $i > j - [L - \frac{N_p}{2} - (N_{\rm ch} - 1)]$, when $1 \leq k \leq j + N_{\rm ch} - 1$, we have

$$k - (L - \frac{N_p}{2}) \leq k \leq j + N_{\rm ch} - 1 - (L - \frac{N_p}{2}) < i$$
$$\longrightarrow k < i + L - \frac{N_p}{2} \tag{2.C.2}$$

and this gives $P^{u}{}_{i,k} = 0$ according to (2.6). Therefore, the first term of (2.C.1),

$$\sum_{k=1}^{j+N_{\rm ch}-1} P^{u}{}_{i,k} H^{m}{}_{k,j} = 0.$$

When $j + N_{\rm ch} \leq k \leq L$, we have $H^{m}{}_{k,j} = 0$ based on (2.10). Therefore, the second term of (2.C.1), $\sum_{k=j+N_{\rm ch}}^{L} P^{u}{}_{i,k} H^{m}{}_{k,j} = 0$, is also proved.

As a result, $A_{i,j} = 0$ is proved, because both its sum terms in (2.C.1) equal zero, when $i > j - [L - \frac{N_p}{2} - (N_{\rm ch} - 1)]$.

References

1 Banelli, P., Buzzi, S., Colavolpe, G. et al. (2014). Modulation formats and waveforms for 5G networks: who will be the heir of OFDM?: An overview of alternative modulation schemes for improved spectral efficiency. *IEEE Signal Processing Magazine* 31 (6): 80–93.

2 Wunder, G., Jung, P., Kasparick, M. et al. (2014). 5GNOW: non-orthogonal, asynchronous waveforms for future mobile applications. *IEEE Communications Magazine* 52 (2): 97–105.

3 Farhang-Boroujeny, B. (2011). OFDM versus filter bank multicarrier. *IEEE Signal Processing Magazine* 28 (3): 92–112.

4 Razavi, R., Xiao, P., and Tafazolli, R. (2015). Information theoretic analysis of OFDM/OQAM with utilized intrinsic interference. *IEEE Signal Processing Letters* 22 (5): 618–622.

5 Michailow, N., Matth, M., Gaspar, I.S. et al. (2014). Generalized frequency division multiplexing for 5th generation cellular networks. *IEEE Transactions on Communications* 62 (9): 3045–3061.

6 Zhong, J., Chen, G., Mao, J. et al. (2018). Iterative frequency domain equalization for MIMO-GFDM systems. *IEEE Access* 6: 19386–19395.

7 Zhang, L., Xiao, P., and Quddus, A. (2016). Cyclic prefix-based universal filtered multicarrier system and performance analysis. *IEEE Signal Processing Letters* 23 (9): 1197–1201.

8 Zaidi, A.A., Luo, J., Gerzaguet, R. et al. (2016). A preliminary study on waveform candidates for 5G mobile radio communications above 6 GHz. 2016 IEEE 83rd Vehicular Technology Conference: VTC2016-Spring 15–18 May 2016, Nanjing, China.

9 Zhang, L., Ijaz, A., Xiao, P., and Tafazolli, R. (2017). Multi-service system: an enabler of flexible 5G air interface. *IEEE Communications Magazine* 55 (10): 152–159.

10 Abdoli, J., Jia, M., and Ma, J. (2015,). Filtered OFDM: a new waveform for future wireless systems. Proceedings of IEEE 16th International Workshop on Signal Processing Advances in Wireless Communications (SPAWC), pp. 66–70.

11 Zhang, X., Jia, M., Chen, L. et al. (2015). Filtered-OFDM – enabler for flexible waveform in the 5th generation cellular networks. Proceedings of IEEE Global Communications Conference (GLOBECOM), pp. 1–6.

12 Xiao, P., Toal, C., Burns, D. et al. (2010). Transmit and receive filter design for OFDM based WLAN systems. Proceedings of IEEE International Conference on Wireless Communications and Signal Processing (WCSP), pp. 1–4.

13 Li, J., Kearney, K., Bala, E., and Yang, R. (2013). A resource block based filtered OFDM scheme and performance comparison. Proceedings of IEEE ICT, May 2013, pp. 1–5.

14 Wu, D., Zhang, X., Qiu, J. et al. (2016). A field trial of f-OFDM toward 5G. 2016 IEEE Globecom Workshops (GC Wkshps), Washington, DC, USA.

15 Guan, P., Zhang, X., Ren, G. et al. (2016). Ultra-low latency for 5G-a lab trial. arXiv preprint arXiv:1610.04362.

16 Weitkemper, P., Bazzi, J., Kusume, K. et al. (2016). On regular resource grid for filtered OFDM. *IEEE Communications Letters* 20 (12): 2486–2489.

17 Guan, P., Wu, D., Tian, T. et al. (2017). 5G field trials: OFDM-based waveforms and mixed numerologies. *IEEE Journal on Selected Areas in Communications* 35 (6): 1234–1243.

18 Yli-Kaakinen, J., Levanen, T., Valkonen, S. et al. (2017). Efficient fast-convolution-based waveform processing for 5G physical layer. *IEEE Journal on Selected Areas in Communications* 35 (6): 1309–1326.

19 Zhang, L., Ijaz, A., Xiao, P. et al. (2018). Filtered OFDM systems, algorithms, and performance analysis for 5G and beyond. *IEEE Transactions on Communications* 66 (3): 1205–1218.

20 3GPP (2019). NR; base station (BS) radio transmission and reception. 3rd Generation Partnership Project, TS 38.104, 09.

21 Zhang, L., Ijaz, A., Xiao, P. et al. (2017). Subband filtered multi-carrier systems for multi-service wireless communications. *IEEE Transactions on Wireless Communications* 16 (3): 1893–1907.

22 Sklar, B. and Harris, F.J. (1988). *Digital Communications: Fundamentals and Applications*, vol. 2001. Englewood Cliffs, NJ: Prentice Hall.

23 Cho, K. and Yoon, D. (2002). On the general BER expression of one-and two-dimensional amplitude modulations. *IEEE Transactions on Communications* 50 (7): 1074–1080.

3

Windowed OFDM for Mixed-Numerology 5G and Beyond Systems

Bowen Yang[1], *Xiaoying Zhang*[2], *Lei Zhang*[1], *Arman Farhang*[3], *Pei Xiao*[4] *and Muhammad Ali Imran*[1]

[1] *School of Engineering, University of Glasgow, UK*
[2] *Institute of Electronic Science, National University of Defense Technology, China*
[3] *Electronic Engineering, Maynooth University, Maynooth, Kildare, Ireland*
[4] *5GIC, University of Surrey, UK*

3.1 Introduction

The main objective of the fifth generation (5G) wireless communication systems is to enable a fully mobile and connected community; this requires the support of a variety of use cases with diverse requirements. To achieve this goal, three main communication scenarios with extremely diverse requirements have been categorized, i.e. enhanced-mobile broadband (eMBB), ultra reliable and low latency communications (URLLC), and massive machine type communications (mMTC) [1, 2]. eMBB is expected to support much higher downlink/uplink peak data rates (up to 100 times) than the long-term evolution (LTE) system, which requires a flexible physical layer configuration, e.g. a large subcarrier spacing (SCS) to counteract the Doppler spread of the fast fading channel, and a small SCS for higher robustness to the long delay spread channel. URLLC caters to applications/services where extremely high reliability and low end-to-end delays are required, e.g. vehicle to vehicle (V2V) services; hence, the SCS needs to be large enough to meet stringent latency requirements. On the other hand, mMTC is designed to support a connection density of up to 10^6 devices/km^2; thus, delay-tolerant devices with long symbol duration (and hence small SCS) is an important precondition [1].

Considering the high cost and deployment complexity, it is not feasible to have separate radio designs for different services. Therefore, mixed numerology, wherein services with different physical signaling formats operate on a common underlying physical layer, is proposed as one of the most promising techniques to support the diverse requirements in 5G. Besides, with mixed numerology, inter-numerology interference (INI) will be generated because different symbols and SCS for different services will adversely affect the system orthogonality [3]. Hence, a key challenge for designing a mixed numerology system is to minimize the performance degradation caused by INI. In comparison with adding a guard band (GB) between services (which will reduce the spectrum utilization efficiency),

Radio Access Network Slicing and Virtualization for 5G Vertical Industries, First Edition.
Edited by Lei Zhang, Arman Farhang, Gang Feng, and Oluwakayode Onireti.

a more practical and attractive way is to design new waveforms with lower out of band emission (OoBE) [4].

A variety of new waveforms have been proposed for the 5G network, such as filter-band multicarrier (FBMC) [5, 6], generalized frequency division multiplexing (GFDM) [7], universal filtered multicarrier (UFMC)[1] [8], filtered orthogonal frequency division multiplexing (F-OFDM) [9, 10], and windowed OFDM (W-OFDM) [11]. All of these waveforms have their advantages and drawbacks, e.g. FBMC applies subcarrier level pulse shaping to achieve the trade-off between time and frequency domain localization. This technique provides the best OoBE among the above waveforms. However, due to the interference between its real and imaginary branches, channel estimation/equalization needs significantly higher computational complexity than OFDM [5]. In addition, it is very hard to combine FBMC with multiple-input multiple-output (MIMO) systems [12]. Similar to FBMC, GFDM also utilizes subcarrier level pulse shaping but in a block-based manner, so that inter-burst tails are avoided [13]. Nevertheless, block-based processing increases the decoding latency of the GFDM system; hence, it may not be an appropriate choice for URLLC services. Moreover, FBMC and GFDM both require different transceiver structures from traditional OFDM systems. This will further reduce the competitiveness of these two waveforms.

Compared with FBMC and GFDM, the filtering/windowing based waveforms, i.e. F-OFDM, UFMC, and W-OFDM, have drawn much more attention from both academia and industry. F-OFDM and UFMC implement subband filtering to suppress the spectrum leakage, while W-OFDM utilizes windowing to smooth the transmission between adjacent symbols and will reduce the OoBE generated from the discontinuous transitions [4]. In W-OFDM based systems, a simple windowing is required for spectral confinement, and it has negligible peak-to-average power ratio (PAPR) overhead compared with the filtering-type techniques [14, 15]. Although W-OFDM cannot reduce the OoBE as much as F-OFDM, it retains the core structure of the traditional OFDM receiver, and the windowing procedure between transmitter and receiver can be transparent to each other, and is aligned with the 3GPP recommendations about 5G waveform candidates [16]. Furthermore, W-OFDM can provide a high robustness against timing and carrier frequency offset in the asynchronous systems [17]. A comparison of the power spectral density (PSD) of the transmit signals of the aforementioned waveforms is illustrated in Figure 3.1. It can be observed that all the new waveforms have much lower OoBE compared with OFDM.

Owing to the aforementioned advantages, W-OFDM has become one of the most discussed candidate waveforms for the physical layer of 5G systems. Great efforts have been made to improve the performance of the W-OFDM system. For instance, a joint windowing scheme is proposed in [18] to eliminate adjacent channel interference (ACI) and reduce OoBE. In [19], a time-asymmetric windowing scheme is proposed to eliminate the inter-symbol interference (ISI) that is introduced due to the insufficient cyclic prefix (CP) length. Nevertheless, analytical investigations about the INI in W-OFDM are missing in the literature. In this chapter, we will introduce our research about W-OFDM [11] in an analytical way. Specifically, W-OFDM system model for single numerology and mixed-numerology scenarios are derived in Sections 3.2.1 and 3.2.2, respectively. Section 3.3 gives the detailed INI models, and the power of INI is investigated analytically

1 Also known as universal filtered OFDM (UF-OFDM).

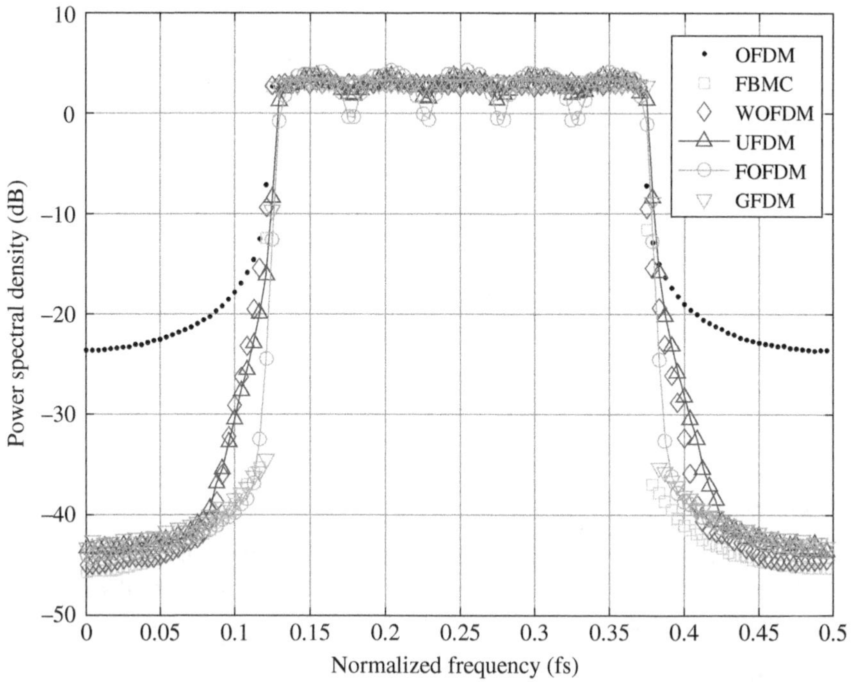

Figure 3.1 Power spectral density comparison of the candidate 5G waveforms.

for different numerologies. Simulation results are provided in Section 3.4 to verify the effectiveness of the derivations and analysis, and Section 3.5 concludes the chapter.

The notations used in this chapter are illustrated as follows. $\{\cdot\}^H$ and $\{\cdot\}^T$ stand for the conjugate transpose and transpose operation, respectively. diag$\{\mathbf{a}\}$ denotes the forming of a diagonal matrix $\mathbf{A}$ based on vector $\mathbf{a}$. Blkdiag($\mathbf{A}, N$) is a block diagonal matrix generated by repeating $\mathbf{A}$ matrix N times. $\lfloor a \rfloor$ denotes the largest integer smaller than or equal to a. $\bar{g(t)}$ denotes the conjugate of the function $g(t)$. $\mathbf{1}_N$ is an $N \times 1$ vector with all its elements being 1. $\mathbf{A}(i,j)$ refers to the i-th row and j-th column of the matrix $\mathbf{A}$. The operator $*$ denotes the linear convolution of two vectors, and $\otimes$ represents Kronecker product.

3.2 W-OFDM System Model

In a W-OFDM system, time-domain windowing operation is performed at both the transmitter and the receiver. Instead of the rectangular window used in the traditional OFDM system, W-OFDM implements windows with soft edges on both sides, which results in much sharper side-lope decay in its frequency domain. Figure 3.2 illustrates the time-domain window processing in a W-OFDM system. It can be seen that although the edges of the window further expand each symbol, the overhead of the signal is not increased because of the transmit roll-off portion (or edge transition region). Unless otherwise stated, the raised cosine (RC) window suggested by Bala et al. [18] will be considered for both transmitter and receiver in this section.

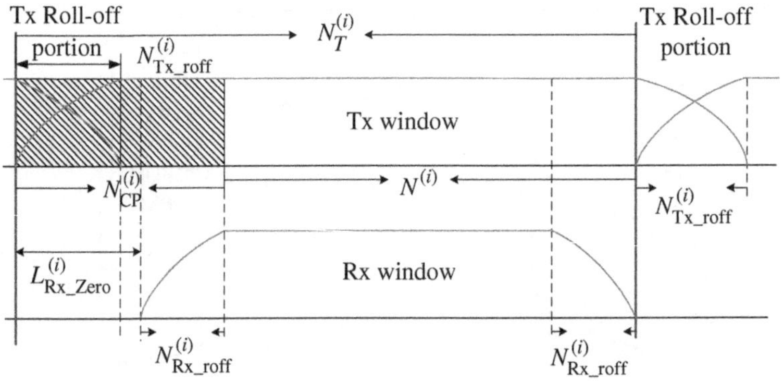

Figure 3.2 Time-domain window processing in transmitter and receiver.

3.2.1 Single Numerology System Model

Let us establish a single numerology W-OFDM system first. Assume that the user n is allocated with $Z^{(i)}$ continuous subcarriers in subband i. Then we can write the discrete-time baseband W-OFDM signal as

$$s^{(i)}(n) = \sum_{u=0}^{Z^{(i)}-1} \sum_{v=-\infty}^{\infty} x_{u,v}^{(i)} g_{u,v}^{(i)}(n), \tag{3.1}$$

where $x_{u,v}^{(i)}$ denotes the v-th independent identically distributed (i.i.d) complex modulation W-OFDM symbol on the u-th subcarrier. In addition, it is assumed that $x_{u,v}^{(i)}$ has zero mean and unit power. Similarly, $g_{u,v}^{(i)}(n)$ is defined as the synthesis function for the u-th subcarrier of the v-th symbol, and can be written as

$$g_{u,v}^{(i)}(n) = \frac{1}{\sqrt{N^{(i)}}} g_u^{(i)}(n - vN_T^{(i)}) e^{j\frac{2\pi}{N^{(i)}}(u+\Delta F^{(i)})(n-N_{\text{CP}}^{(i)}-vN_T^{(i)})}, \tag{3.2}$$

where $N_T^{(i)} = N_{\text{CP}}^{(i)} + N^{(i)}$ is the transmitted symbol duration by taking the CP length $N_{\text{CP}}^{(i)}$ into account, and $\Delta F^{(i)}$ denotes the frequency offset of the subband. Let us define $g_u^{(i)}(n)$ for $0 \leq n < N_w^{(i)}$ as the transmit window (as can be seen in Figure 3.2), and assume that the window has a unit response in the middle $N_T^{(i)} - N_{\text{Tx\_roff}}^{(i)}$ samples and smooth roll-off portions at the first and last $N_{\text{Tx\_roff}}^{(i)}$ samples. Thus, the first and last $N_{\text{Tx\_roff}}^{(i)}$ samples of the transmit symbol will be weighted by the roll-off portions of the window [20], which results in the smooth transitions between adjacent symbols, thus reducing the OoBE. Note that if we set $N_{\text{Tx\_roff}}^{(i)} = 0$, the W-OFDM system degenerates to the normal OFDM system.

Assume that the time invariant channel for subband i has length $L_{\text{CH}}^{(i)}$, and the CIR is $\mathbf{h}^{(i)} = [h^{(i)}(0), \ldots, h^{(i)}(L_{\text{CH}}^{(i)} - 1)]^T$. The received signal can be written as

$$y^{(i)}(n) = \sum_{l=0}^{L_{\text{CH}}^{(i)}-1} h^{(i)}(l) s^{(i)}(n-l) + \tilde{w}^{(i)}(n), \tag{3.3}$$

where $\tilde{w}^{(i)}(n)$ is the additive white Gaussian noise (AWGN). By defining the analysis function as

$$q_{k,m}^{(i)}(n) = \frac{1}{\sqrt{N^{(i)}}} q_k^{(i)}(n - mN_T^{(i)}) e^{j\frac{2\pi}{N^{(i)}}(k+\Delta F^{(i)})(n-N_{\mathrm{CP}}^{(i)}-mN_T^{(i)})}, \tag{3.4}$$

the demodulated m-th symbol on subcarrier k can be written as

$$\hat{y}_{k,m}^{(i)} = \sum_n y^{(i)}(n) \overline{q_{k,m}^{(i)}}(n). \tag{3.5}$$

In (3.4), $q_k^{(i)}(n)$, $0 \le n < N_T^{(i)}$ denotes the receiver window, which is similar to the transmitter one, but with a shorter window length (as can be seen in Figure 3.2). Additionally, to maintain the system orthogonality, the first $L_{\mathrm{Rx\_Zero}}^{(i)}$ elements of $q_k^{(i)}(n)$ are set to zero; thus ISI can be avoided if the following condition is met:

$$L_{\mathrm{Rx\_Zero}}^{(i)} - N_{\mathrm{Tx\_roff}}^{(i)} \ge L_{\mathrm{CH}}^{(i)} - 1. \tag{3.6}$$

Assume that (3.6) is satisfied and the receiver window has vestigial symmetry, i.e.

$$q_k^{(i)}(n) + q_k^{(i)}(n + N^{(i)}) = 1, 0 \le n \le N_{\mathrm{CP}}^{(i)} - 1, \tag{3.7}$$

we have

$$\frac{1}{N^{(i)}} \sum_{n=0}^{N_T^{(i)}-1} q_k^{(i)}(n) e^{-j\frac{2\pi}{N^{(i)}}(k-u)(n-N_{\mathrm{CP}}^{(i)})} = \delta(k-u), \tag{3.8}$$

where $\delta(\cdot)$ denotes the Dirac delta function. By substituting (3.1) and (3.3) into (3.5), (3.5) can be simplified as (derivation is provided in Appendix A):

$$\hat{y}_{k,m}^{(i)} = x_{k,m}^{(i)} H^{(i)}(k) + w^{(i)}(k), \tag{3.9}$$

where $H^{(i)}(k) = \sum_{l=0}^{L_{\mathrm{CH}}^{(i)}-1} h^{(i)}(l) e^{-j\frac{2\pi}{N^{(i)}}(k+\Delta F^{(i)})(l)}$ is the channel frequency response (CFR) of the k-th subcarrier in subband i, and $w^{(i)}(k)$ is the corresponding AWGN with the variance $\sigma_{(i)}^2(k)$. Moreover, let us assume that the channel is flat for each subcarrier, such that there is no ISI and inter-channel interference (ICI) in the underlying W-OFDM systems; we can then focus on the INI investigation.

The above signal modeling in (3.1)–(3.9) can be reformulated in matrix form. We can express the information symbols as an $N_s^{(i)} Z^{(i)} \times 1$ vector:

$$\begin{aligned} \mathbf{x}_{N_s^{(i)}}^{(i)} = [&x_{0,0}^{(i)}, x_{1,0}^{(i)}, \dots, x_{Z^{(i)}-1,0}^{(i)}, x_{0,1}^{(i)}, \dots, x_{Z^{(i)}-1,1}^{(i)}, \\ &x_{0,N_s^{(i)}-1}^{(i)}, \dots, x_{Z^{(i)}-1,N_s^{(i)}-1}^{(i)}]^T \end{aligned}. \tag{3.10}$$

The received signal $\mathbf{y}_{N_s^{(i)}}^{(i)} \in \mathbb{C}^{N_s^{(i)} N_T^{(i)} \times 1}$ can be written as

$$\mathbf{y}_{N_s^{(i)}}^{(i)} = \mathbf{\Psi}_{N_s^{(i)}}^{(i)} \mathbf{G}_{N_s^{(i)}}^{(i)} \mathbf{x}_{N_s^{(i)}}^{(i)} + \tilde{w}^{(i)}, \tag{3.11}$$

with

$$\begin{aligned} \mathbf{G}_{N_s^{(i)}}^{(i)} = [&\mathbf{g}_{0,0}^{(i)}\ \mathbf{g}_{1,0}^{(i)}, \dots, \mathbf{g}_{Z^{(i)}-1,0}^{(i)}, \mathbf{g}_{0,1}^{(i)}\ \mathbf{g}_{1,1}^{(i)}, \dots, \\ &\mathbf{g}_{0,N_s^{(i)}-1}^{(i)}, \dots, \mathbf{g}_{Z^{(i)}-1,N_s^{(i)}-1}^{(i)}], \end{aligned} \tag{3.12}$$

where $\mathbf{G}^{(i)}_{N_s^{(i)}}$ is the $L^{(i)} \times N_s^{(i)} Z^{(i)}$ transmit matrix and $L^{(i)} = N_s^{(i)} N_T^{(i)} + N^{(i)}_{\text{Tx\_roff}}$ is the total signal length. In addition, $\mathbf{g}^{(i)}_{u,v}$, for $0 \leq u < Z^{(i)}$ and $0 \leq v < N_s^{(i)}$, is an $L^{(i)} \times 1$ vector with its n-th element being $\{g^{(i)}_{u,v}(n)\}_{n=0}^{(L^{(i)}-1)}$. The $L^{(i)}$-dimensional Toeplitz matrix of the channel $\mathbf{h}^{(i)}$ is constructed with $[h^{(i)}(0), \ldots, h^{(i)}(L^{(i)}_{\text{CH}} - 1), \mathbf{0}_{1\times(N_s^{(i)} N_T^{(i)} - L^{(i)}_{\text{CH}})}]T \in \mathbb{C}^{N_s^{(i)} N_T^{(i)} \times 1}$ as the first row and $[h^{(i)}(0),\ \mathbf{0}_{1\times(L^{(i)}-1)}] \in \mathbb{C}^{1\times L^{(i)}}$ as the first column. Since the last $N^{(i)}_{\text{Tx\_roff}}$ samples of the transmitted symbols are ignored, the equivalent channel matrix $\mathbf{\Psi}^{(i)}_{N_s^{(i)}} \in \mathbb{C}^{N_s^{(i)} N_T^{(i)} \times L^{(i)}}$ can be constructed by taking the first $N_s^{(i)} N_T^{(i)}$ rows of the aforementioned Toeplitz matrix. The demodulated signal can be written as

$$\mathbf{y}^{(i)}_{N_s^{(i)}} = \mathbf{Q}^{(i)}_{N_s^{(i)}} \mathbf{y}^{(i)}_{N_s^{(i)}} + \mathbf{w}^{(i)}, \tag{3.13}$$

where $\hat{\mathbf{y}}^{(i)}_{N_s^{(i)}} = [\hat{y}^{(i)}_{0,0}, \hat{y}^{(i)}_{1,0}, \ldots, \hat{y}^{(i)}_{Z^{(i)}-1,0}, \hat{y}_0, 1(i), \ldots, \hat{y}_{Z(i)} - 1, 1(i), \hat{y}^{(i)}_{0,N_s^{(i)}-1}, \ldots, \hat{y}^{(i)}_{Z^{(i)}-1,N_s^{(i)}-1}]^T$ is an $N_s^{(i)} Z^{(i)} \times 1$ demodulated signal and $\mathbf{w}^{(i)} \in \mathbb{C}^{N_s^{(i)} Z^{(i)} \times 1}$ is the corresponding Gaussian noise vector. The receiver window $\mathbf{Q}^{(i)}_{N_s^{(i)}} \in \mathbb{C}^{N_s^{(i)} Z^{(i)} \times N_s^{(i)} N_T^{(i)}}$ is defined by

$$\begin{aligned}\mathbf{Q}^{(i)}_{N_s^{(i)}} = [&\mathbf{q}^{(i)}_{0,0},\ \mathbf{q}^{(i)}_{1,0}, \ldots, \mathbf{q}^{(i)}_{Z^{(i)}-1,0}, \mathbf{q}^{(i)}_{0,1},\ \mathbf{q}^{(i)}_{1,1}, \ldots, \\ &\mathbf{q}^{(i)}_{0,N_s^{(i)}-1}, \ldots, \mathbf{q}^{(i)}_{Z^{(i)}-1,N_s^{(i)}-1}\quad]^T \quad ,\end{aligned} \tag{3.14}$$

where $\mathbf{q}^{(i)}_{k,m} \in \mathbb{C}^{(N_s^{(i)} N_T^{(i)})\times 1}$, for $0 \leq k < Z^{(i)}$ and $0 \leq m < N_s^{(i)}$, is the demodulation vector with its n-th element equal to $\{q^{(i)}_{k,m}(n)\}_{n=0}^{(N_s^{(i)} N_T^{(i)}-1)}$. According to (3.9), (3.13) can be simplified to

$$\mathbf{y}^{(i)}_{N_s^{(i)}} = \mathbf{H}^{(i)}_{N_s^{(i)}} \mathbf{x}^{(i)}_{N_s^{(i)}} + \mathbf{w}^{(i)}, \tag{3.15}$$

where $\mathbf{H}^{(i)}_{N_s^{(i)}} = \text{Blkdiag}(\mathbf{H}^{(i)}, N_s^{(i)})$ and $\mathbf{H}^{(i)} = \text{diag}\{[H^{(i)}(0), \ldots, H^{(i)}(Z^{(i)} - 1)]^T\}$ is an $N_s^{(i)} Z^{(i)}$ dimensional CFR matrix. The noise variance matrix is $\mathbf{\Gamma}^{(i)}_{N_s^{(i)}} = \text{Blkdiag}(\mathbf{\Gamma}^{(i)}, N_s^{(i)})$ and $\mathbf{\Gamma}^{(i)} = \text{diag}\{[\sigma^2_{(i)}(0), \ldots, \sigma^2_{(i)}(Z^{(i)} - 1)]^T\}$.

3.2.2 System Model for Mixed Numerologies

When it comes to mixed numerologies transmission, the system bandwidth is divided into several subbands where each of them is assigned with a different set of numerologies. A commonly accepted relationship between the SCS and/or CP length for different numerologies [21–23] can be defined as

$$\Delta f^{(i)} = n_i \Delta f^{(i-1)}, N^{(i)}_{\text{CP}} = N^{(i-1)}_{\text{CP}} / n_i, n_i = 2^{\kappa}, \tag{3.16}$$

where $\Delta f^{(i)}$ is the SCS for the subband i, and κ is an integer. Figure 3.3 depicts a dual numerologies system with a GB, F_{GB}, between them. Without loss of generality, we assume that the two subbands have the same sampling rate. Let us consider that the W-OFDM system is in a generalized synchronized scenario [4], where an integral least common multiplier (LCM) symbol consists of $N_T^{(1)}$ samples. In addition, the subframe structures of different numerologies in the system are aligned; thus, we have

$$N_T^{(1)} = N^{(1)} + N^{(1)}_{\text{CP}} = n_2(N^{(2)} + N^{(2)}_{\text{CP}}) = n_2 N_T^{(2)}. \tag{3.17}$$

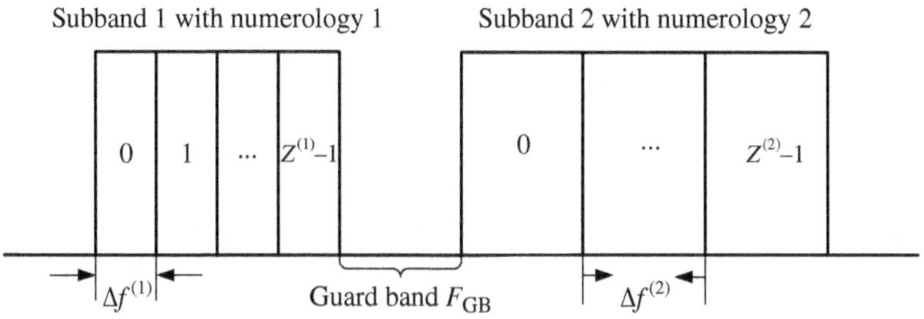

Figure 3.3 An example of mixed numerologies transmission with two subbands.

Using (3.3), the received signal can be written as

$$\begin{aligned} y(n) &= y^{(1)}(n) + y^{(2)}(n) \\ &= \rho^{(1)} \sum_{l=0}^{L_{\mathrm{CH}}^{(1)}-1} h^{(1)}(l) s^{(1)}(n-l) \\ &\quad + \rho^{(2)} \sum_{l=0}^{L_{\mathrm{CH}}^{(2)}-1} h^{(2)}(l) s^{(2)}(n-l) + \tilde{w}(n), \end{aligned} \tag{3.18}$$

where $\rho^{(i)}$ is the factor for power adjustment in subband i, and $\tilde{w}(n)$ is the white noise. If the two subband signals have the same transmit power, then we have

$$\rho^{(i)} = \sqrt{\frac{N^{(i)} L^{(i)}}{N_w^{(i)} N_s^{(i)} \sum_{u=0}^{Z^{(i)}-1} P_{g_u}^{(i)}}}, \tag{3.19}$$

where $P_{g_u}^{(i)}$ is the average power of the transmit window $g_u^{(i)}(n)$. Considering similar steps as (3.1)–(3.9), the demodulated m-th W-OFDM symbol on subcarrier k can be written as

$$\begin{aligned} \hat{y}_{k,m}^{(i)} &= \sum_n y(n) q_{k,m}^{(i)^{-}}(n) \\ &= \rho^{(i)} x_{k,m}^{(i)} H^{(i)}(k) + \rho^{(j)} \sum_{u=0}^{Z^{(j)}-1} \sum_{v=-\infty}^{\infty} x_{u,v}^{(j)} \sum_{l=0}^{L_{\mathrm{CH}}^{(j)}-1} h^{(j)}(l) \\ &\quad \times \sum_n g_{u,v}^{(j)}(n-l) q_{k,m}^{(i)^{-}}(n) + w^{(i)}(k). \end{aligned} \tag{3.20}$$

The three terms in (3.20) are the desired signal, INI caused by numerology j ($j \neq i$), and AWGN, respectively.

By setting $N_s^{(1)} = 1$, $N_s^{(2)} = n_2$, the received mixed signal during one LCM symbol can be written as

$$\mathbf{y} = \rho^{(1)} \mathbf{\Psi}_1^{(1)} \mathbf{G}_1^{(1)} \mathbf{x}_1^{(1)} + \rho^{(2)} \mathbf{\Psi}_{n_2}^{(2)} \mathbf{G}_{n_2}^{(2)} \mathbf{x}_{n_2}^{(2)} + \mathbf{w}, \tag{3.21}$$

where $\mathbf{\Psi}_1^{(1)} \in \mathbb{C}^{N_T^{(1)} \times L^{(1)}}$, $\mathbf{G}_1^{(1)} \in \mathbb{C}^{L^{(1)} \times Z^{(1)}}$, $\mathbf{x}_1^{(1)} \in \mathbb{C}^{Z^{(1)} \times 1}$ and $\mathbf{\Psi}_{n_2}^{(2)} \in \mathbb{C}^{n_2 N_T^{(2)} \times L^{(2)}}$, $\mathbf{G}_{n_2}^{(2)} \in \mathbb{C}^{L^{(2)} \times n_2 Z^{(2)}}$, $\mathbf{x}_{n_2}^{(2)} \in \mathbb{C}^{n_2 Z^{(2)} \times 1}$ are the channel matrices, transmit matrices, and information symbols of subband 1 and subband 2, respectively, and $\mathbf{w} \in \mathbb{C}^{N_T^{(1)} \times 1}$ denotes the Gaussian noise. Based on

(3.13), (3.14), and (3.15), the demodulated signal from numerology 1 can be written as

$$\mathbf{y}_1^{(1)} = \mathbf{Q}_1^{(1)}\mathbf{y} = \rho^{(1)}\mathbf{H}_1^{(1)}\mathbf{x}_1^{(1)} + \mathbf{E}_{(1,2)}\mathbf{x}_{n_2}^{(2)} + \mathbf{w}^{(1)}, \tag{3.22}$$

where $\rho^{(1)}\mathbf{H}_1^{(1)}\mathbf{x}_1^{(1)}$ is the desired signal with $\mathbf{H}_1^{(1)} \in \mathbb{C}^{Z^{(1)}\times Z^{(1)}}$ being the diagonal CFR matrix. Moreover, $\mathbf{E}_{(1,2)}\mathbf{x}_{n_2}^{(2)}$ denotes the INI that is caused by the system with numerology 2, and $\mathbf{E}_{(1,2)} = \rho^{(2)}\mathbf{Q}_1^{(1)}\mathbf{\Psi}_{n_2}^{(2)}\mathbf{G}_{n_2}^{(2)}$ is the $Z^{(1)} \times n_2Z^{(2)}$ interference matrix. $\mathbf{w}^{(1)}$ in (3.22) is the Gaussian noise term with the variance vector being $\gamma^{(1)} = [\sigma_{(1)}^2(0), \dots, \sigma_{(1)}^2(Z^{(1)}-1)]^T$. Similarly, the demodulated numerology 2 signal $\hat{y}_{n_2}^{(2)} \in \mathbb{C}^{n_2Z^{(2)}\times 1}$ can be written as

$$\mathbf{y}_{n_2}^{(2)} = \mathbf{Q}_{n_2}^{(2)}\mathbf{y} = \rho^{(2)}\mathbf{H}_{n_2}^{(2)}\mathbf{x}_{n_2}^{(2)} + \mathbf{E}_{(2,1)}\mathbf{x}_1^{(1)} + \mathbf{w}^{(2)}, \tag{3.23}$$

where $\mathbf{H}_{n_2}^{(2)} = \mathrm{BlkDiag}(\mathbf{H}_1^{(2)}, n_2)$ and $\mathbf{H}_1^{(2)} = \mathrm{diag}\{[H^{(2)}(0), \dots, H^{(2)}(Z^{(2)}-1)]^T\}$. $\mathbf{E}_{(2,1)}\mathbf{x}_1^{(1)}$ denotes the INI coming from subband 1, and $\mathbf{E}_{(2,1)} = \rho^{(1)}\mathbf{Q}_{n_2}^{(2)}\mathbf{\Psi}_1^{(1)}\mathbf{G}_1^{(1)}$ is the $n_2Z^{(2)} \times Z^{(1)}$ interference matrix. The variance vector for noise $\mathbf{w}^{(2)}$ is $\gamma_{n_2}^{(2)} = \mathbf{1}_{n_2} \otimes \gamma^{(2)}$ with $\gamma^{(2)} = [\sigma_{(2)}^2(0), \dots, \sigma_{(2)}^2(Z^{(2)}-1)]^T$.

3.3 Inter-numerology Interference Analysis

In this section, we will discuss the INI generated between two adjacent subbands, wherein different numerologies are allocated.

3.3.1 Inter-numerology Interference Analysis for Numerology 1

By setting $i = 1$ and $j = 2$, the INI suffered by the m-th W-OFDM symbol in subcarrier k of numerology 1, i.e. the interference term in (3.20), can be written as

$$I^{(1)}(k,m) = \sum_{u=0}^{Z^{(2)}-1} I_u^{(1)}(k,m), \tag{3.24}$$

where

$$I_u^{(1)}(k,m) = \rho^{(2)} \sum_{v=-\infty}^{\infty} x_{u,v}^{(2)} \sum_{l=0}^{L_{\mathrm{CH}}^{(2)}-1} h^{(2)}(l) \sum_n g_{u,v}^{(2)}(n-l)\overline{q_{k,m}^{(1)}}(n). \tag{3.25}$$

The term $I_u^{(1)}(k,m)$ in (3.25) is the INI caused by the u-th subcarrier of numerology 2. As we assumed the transmit signal is i.i.d, the power of $I_u^{(1)}(k,m)$ can be calculated as

$$P_u^{(1)}(k,m) = \sum_{v=-\infty}^{\infty} \left| \rho^{(2)} \sum_{l=0}^{L_{\mathrm{CH}}^{(2)}-1} h^{(2)}(l) \sum_n g_{u,v}^{(2)}(n-l)\overline{q_{k,m}^{(1)}}(n) \right|^2 . \tag{3.26}$$

Accordingly, we can calculate the power of $I^{(1)}(k,m)$ as

$$P^{(1)}(k,m) = \sum_{u=0}^{Z^{(2)}-1} P_u^{(1)}(k,m). \tag{3.27}$$

Without loss of generality, we set $\Delta F^{(1)} = 0$ and assume that the channel delay spread is smaller than the symbol duration. By substituting (3.2) and (3.4) into (3.26), we can have the following expression (derivations can be found in Appendix B):

$$P_u^{(1)}(k,m) \approx \left(\frac{\rho^{(2)}}{\sqrt{N^{(1)}N^{(2)}}}\right)^2 |H^{(2)}(u)|^2 \times \sum_{d=-1}^{n_2-1} \left| \sum_n \Omega_{g_u^{(2)},dN_T^{(2)}}^{q_k^{(1)},0}(n) e^{-j\frac{2\pi}{N^{(1)}}\Delta k^{(1)} n} \right|^2, \tag{3.28}$$

where $\Omega_{g,\zeta}^{q,\tau}(n)\Delta = q(n-\tau)g(n-\zeta)$ denotes the equivalent window response taking both numerology 1 transmitter window, i.e. $g(n-\zeta)$, and numerology 2 receiver window, i.e. $q(n-\tau)$, into account. Moreover, $\Delta k^{(1)} = k - \hat{k}(u)$ is the spectral distance between subcarrier k and $\hat{k}(u)$, where $\hat{k}(u) = n_2(u + \Delta F^{(2)})$, for $0 \le u < Z^{(2)}$, represents the subcarrier indexes of numerology 2 when its subcarriers are observed with a granularity of $\Delta f^{(1)}$. Furthermore, $H^{(2)}(u) = \sum_{l=0}^{L_{CH}^{(2)}-1} h^{(2)}(l) e^{-j\frac{2\pi}{N^{(2)}}(u+\Delta F^{(2)})l}$ is the CFR on the u-th subcarrier of numerology 2. From (3.28), we can find that the INI power is a function of the overlapping window, the spectral distance between the victim and interfering subcarrier, and the CFR of the interfering subband.

In Figure 3.4, we depict the overlapping windows in the multi numerologies W-OFDM system, where n_2 is set to be 2. By taking into account the transmitter window of the previous LCM symbol for numerology 2, i.e. $g_u^{(2)}[n-(n_2m-1)N_T^{(2)}]$, there are n_2+1 transmitter windows that may overlap with the receiver window of numerology 1. It can be observed that when the two subbands have the same numerology, the overlapping window is equivalent to the numerology 1 receiver window, and can satisfy the condition in (3.8). In this case, there is no INI generated. In addition, as indicated by (3.28), the value of $P_u^{(1)}(k,m)$ is not related to the symbol index m under generalized synchronization scenario; thus, $P_u^{(1)}(k)$ is used instead of $P_u^{(1)}(k,m)$ in the following discussions.

As a special case of W-OFDM, the overlapping window of the CP-OFDM system is shown in Figure 3.5, where $g_u^{(2)}(n)$ and $q_k^{(1)}(n)$ are changed to rectangular windows.

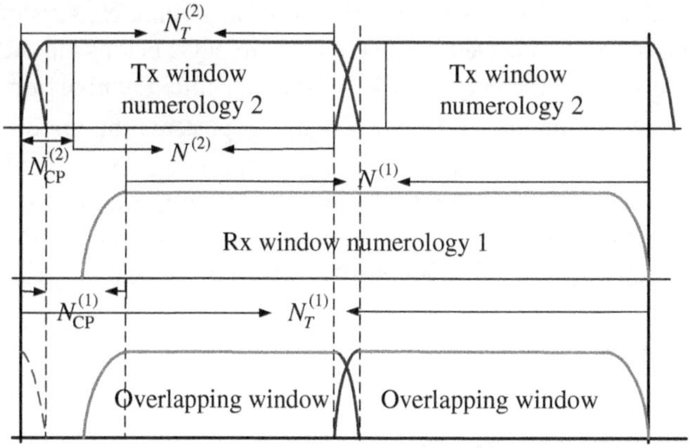

Figure 3.4 INI analysis for subband when using numerology 1.

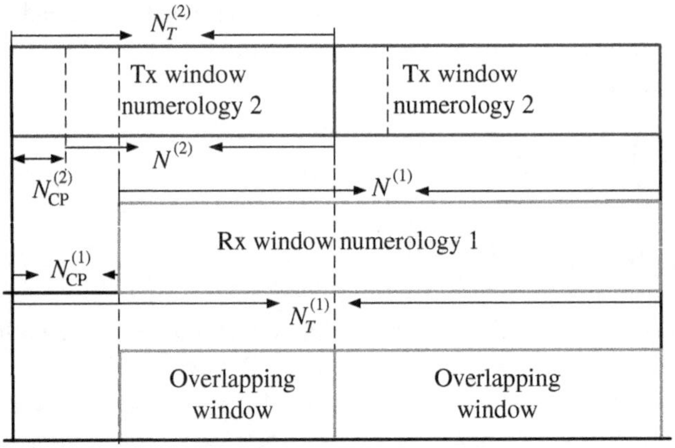

Figure 3.5 INI analysis for numerology 1 in the CP-OFDM case.

The difference between W-OFDM and CP-OFDM can be observed by comparing Figure 3.4 and 3.5, i.e. overlapping windows of CP-OFDM are rectangular windows with sharp transitions, while in W-OFDM system, the overlapping windows have smooth transition edges. As the level of INI depends on the discrete Fourier transform (DFT) magnitude of overlapping windows, the smooth time-domain transitions between symbols result in faster decay of the spectrum magnitude and hence a reduced INI.

3.3.2 Inter-numerology Interference Analysis for Numerology 2

By performing similar derivations as we did for numerology 1, the INI power level of numerology 2 can be expressed as

$$P_u^{(2)}(k,m) = \sum_{v=-\infty}^{\infty} \left| \rho^{(1)} \sum_{l=0}^{L_{CH}^{(1)}-1} h^{(1)}(l) \sum_n g_{u,v}^{(1)}(n-l) q_{k,m}^{(2)^-}(n) \right|^2. \tag{3.29}$$

Consider that in the generalized synchronous scenario [4], one LCM symbol consists of $N_T^{(1)}$ samples. If we assume that the m-th received symbol of numerology 2 belongs to the $\tilde{m}$-th LCM symbol, it may be interfered by the $\tilde{m}$-th and $(\tilde{m}-1)$-th transmitted symbol from numerology 1, where $\tilde{m} = \left\lfloor \frac{m}{n_2} \right\rfloor$. We define $m = \tilde{m}n_2 + d$, where d represents the relative position index for the m-th receiver window of numerology 2 within one LCM symbol, and $d = 0, 1, \ldots, n_2 - 1$. Again, we consider a channel with a smaller delay spread than the symbol duration. By substituting (3.2) and (3.4) into (3.29), the power of component INI can be calculated as (derivations are provided in Appendix C)

$$P_u^{(2)}(k,m) \approx \left(\frac{\rho^{(1)}}{\sqrt{N^{(2)}N^{(1)}}} \right)^2 |H^{(1)}(u)|^2 \times \sum_{j=0}^{1} \left| \sum_n \Omega_{g_u^{(1)},-jN_T^{(1)}}^{q_k^{(2)},dN_T^{(2)}}(n) e^{-j\frac{2\pi}{N^{(1)}}\Delta k^{(2)} n} \right|^2, \tag{3.30}$$

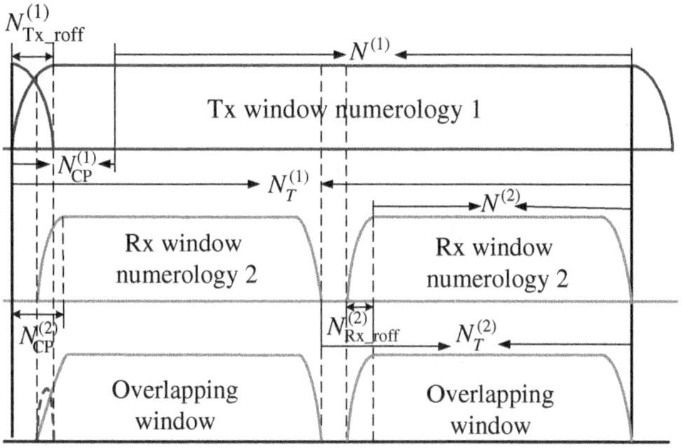

Figure 3.6 INI analysis for subband using numerology 2.

where $\Omega^{q_k^{(2)}, dN_T^{(2)}}_{g_u^{(1)}, -jN_T^{(1)}}(n)$ is the overlapping window generated by $g_u^{(1)}(n + jN_T^{(1)})$ and $q_k^{(2)}(n - dN_T^{(2)})$ (as illustrated in Figure 3.6). $\Delta k^{(2)} = \hat{\text{k}}(k) - u$ denotes the spectral distance between subcarrier $\hat{\text{k}}(k)$ and the u-th subcarrier of numerology 1, where $\hat{\text{k}}(k) = n_2(k + \Delta F^{(2)})$ represents the subcarrier indexes of numerology 2 when its k-th subcarrier is observed with a granularity of $\Delta f^{(1)}$. In addition, $H^{(1)}(u) = \sum_{l=0}^{L_{\text{CH}}^{(1)}-1} h^{(1)}(l) e^{-j\frac{2\pi}{N^{(1)}}ul}$ is the CFR for the u-th subcarrier of numerology 1. Note that the interference from the $(\tilde{m} - 1)$-th transmitted symbol of numerology 1 is considered, which is shown as the dashed part in Figure 3.6.

As shown in Figure 3.6, the first received numerology 2 symbol may be interfered by the numerology 1 transmitter window of the previous LCM symbol duration. Thus, for different receiver windows of numerology 2, the INI power within one LCM symbol might be slightly different. Hence, we define the average power of INI to measure the INI suffered by subcarrier k of numerology 2 as

$$P_u^{(2)}(k) = \frac{1}{n_2} \sum_{m=\tilde{m}n_2}^{\tilde{m}n_2+n_2-1} P_u^{(2)}(k, m). \tag{3.31}$$

Likewise, the overlapping window in the CP-OFDM system is illustrated in Figure 3.7 for comparison.

With (3.28) and (3.31), for the k-th subcarrier of numerology i, the component signal to interference ratio (SIR) caused by the u-th subcarrier of numerology j can be calculated as

$$R_u^{(i)}(k) = \frac{P_D^{(i)}(k)}{P_u^{(i)}(k)}, 0 \le k < Z^{(i)}, 0 \le u < Z^{(j)}, \tag{3.32}$$

where $P_D^{(i)}(k) = |\rho^{(i)} H^{(i)}(k)|^2$ is the power of the desired signal based on (3.20). Furthermore, the total SIR of numerology i can be written as

$$R_T^{(i)}(k) = \frac{P_D^{(i)}(k)}{P_I^{(i)}(k)}, 0 \le k < Z^{(i)}, \tag{3.33}$$

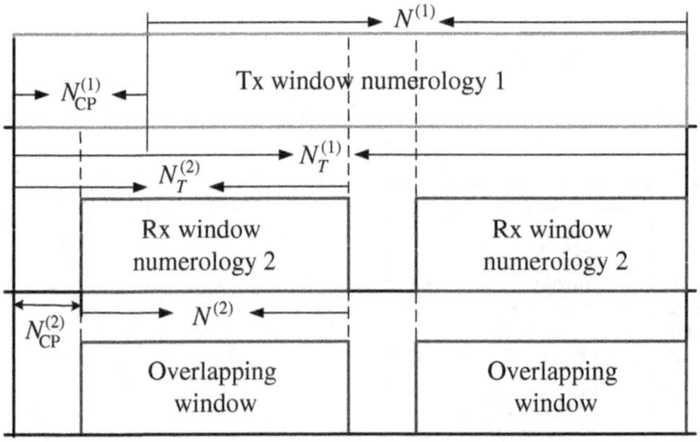

Figure 3.7 INI analysis for numerology 2 in CP-OFDM case.

where $P_I^{(i)}(k) = \sum_{u=0}^{Z^{(j)}-1} P_u^{(i)}(k)$ is the entire INI power generated from numerology j. The quantitative comparison of SIRs between CP-OFDM and W-OFDM will be provided in Section 3.4.

3.4 Numerical Results and Discussion

In this section, simulation results are provided to evaluate the effectiveness of the derived INI power for W-OFDM systems. In the simulations, we consider that two different numerologies are assigned to adjacent subbands as illustrated in Figure 3.3. The first numerology is with $\Delta f^{(1)} = 15\,\text{kHz}$ SCS and 1024-points DFT, while the second adopts $\Delta f^{(2)} = 30\,\text{kHz}$ with 512-points DFT (or $\Delta f^{(2)} = 60\,\text{kHz}$ with 256-points DFT). We define $\mu_{\text{CP}} = N_{\text{CP}}^{(i)}/N^{(i)}$, $\mu_{\text{Tx}}^{(i)} = N_{\text{Tx\_roff}}^{(i)}/N^{(i)}$, and $\mu_{\text{Rx}}^{(i)} = N_{\text{Rx\_roff}}^{(i)}/N^{(i)}$ $(i = 1, 2)$ as the normalized length of CP, and the roll-off lengths of the transmitter window and the receive window, respectively. Specifically, μ_{CP} is set to 7%, which means the CP length is 72 OFDM samples. As mentioned before, we apply RC windows in both the transmitter and the receiver.

In Figure 3.8, the simulated SIR and analytical SIR are compared with different SCS. We set the GB between the two numerologies as 0, $\Delta f^{(2)}$, and $2\Delta f^{(2)}$, respectively. Here, we assume $\Delta f^{(2)} = 30\,\text{kHz}$ and one resource block (RB) is allocated to each numerology. $\mu_{\text{Tx}}^{(i)}$ and $\mu_{\text{Rx}}^{(i)}$ are assumed to be 4% and 1%, respectively. In addition, quadrature phase shift keying (QPSK) modulation scheme is used for both numerologies, and the Pedestrian-A channel model [24] is considered.

As we can see in Figure 3.8, the theoretical and numerical SIR curves match perfectly, which verifies the effectiveness of the analysis in Section 3.2.2. It is also observed that larger GB results in better SIR for both numerologies. Generally, the subcarriers that have smaller spectral distance from the interfering numerology have lower SIR; this can be observed in Figure 3.8 as well; e.g. the subcarrier with index 13 (in subband 2) is the nearest one to subband 1 and has the lowest SIR.

In Figure 3.9, we compare the SIR for OFDM system and W-OFDM system. In this simulation, the GB is assumed to be 0 and both numerologies have 20 RBs. As can be seen from

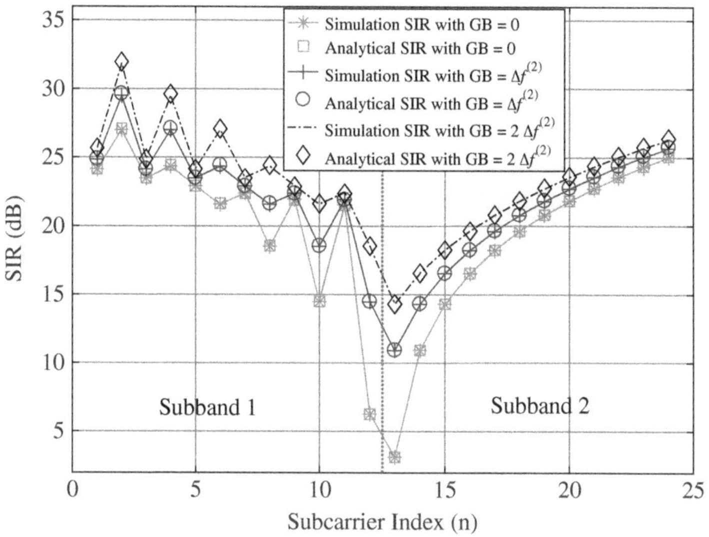

Figure 3.8 SIR for the adjacent subbands.

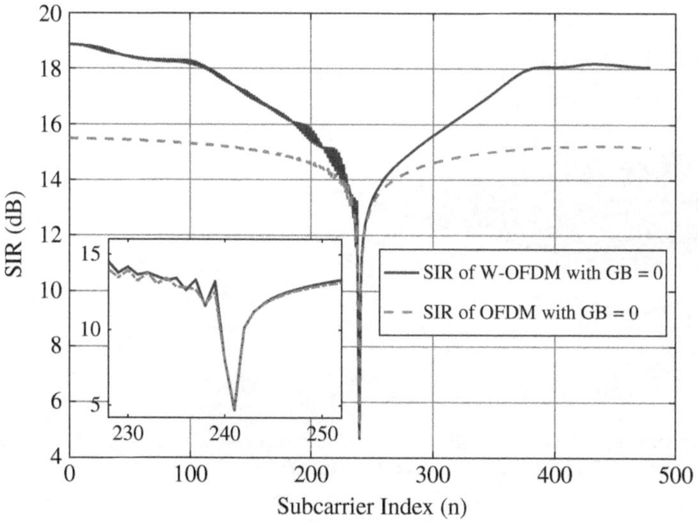

Figure 3.9 SIR comparison for W-OFDM and OFDM systems.

Figure 3.9, except the edge subcarriers between the two numerologies, the SIR of W-OFDM is significantly larger than that of OFDM. The reason is that the smooth transition edges of overlapping windows in W-OFDM system lead to more rapid decrease of INI compared to OFDM. However, we can find that the SIR in edge subcarriers of W-OFDM and OFDM is almost the same, which reveals that in small GB cases, the spectral distance is the dominant factor of the INI level compared with the roll-off length of the window.

It was mentioned in [18, 23] that a larger transmitter window roll-off portion offers better spectral containment, and a larger receiver window roll-off portion brings benefit

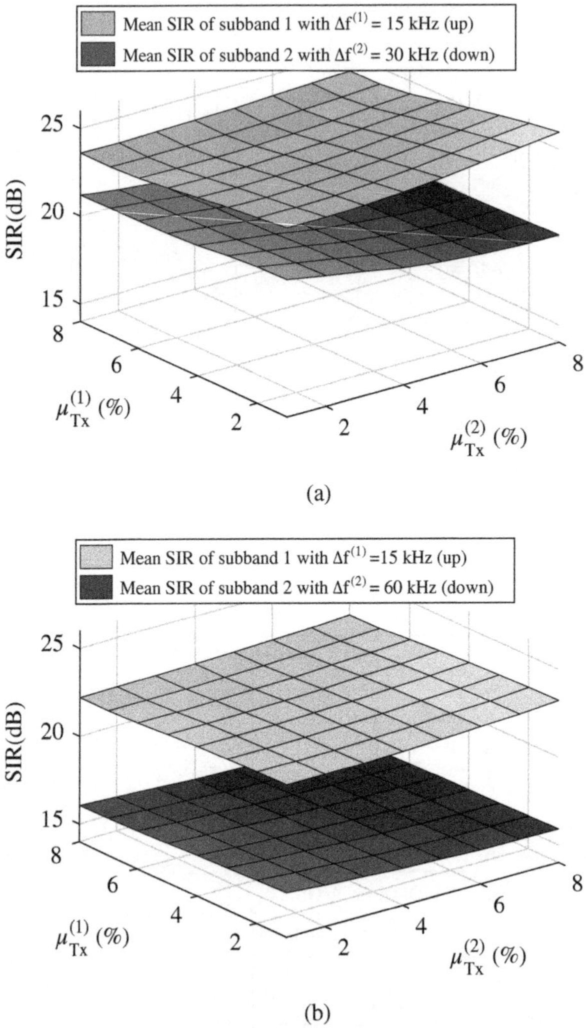

Figure 3.10 SIR for different transmit window roll-off lengths and receiver window roll-off lengths. (a) $\Delta f^{(1)} = 15$ kHz, $\Delta f^{(2)} = 30$ kHz and (b) $\Delta f^{(1)} = 15$ kHz, $\Delta f^{(2)} = 60$ kHz.

for reducing ACI. Nevertheless, to avoid ISI and keep a fixed CP overhead, increasing the roll-off length of a transmit window leads to a decrease in receiver window roll-off length. In Figure 3.10, we evaluate the average SIR of two numerologies with different values of $\mu_{\text{Tx}}^{(1)}$ and $\mu_{\text{Tx}}^{(2)}$, in order to find how Tx and Rx roll-off lengths will affect the INI level. In this study, GB = 60 kHz is considered, and $\mu_{\text{Tx}}^{(i)} + \mu_{\text{Rx}}^{(i)} = 7\%$ is set for numerology 1 and 2. In Figure 3.10a, we set $\Delta f^{(1)} = 15$ kHz, $\Delta f^{(2)} = 30$ kHz, while $\Delta f^{(1)} = 15$ kHz, $\Delta f^{(2)} = 60$ kHz are adopted in Figure 3.10b. As shown in Figure 3.10, both larger $\mu_{\text{Tx}}^{(i)}$ for interferer and larger $\mu_{\text{Rx}}^{(i)}$ for victim offer SIR gain for numerology 1. In Figure 3.10a, when $\mu_{\text{Tx}}^{(2)} = 0$ in subband 2, increasing $\mu_{\text{Rx}}^{(1)}$ from 0% to 7% (i.e. decreasing $\mu_{\text{Tx}}^{(1)}$ from 7%

to 0%) brings 1.2 dB SIR gain for numerology 1 whereas when $\mu_{\text{Rx}}^{(1)} = 0\%$ in subband 1, increasing $\mu_{\text{Tx}}^{(2)}$ from 0% to 7% leads to a 0.4 dB gain for the average SIR of numerology 1. A similar phenomenon can be observed in Figure 3.10b. However, no apparent SIR gain can be found for numerology 2 when $\mu_{\text{Tx}}^{(1)}$ is increasing, which is probably because of the two following factors. Firstly, although the increase of $\mu_{\text{Tx}}^{(1)}$ may decrease the sidelobes of numerology 1, it also generates larger overlapping of adjacent symbols, which leads to extra INI from previous LCM symbol for numerology 2 (as shown in Figure 3.6). Secondly, for numerology 2, the smooth transition edges of overlapping windows are mostly inherited from the roll-off parts of the victim's receive window.

3.5 Conclusions

To conclude, a promising low OoBE radiation waveform, i.e. W-OFDM, is discussed analytically in this chapter. Specifically, the INI model of W-OFDM system is investigated in the scenario of mixed numerologies transmission. Analytical expressions of the INI power are derived, which explains how the transmitter and receiver windows affect the INI level. Simulations are performed to verify the effectiveness of the derivations. According to the analysis in this chapter, it can be observed that the INI level in a mixed numerologies system depends on the spectral distance between the numerologies, the CFR of interferer, and the overlapping windows generated by the aggressor's transmitter windows and victim's receiver window. Compared to the normal OFDM, significant SIR gain can be achieved for the subcarriers with relatively large spectral distance by applying W-OFDM. The analytical expressions in this chapter can be easily extended to other OFDM-based systems and can be considered as the guideline for the design and waveform optimization of systems of 5G and beyond.

3.A Appendix

3.A.1 Derivation of (3.9)

Substitute (3.3) and (3.1) into (3.5) and assume that the transmitter window and the receiver window satisfy conditions (3.6) and (3.7); we have

$$
\begin{aligned}
\hat{y}_{k,m}^{(i)} &= \sum_{n} \sum_{l=0}^{L_{\text{CH}}^{(i)}-1} h^{(i)}(l) \sum_{u=0}^{Z^{(i)}-1} \sum_{v=-\infty}^{\infty} x_{u,v}^{(i)} g_{u,v}^{(i)}(n-l) \overline{q_{k,m}^{(i)}}(n) + w^{(i)}(k) \\
&= \frac{1}{N^{(i)}} \sum_{u=0}^{Z^{(i)}-1} x_{u,v}^{(i)} \sum_{l=0}^{L_{\text{CH}}^{(i)}-1} h^{(i)}(l) e^{-j\frac{2\pi}{N^{(i)}}(u+\Delta F^{(i)})l} \sum_{n} g_{u}^{(i)}(n-l) \\
&\quad \times q_{k}^{(i)}(n) e^{-j\frac{2\pi}{N^{(i)}}(k-u)(n-N_{\text{CP}}^{(i)})} + w^{(i)}(k) \\
&= \frac{1}{N^{(i)}} \sum_{u=0}^{Z^{(i)}-1} x_{u,m}^{(i)} H^{(i)}(u) \sum_{n'=0}^{N_{T}^{(i)}-1} q_{k}^{(i)}(n') e^{j\frac{2\pi}{N^{(i)}}(u-k)(n'-N_{\text{CP}}^{(i)})} \\
&\quad + w^{(i)}(k).
\end{aligned}
\tag{3.A.1}
$$

Then, (3.9) can be obtained by substituting (3.8) in (3.A.1). In addition, $\gamma_i(q, n, e, d)$ is defined as

$$\gamma_i(q,n,e,d) = \sum_{l_1=0}^{L_{\text{SYM}}-1} \sum_{l_2=0}^{l_1} \sum_{r=l_1-l_2}^{L_{\text{SYM}}-1} \beta_i(n,e,l_1,l_2)\eta_i(q,r,r-l_1+l_2)\lambda_i$$
$$+ \sum_{l_1=0}^{L_{\text{SYM}}-1} \sum_{l_2=l_1}^{L_{\text{SYM}}-1} \sum_{r=l_1-l_2}^{L_{\text{SYM}}-1-(l_2-l_1)} \beta_i(n,e,l_1,l_2)\eta_i(q,r,r-l_1+l_2)\lambda_i, \tag{3.A.2}$$

where $\lambda_i = R_i(r - l_1 - dL_{\text{SYM}} + \tau_k(i) + \mu_3 L_s)$, $\beta_i(n,e,l_1,l_2) = \alpha_i(n,e,l_1)\alpha_i(n,e,l_2)^*$with $\alpha_i(n,e,l) = Q_i(n,n)\sum_{m=0}^{\mu_1-1} e^{j\cdot\frac{2\pi(m-\mu_2)[(i-1)M+n]}{N}} a_i(l-m-eL_{\text{SYM}})$, and $\eta_i(q,r_1,r_2) = \theta_i(q,r_1)\theta_i(q,r_2)^*$ with $\theta_i(q,r) = \sum_{p=0}^{N-1} e^{-j\cdot 2\pi p[(i-1)M+q-1]/N} c_i(p - r + \mu_3(L_F - L_s 1) + \mu_2)$. For the ACI and ISI calculation, $\mu_1 = L_{\text{SYM}}$, $\mu_2 = L_{\text{CP}}$ and $\mu_3 = 1$, while $\mu_1 = L_O - L_{\text{CP}}$, $\mu_2 = L_O$, and $\mu_3 = 0$ for ICI calculation.

3.B Appendix

3.B.1 Derivations of (3.28)

Substitute (3.2) and (3.4) into (3.26) and define $\rho_N^{(2)} = \left(\frac{\rho^{(2)}}{\sqrt{N^{(1)}N^{(2)}}}\right)^2$; we have

$$P_u^{(1)}(k,m) = \rho_N^{(2)} \sum_{v=-\infty}^{\infty} \Bigg| \sum_{l=0}^{L_{\text{CH}}^{(2)}-1} h^{(2)}(l) e^{-j\frac{2\pi}{N^{(2)}}(u+\Delta F^{(2)})(l)}$$
$$\times \sum_n g_u^{(2)}(n - vN_T^{(2)} - l) q_k^{(1)}(n - mN_T^{(1)})$$
$$\times e^{j\frac{2\pi(u+\Delta F^{(2)})}{N^{(2)}}(n - N_{\text{CP}}^{(2)} - vN_T^{(2)})} e^{-j\frac{2\pi k}{N^{(1)}}(n - N_{\text{CP}}^{(1)} - mN_T^{(1)})} \Bigg|^2$$
$$\overset{(a)}{\approx} \rho_N^{(2)} \sum_{v=-\infty}^{\infty} \Bigg| H^{(2)}(u) \sum_n g_u^{(2)}(n - vN_T^{(2)}) q_k^{(1)}(n - mN_T^{(1)})$$
$$\times e^{-j\frac{2\pi}{N^{(1)}}[k - n_2(u+\Delta F^{(2)})](n)} \Bigg|^2$$
$$\overset{(b)}{=} \rho_N^{(2)} |H^{(2)}(u)|^2 \sum_{d=-1}^{n_2-1} \Bigg| \sum_n g_u^{(2)}(n - mN_T^{(1)} + N_T^{(2)})$$
$$\times q_k^{(1)}(n - mN_T^{(1)}) e^{-j\frac{2\pi}{N^{(1)}}[k - \text{k}(u)](n)} \Bigg|^2$$
$$= \rho_N^{(2)} |H^{(2)}(u)|^2 \sum_{d=-1}^{n_2-1} \Bigg| \sum_n g_u^{(2)}(n - dN_T^{(2)}) q_k^{(1)}(n) e^{-j\frac{2\pi}{N^{(1)}}\Delta k^{(1)} n} \Bigg|^2$$
$$= \rho_N^{(2)} |H^{(2)}(u)|^2 \sum_{d=-1}^{n_2-1} \Bigg| \sum_n \Omega_{g_u^{(2)}, dN_T^{(2)}}^{q_k^{(1)}, 0}(n) e^{-j\frac{2\pi}{N^{(1)}}\Delta k^{(1)} n} \Bigg|^2. \tag{3.B.1}$$

In (3.B.1), approximate equality (a) follows the assumption that the channel is weakly or mildly time-dispersive, and thus $g_u^{(2)}(n - vN_T^{(2)} - l) \approx g_u^{(2)}(n - vN_T^{(2)})$; equality (b) holds due to $n_2 N_T^{(2)} = N_T^{(1)}$ and only those transmit windows $g_k^{(2)}(n - vN_T^{(2)})$ with $v = mn_2 - 1, mn_2, mn_2 + 1, \ldots, mn_2 + n_2 - 1$, may possibly overlap with the given receive window $q_k^{(1)}(n - mN_T^{(1)})$ for numerology 1.

3.C Appendix

3.C.1 Derivations of (3.30)

Substitute (3.2) and (3.4) into (3.29) and define $\rho_N^{(1)} = \left(\frac{\rho^{(1)}}{\sqrt{N^{(1)}N^{(2)}}}\right)^2$; we have

$$
\begin{aligned}
P_u^{(2)}(k,m) &= \rho_N^{(1)} \sum_{v=-\infty}^{\infty} |\sum_n \sum_{l=0}^{L_{\rm CH}^{(1)}-1} h^{(1)}(l) g_u^{(1)}(n-l-vN_T^{(1)}) \\
&\quad \times e^{j\frac{2\pi}{N^{(1)}}u(n-l-N_{\rm CP}^{(1)}-vN_T^{(1)})} \\
&\quad \times q_k^{(2)}(n-mN_T^{(2)}) e^{-j\frac{2\pi}{N^{(2)}}(k+\Delta F^{(2)})(n-N_{\rm CP}^{(2)}-mN_T^{(2)})}|^2 \\
&\overset{(d)}{\approx} \rho_N^{(1)} \sum_{v=-\infty}^{\infty} | \sum_{l=0}^{L_{\rm CH}^{(1)}-1} h^{(1)}(l) e^{-j\frac{2\pi}{N^{(1)}}ul} \sum_n g_u^{(1)}(n-vN_T^{(1)}) \\
&\quad \times e^{j\frac{2\pi}{N^{(1)}}u(n-N_{\rm CP}^{(1)}-vN_T^{(1)})} \\
&\quad \times q_k^{(2)}(n-mN_T^{(2)}) e^{-j\frac{2\pi}{N^{(2)}}(k+\Delta F^{(2)})(n-N_{\rm CP}^{(2)}-mN_T^{(2)})}\Big|^2 \\
&\overset{(e)}{=} \rho_N^{(1)} |H^{(1)}(u)|^2 \sum_{j=0}^{1} \Big| \sum_n q_k^{(2)}(n-\tilde{m}n_2N_T^{(2)}-dN_T^{(2)}) \\
&\quad \times q_k^{(2)}(n-\tilde{m}n_2N_T^{(2)}-dN_T^{(2)}) e^{-j\frac{2\pi}{N^{(1)}}[n_2(k+\Delta F^{(2)})-u]n}\Big|^2 \\
&= \rho_N^{(1)} |H^{(1)}(u)|^2 \sum_{j=0}^{1} | \sum_n g_u^{(1)}(n-jN_T^{(1)}) q_k^{(2)}(n-dN_T^{(2)}) \\
&\quad \times e^{-j\frac{2\pi}{N^{(1)}}\Delta k^{(2)}(n)}|^2 \\
&= \rho_N^{(1)} |H^{(1)}(u)|^2 \sum_{j=0}^{1} \Big| \sum_n \Omega_{g_u^{(1)},-jN_T^{(1)}}^{q_k^{(2)},dN_T^{(2)}}(n) e^{-j\frac{2\pi}{N^{(1)}}\Delta k^{(2)}n}\Big|^2
\end{aligned} \tag{3.C.1}
$$

In (3.C.1), approximate equality (d), similar to approximate equality (a) in (3.B.1), follows the assumption that the channel is weakly or mildly time-dispersive, and thus $g_u^{(1)}(n-l-vN_T^{(1)}) \approx g_u^{(1)}(n-vN_T^{(1)})$; equality (e) holds due to the fact of the m-th received symbol of subband 2 lies in the $\tilde{m} = m/n_2\rfloor$-th LCM symbol; using $m = \tilde{m}n_2 + d$, with $d = 0, 1, \dots, n_2 - 1$, the m-th received symbol may possibly be interfered by the $\tilde{m}$-th and $\tilde{m}-1$-th transmitted symbol of subband 1.

References

1 Andrews, J.G., Buzzi, S., Choi, W. et al. (2014). what will 5G be? *IEEE Journal on Selected Areas in Communications* 32 (6): 1065–1082.

2 Ijaz, A., Grau, M., Zhang, L. et al. (2016). Enabling massive IoT in 5G and beyond systems: PHY radio frame design considerations. *IEEE Access* 4: 3322–3339.

3 Zhang, L., Ijaz, A., Xiao, P., and Tafazolli, R. (2017). Channel equalization and interference analysis for uplink narrowband internet of things (NB-IoT). *IEEE Communications Letters* 21 (10): 2206–2209.

4 Zhang, L., Ijaz, A., Xiao, P., and Tafazolli, R. (2017). Subband filtered multi-carrier systems for multi-service wireless communications. *IEEE Transactions on Wireless Communications* 16 (3): 1893–1907.

5 Zhang, L., Xiao, P., Zafar, A. et al. (2016). FBMC system: an insight into doubly dispersive channel impact. *IEEE Transactions on Vehicular Technology* PP (99): 1–14.

6 Taheri, S., Ghoraishi, M., Xiao, P. et al. (2018). Square-root nyquist filter design for QAM-based filter bank multicarrier systems. *IEEE Transactions on Vehicular Technology* 67 9: 9006–9010.

7 Fettweis, G., Krondorf, M., and Bittner, S. (2009). GFDM - generalized frequency division multiplexing. VTC Spring 2009 - IEEE 69th Vehicular Technology Conference, Barcelona, pp. 1–4.

8 Zhang, L., Xiao, P., and Quddus, A. (2016). Cyclic prefix based universal filtered multi-carrier system and performance analysis. *IEEE Signal Processing Letters* 23 (9): 1197–1201.

9 Zhang, L., Ijaz, A., Xiao, P. et al. (2018). Filtered OFDM systems, algorithms and performance analysis for 5G and beyond. *IEEE Transactions on Communications* 66 (3): 1205–1218.

10 Zhang, L., Ijaz, A., Xiao, P., and Tafazolli, R. (2017). Multi-service system: an enabler of flexible 5G air interface. *IEEE Communications Magazine* 55 (10): 152–159.

11 Zhang, X., Zhang, L., Xiao, P. et al. (2015). Mixed numerologies interference analysis and inter-numerology interference cancellation for windowed OFDM systems. *IEEE Transactions on Vehicular Technology* 64 (11): 5070–5082.

12 Ikhlef, A. and Louveaux, J. (2009). Per subchannel equalization for MIMO FBMC/OQAM systems. IEEE Pacific Rim Conference on Communications, Computers and Signal Processing, August 2009, pp. 559–564. doi: https://doi.org/10.1109/PACRIM.2009.5291308.

13 Michailow, N., Matthe, M., Gaspar, I. et al. (2014). Generalized frequency division multiplexing for 5th generation cellular networks. *IEEE Transactions on Communications* 62 (9): 3045–3061.

14 Lien, S.-Y., Shieh, S.-L., Huang, Y. et al. (2017). 5G new radio: waveform, frame structure, multiple access, and initial access. *IEEE Communications Magazine* 55 (6): 64–71.

15 Liu, X., Zhang, X., Zhang, L. et al. (2020). PAPR reduction using iterative clipping/filtering and ADMM approaches for OFDM-based mixed-numerology systems. *IEEE Transactions on Wireless Communications*: 1–1. doi: https://doi.org/10.1109/TWC.2020.2966600.

16 RAN1 Chairman' s Notes (2016). Document 3GPP TSG RAN WG1 Meeting #86.

17 Zayani, R., Medjahdi, Y., Shaiek, H., and Roviras, D. (2016) WOLA-OFDM: a potential candidate for asynchronous 5G. Proceedings of IEEE Globecom Workshops (GC Wkshps), Washington, DC.

18 Bala, E., Li, J., and Yang, R. (2013). Shaping spectral leakage: a novel low complexity transceiver architecture for cognitive radio. *IEEE Vehicular Technology Magazine* 8 (3): 38–46.

19 Güvenkaya, E., Bala, E., Yang, R., and Arslan, H. (2015). Time-asymmetric and subcarrier-specific pulse shaping in OFDM-based waveforms. *IEEE Transactions on Vehicular Technology* 64 (11): 5070–5082.

20 Waveform Candidates (2016). 3GPP Documant R1-162199, Qualcomm Inc.

21 Guan, P., Wu, D., and Tian, T. et al. (2017). 5G field trials: OFDM-based waveforms and mixed numerologies. *IEEE Journal on Selected Areas in Communications* 35 (6): 1234–1243.

22 Way Forward on Assumptions for Waveform Evaluation (2016). 3GPP Documant R1-163558, Huawei, Hisilicon.

23 Zaidi, A.A., Baldemair, R., Tullberg, H. et al. (2016). Waveform and numerology to support 5G services and requirements. *IEEE Communications Magazine* 54 (11): 90–98.

24 Evolved Universal Terrestrial Radio Access (E-UTRA) (2020). User Equipment (UE) Radio Transmission and Reception, 3GPP Specification #: 36.101, Technical Specification Group Radio Access Network.

4

Generalized Frequency Division Multiplexing: Unified Multicarrier Framework

Ahmad Nimr[1], Zhongju Li[1], Marwa Chafii[2] and Gerhard Fettweis[1]

[1] Vodafone Chair Mobile Communication Systems, Technische Universität Dresden, Germany
[2] ETIS UMR 8051, CY Paris Université, ENSEA, CNRS, France

4.1 Overview of Multicarrier Waveforms

The original motivation behind multicarrier modulation (MCM) is to tackle the frequency selectivity of the wireless channel for broadband communications [1]. The idea is to transmit high data rate using parallel streams; each stream has a lower data rate and occupies narrower bandwidth. The name carrier is inherited from the concept of frequency division multiplexing (FDM) [2], where the available bandwidth is divided into several narrow bands centered around subcarriers. Filters are used to separate the bands to eliminate the cross talk (inter-carrier-interference (ICI)). The practical subband filters can be generated from a prototype filter with frequency shift equal to the subcarrier spacing as in filtered multitone (FMT) [3]. ICI depends on the overlap between adjacent filters. To eliminate ICI and retain the orthogonality, the subcarrier spacing can be set larger than the filter bandwidth. With that, the modulation becomes orthogonal FDM. However, spectral efficiency in this case is reduced. In order to preserve both orthogonality and spectral efficiency, two approaches emerged during the past years; the first leads to offset quadrature amplitude modulation (OQAM)–orthogonal frequency division multiplexing (OFDM) and the other leads to discrete Fourier transform (DFT)-OFDM [4]. In OQAM–OFDM, the prototype filter is designed to have overlap with the adjacent subcarriers. The orthogonality is – only – maintained at the in-phase and quadrature components of the complex data symbols. This is done by delaying the in-phase component by half the symbol period in one channel and vice versa on the adjacent channel [5]. This approach is revised and reintroduced under the name filter bank multicarrier (FBMC) [6]. In the beginning of DFT-OFDM, the complex data symbols are transformed with DFT, and the real part is passed through low-pass filter [7]. Later, similar approach is used with the consideration of complex signal and using inverse discrete Fourier transform (IDFT) under the name discrete multitone (DMT) [8], which is nowadays known as OFDM. The prototype filter used in OFDM has a sinc response, which overlaps with more than one adjacent subcarriers. With the introduction of cyclic prefix (CP) to OFDM, which was actually proposed prior to the well-known form of OFDM [9], the frequency-domain processing of CP-OFDM eliminates the inter-symbol-interference (ISI) introduced between successive OFDM symbols

Radio Access Network Slicing and Virtualization for 5G Vertical Industries, First Edition.
Edited by Lei Zhang, Arman Farhang, Gang Feng, and Oluwakayode Onireti.

as a result of the channel delay spread. Considering the frequency domain equalization (FDE) of a single carrier system, which was first proposed in [10], it can be seen as precoded OFDM, where the precoding matrix is the DFT matrix as noted in [11]. The idea of precoded-OFDM is extended to cover several OFDM symbols. Another famous precoding is DFT-spread-OFDM, which is exploited for the uplink multiple access under the name single carrier-frequency division multiple access (SC-FDMA) [12, 13].

Another family of waveforms uses periodic pulse shaping and windowing, which leads to circular filtering. It is worth noting that OFDM also follows this principle. To generalize the concept of circular filtering, generalized frequency division multiplexing (GFDM) [14] was proposed as an extension of OFDM using general circular filtering [15]. At the beginning, GFDM is introduced as a system of parallel single carrier (SC)-CPs, which are filtered to reduce the ICI. The tail-biting idea, which is equivalent to circular filtering, is used to reduce the CP length due to the filtering. In this scheme described in [14], a finite number of symbols per subcarrier, denoted as subsymbols, are circularly filtered with a prototype filter and shifted to the corresponding channel. Moreover, the modulated blocks can be isolated by means of guard interval that is longer than the channel delay to prevent inter-block interference [14]. Circular filtering is also introduced for OQAM–OFDM to facilitate the CP insertion [16].

The common thread among all multicarrier techniques is the transmission of data in parallel streams, which are then superimposed to formulate the final signal. Each stream can be seen as a single-carrier modulation with a specific pulse shape. In the state of the art multicarrier systems, pulse shapes are generated by a shift of the prototype filter[1] in the frequency-domain (FD) and each stream is denoted as a subcarrier. The subsymbol concept, which is introduced by GFDM, corresponds to a shift in the time-domain (TD). Therefore, a general time–frequency description of waveforms is useful to put different waveforms in a unified framework.

4.1.1 Time–Frequency Representation

In SC waveform, a stream of data symbols d_i with the symbol spacing ΔT modulates a pulse shape $g(t)$, where the baseband signal can be expressed as

$$x_{\text{SC}}(t) = \sum_{i\in\mathbb{Z}} d_i g(t - i\Delta T) = g(t) * \sum_{i\in\mathbb{Z}} d_i \delta(t - i\Delta T). \tag{4.1}$$

Here, i denotes the symbol index, $\mathbb{Z}$ in the set of integer numbers, and $*$ represents linear convolution operation. This is also interpreted as filtering of the random process $\sum_{i\in\mathbb{Z}} d_i \delta(t - i\Delta T)$, where d_i is a random variable and $\delta(t)$ is the Dirac function. A multicarrier waveform is a superposition of multiple SC signals, and each uses a pulse $g_k(t)$. Thus, the multicarrier waveform can be expressed as

$$x_{\text{MC}}(t) = \sum_{i\in\mathbb{Z}} \sum_{k\in\mathcal{K}_{\text{on}}} d_{k,i} g_k(t - i\Delta T), \tag{4.2}$$

where $\mathcal{K}_{\text{on}}$ is a finite set used for the subcarriers indexing. The original idea behind multicarrier comes from FDM [2], where the pulses $\{g_k(t)\}$ are designed to occupy disjoint narrow bands. The idea is extended to allow overlapping among the subcarriers as in FMT [3].

1 Prototype pulse and prototype filter are used exchangeably in this chapter.

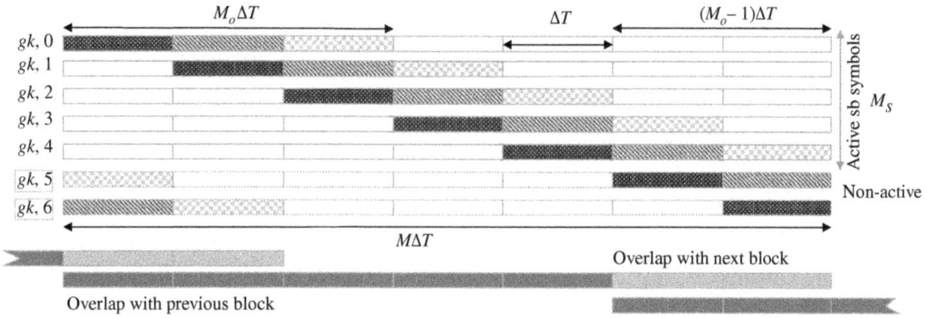

Figure 4.1 Subsymbol concept. In this example $M = 7$, $M_o = 3$, and $M_s = 5$.

Accordingly, $g_k(t)$ is generated by means of frequency shift of a prototype pulse $g(t)$, such that $g_k(t) = g(t)e^{j2\pi k\Delta ft}$, where Δf is the subcarrier spacing and $\mathcal{K}_{\text{on}}$ becomes the set of active subcarriers. Moreover, by expressing the symbol index using two indices (m, b) and an integer $M_s \geq 1$, such that $i = m + bM_s$, we get

$$x_{\text{MC}}(t) = \sum_b \sum_{k\in\mathcal{K}_{\text{on}}} \sum_{m=0}^{M_s-1} d_{k,m+bM_s} g_k(t - [m + bM_s]\Delta T)$$
$$= \sum_b \sum_{k\in\mathcal{K}_{\text{on}}} \sum_{m=0}^{M_s-1} d_{k,m,b} g_{k,m}(t - bT_s), \tag{4.3}$$

where $g_{k,m}(t) = g_k(t - m\Delta T)$, $d_{k,m,b} = d_{k,m+bM_s}$, and $T_s = M_s\Delta T$. The index m refers to the subsymbol index, and accordingly, ΔT defines the subsymbol spacing, and M_s represents the number of active subsymbols. Therefore, $g_{k,m}(t) = g(t - m\Delta T)e^{j2\pi k\Delta f(t-m\Delta T)}$. In practice, the pulse $g(t)$ is finite and causal. As shown in Figure 4.1, let M_o be the number of subsymbol spacing that $g(t)$ spans, such that $g(t) = 0, t \notin [0, M_o\Delta T]$. Thus, $\{g_{k,m}(t)\}$ are also finite and causal, with $g_{k,m}(t) = 0, t \notin [0, M\Delta T]$, where the parameter $M = M_o + M_s - 1$ denotes the number of subsymbol spacing that the pulses $\{g_{k,m}(t)\}$ span. Based on that, we define the periodic pulse $g_T(t)$, where $g_T(t) = g(t), t \in [0, T]$, and $T = M\Delta T$. As a result,

$$g_{k,m}(t) = u_T(t)g_T(t - m\Delta T)e^{j2\pi k\Delta f(t-m\Delta T)}, m \in \mathcal{M}_{\text{on}}. \tag{4.4}$$

Here, $u_T(t) = 1, t \in [0, T]$ is a rectangular window, and the set $\mathcal{M}_{\text{on}} = \{0, M_s - 1\}$ defines the set of active subsymbols, where $d_{k,m,b} = 0$, if $m \notin \mathcal{M}_{\text{on}}$. Therefore,

$$x_{\text{MC}}(t) = \sum_b x_b(t - bT_s), \quad x_b(t) = \sum_{k\in\mathcal{K}_{\text{on}}} \sum_{m\in\mathcal{M}_{\text{on}}} d_{k,m,b} g_{k,m}(t), \tag{4.5}$$

where $x_b(t)$, which is denoted as the modulation block, is a finite signal of length T. The multicarrier signal is a results of multiplexing the modulated blocks, where T_s is the spacing between the blocks, and $T_o = T - T_s = (M_o - 1)\Delta T$ is the overlapping among blocks.

4.1.1.1 Discrete-Time Representation

We consider $N = MK$ samples per block, such that $T = \frac{N}{F_s}$, where F_s is the sampling frequency. By sampling (4.4), we get

$$g_{k,m}\left(\frac{n}{F_s}\right) = g_T\left(\frac{1}{F_s}(n - mF_s\Delta T)\right) e^{j2\pi k\frac{\Delta f}{F_s}(n-mF_s\Delta T)}.$$

For practical implementation, we assume that $P = \Delta T F_s$, $Q = \frac{\Delta f}{F_s} N = \Delta f T$ are integer numbers. P is the size of subsymbol spacing measured in time samples, and Q is the subcarrier spacing in number of frequency samples, where a frequency sample corresponds to $\frac{F_s}{N}$. Therefore, the discrete pulse can be expressed as

$$g_{k,m}[n] = e^{j\phi_{k,m}} g_T[\langle n - mP\rangle_N] e^{j2\pi \frac{nkQ}{N}}, \phi_{k,m} = -2\pi \frac{kmPQ}{N}. \tag{4.6}$$

Here, $\langle\cdot\rangle N$ is the modulo-N operator, which is used due to the periodicity of $g_T(t)$. The phase term can be considered as part of the data symbols. The frequency domain representation is achieved by means of N-DFT, where

$$\tilde{g}_{k,m}[s] = \tilde{g}[\langle s - kQ\rangle_N] e^{-j2\pi \frac{smP}{N}}, \tilde{g}[s] = \sum_{n=0}^{N-1} g[n] e^{-j2\pi \frac{ns}{N}}. \tag{4.7}$$

4.1.1.2 Relation to Gabor Theory

The pulses in (4.6) belong to Gabor time–frequency lattices [17], with $g_{k\alpha,m\beta}[n] = g_T[\langle n - m\beta\rangle_N] e^{j2\pi nk\alpha}$, $\alpha = Q/N$, $\beta = P$. To uniquely demodulate the data symbols $d_{k,m}$ from a given $x[n]$, Wexler–Raz duality condition [18] must be satisfied. Thus, there exists a pulse $h[n]$ that attains $\langle h, g_{k/\beta,m/\alpha}\rangle = \alpha\beta\delta_{k,0}\delta_{m,0}$, where $g_{k/\beta,m/\alpha}$ is the dual Gabor lattice, δ_{ij} refers to Kronecker delta, and $\langle\cdot\rangle$ denotes the inner product.[2] Furthermore, h is the demodulator prototype pulse such that $d_{k,m} = \langle x, h_{k\alpha,m\beta}\rangle$. It has been rigorously proved in [19] that Wexler–Raz duality condition cannot be fulfilled if $\alpha\beta > 1$. Based on that, the choice of P and Q is influenced by

- $QP \leq N$, which is necessary to achieve Wexler–Raz duality condition.
- $\Delta f K \leq F_s = \frac{N}{T} \Rightarrow QK \leq N$, which is necessary to confine the signal within a bandwidth $B \leq F_s$.

Consequently, the design needs to fulfill the conditions $Q \leq M$, $P \leq K$.

4.1.2 GFDM As a Flexible Waveform

The case $P = K$ and $Q = M$, i.e. $PQ = N$, in (4.8) corresponds to a critically sampled system, where $\Delta f \Delta T = \frac{QP}{N} = 1$, which is actually GFDM [14]:

$$g_{k,m}[n] = g_T[\langle n - mK\rangle_N] e^{j2\pi \frac{nk}{K}}, m \in \mathcal{M}_{\text{on}}, \text{and } k \in \mathcal{K}_{\text{on}}. \tag{4.8}$$

GFDM structure enables efficient implementation, as discussed in Section 4.2.2 and the related parameters (Table 4.1), which can be customized to produce different waveforms. For example, OFDM corresponds to the case where $M = 1$ and $g[n]$ is a rectangular pulse. SC is achieved with $K = 1$ and $g[n] = \delta[n]$. FMT can be generated with proper configuration of $\mathcal{M}_{\text{on}}$ (see Figure 4.1).

---

2 In simple words, let $\boldsymbol{x} = \boldsymbol{A}\boldsymbol{d}$, $\boldsymbol{A}[k + mK, n] = g_{k\alpha,m\beta}[n]$, $\boldsymbol{d}[k + mK] = d_{k,m}$, and $\boldsymbol{B}$, where $\boldsymbol{d} = \boldsymbol{B}^{\text{H}}\boldsymbol{x}$. Therefore $\boldsymbol{B}^{\text{H}}\boldsymbol{A} = \boldsymbol{I}$. Wexler–Raz duality condition states that there exist h such that $\boldsymbol{B}[k + mK, n] = h_{k\alpha,m\beta}[n]$, $h_{k\alpha,m\beta}[n] = h_T[\langle n - m\beta\rangle_N] e^{j2\pi nk\alpha}$.

Table 4.1 GFDM-based waveform parameters.

Number of subcarriers	K
Number of subsymbols	M
Block length	$N = KM$
Active subcarriers	$\mathcal{K}_{\text{on}} \subseteq \{0, \cdots K-1\}$
Active subsymbols	$\mathcal{M}_{\text{on}} \subseteq \{0, \cdots M-1\}$
Active symbols	$\mathcal{N}_{\text{on}} = \{k + mK, d_{k,m} \neq 0\}$
Prototype pulse	$g[n]$
Pulse shapes	$g_{k,m}[n] = g[\langle n - mK \rangle_N] e^{j2\pi \frac{nk}{K}}$
GFDM block	$x[n] = \sum_{k \in \mathcal{K}_{\text{on}}} \sum_{m \in \mathcal{M}_{\text{on}}} d_{k,m} g_{k,m}[n]$

4.1.2.1 GFDM with Multiple Prototype Pulses

The case $PQ < N$ corresponds to an over-sampled system where $\Delta f \Delta T < 1$ does not fit in the GFDM representation. However, by managing the sampling frequency, the block length, and properly defining the active sets, it can be represented as a superposition of two or more GFDM blocks. For example, the block length and the sampling frequency can be adjusted to fulfill the conditions $Q = T\Delta f = M$ and $P = \Delta T F_s = \frac{K}{L}$, where L is a positive integer. The set of active subcarriers $\mathcal{K}_{\text{on}}$ is configured with respect to the available bandwidth $B \leq F_s$. The modulation pulses (4.8) can be redefined in the form

$$g_{k,mL+l}[n] = g_T[\left\langle n - mK - l\frac{K}{L} \right\rangle_N] e^{j2\pi \frac{nkM}{N}} = g^{(l)}[n - mK] e^{j2\pi \frac{nk}{K}},$$

where $g^{(l)}[n] = g_T[\left\langle n - l\frac{K}{L} \right\rangle_N]$, $l = 0, \ldots, L-1$. Each pulse shape is used to generate a subset of pulse shapes defined by the active subsymbol set

$$\mathcal{M}_{\text{on}}{}^{(l)} = \{mL + l,\ l \leq mL + l < M\}.$$

Therefore, the final block is a superposition of L GFDM blocks, such that

$$x_b[n] = \sum_{l=0}^{L-1} x^{(l)}[n],\ x_b^{(l)}[n] = \sum_{k \in \mathcal{K}_{\text{on}}} \sum_{m \in \mathcal{M}_{\text{on}}{}^{(l)}} d_{k,m,b} g^{(l)}[\langle n - mK \rangle_N] e^{j2\pi \frac{nk}{K}}. \tag{4.9}$$

An alternative reformulation can be achieved in the FD representation with respect to $\tilde{g}_{k,m}[q]$ (4.7). Here, the design is adjusted such that $P = \Delta T F_s = K$ and $Q = \frac{M}{L}$ resulting in the prototype pulses $\tilde{g}^{(l)}[s] = \tilde{g}[\left\langle s - l\frac{M}{L} \right\rangle_N]$ with the sets $\mathcal{K}_{\text{on}}{}^{(l)} = \{kL + l,\ l \leq kL + l < K\}$.

General Multiple Prototype Pulses

This approach can be generalized to design a waveform with L prototype pulses $\{g^{(l)}[n]\}$. Each prototype pulse is associated with the sets $\mathcal{K}_{\text{on}}{}^{(l)}$ and $\mathcal{M}_{\text{on}}{}^{(l)}$, as shown in Figure 4.2. Thus, for $(k, m) \in \mathcal{K}_{\text{on}}{}^{(l)} \times \mathcal{M}_{\text{on}}{}^{(l)}$, $g_{k,m}[n] = g^{(l)}(\langle n - mK \rangle_N) e^{j2\pi \frac{nk}{K}}$, the input data symbol can be further preprocessed prior to modulation, for instance, to produce OQAM [20].

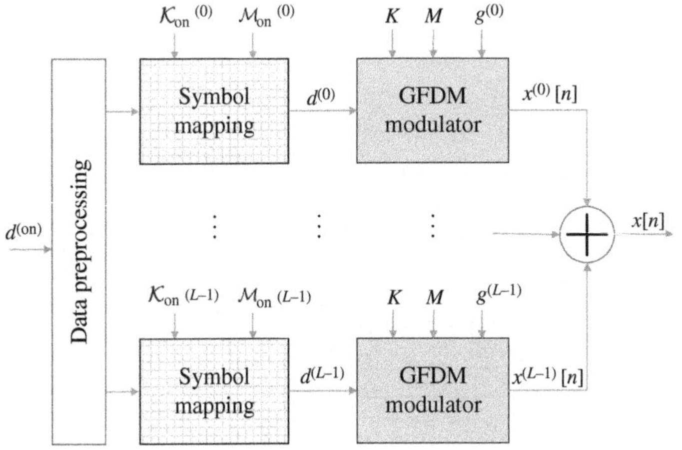

Figure 4.2 GFDM with multiple prototype pulses.

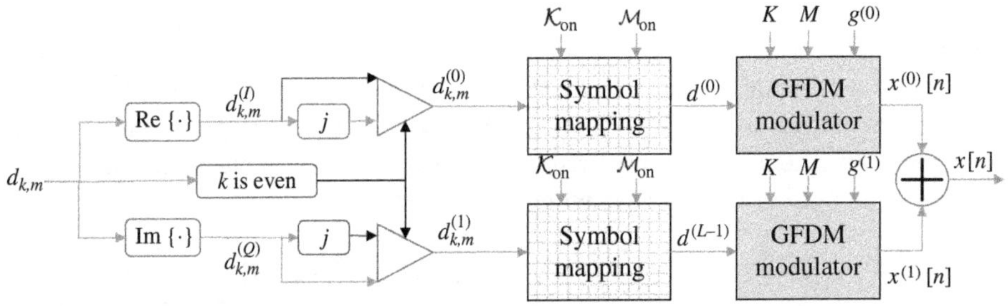

Figure 4.3 OQAM preprocessing.

Example: OQAM-OFDM In this waveform, $Q = M$ and $P = K/2$; first, the complex QAM data symbols $d_{k,m} = d_{k,m}^{(I)} + jd_{k,m}^{(Q)}$ are split into two OQAM precoded streams, $d_{k,m}^{(0)} = \theta_{0,k} d_{k,m}^{(I)}$, $d_{k,m}^{(1)} = \theta_{1,k} d_{k,m}^{(Q)}$, as illustrated in Figure 4.3, where

$$\theta_{0,k} = \begin{Bmatrix} j, & k \text{ is even} \\ 1, & k \text{ is odd} \end{Bmatrix}, \theta_{1,k} = \begin{Bmatrix} 1, & k \text{ is even} \\ j, & k \text{ is odd} \end{Bmatrix}.$$

The streams are then fed to two GFDM modulators with the parameters $g^{(0)}[n] = g[n]$, $g^{(1)}[n] = g[< n - K/2 >_N]$, $\mathcal{K}_{on}{}^{(0)} = \mathcal{K}_{on}{}^{(1)}$, and $\mathcal{M}_{on}{}^{(0)} = \mathcal{M}_{on}{}^{(1)} = \{0, \dots, M_s - 1\}$, where $M_o = M - M_s + 1$ is the overlapping factor of the prototype filter $g[n]$ as shown in Figure 4.1. The output block is a superposition of both GFDM blocks i.e. $x_b[n] = x_b^{(0)}[n] + x_b^{(1)}[n]$. The blocks are multiplexed with spacing $N_s = M_s K$, and thus, $x_{oqam}[n] = \sum_b x_b[n - N_s]$.

4.1.3 Generalized Block-Based Multicarrier

In addition to the pulse shaping, further processing can be performed on top of the modulated block, including CP/cyclic suffix (CS) insertion, windowing, and filtering. Accordingly, we introduce a generic multicarrier framework.

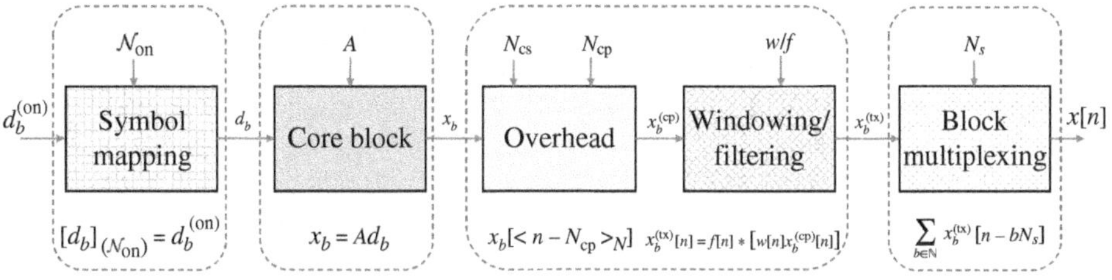

Figure 4.4 Stages of multicarrier waveforms generator.

4.1.3.1 Transmitter

The waveform generation can be achieved in four steps as illustrated in Figure 4.4:

1. *Data symbol mapping:* The data symbols are mapped to the active set of symbols $\mathcal{N}_{\text{on}} = \{n = k + mK\} \subset \mathcal{K}_{\text{on}} \times \mathcal{M}_{\text{on}}$, such that $d_{k,m} = 0, k + mK \notin \mathcal{N}_{\text{on}}$. Therefore, an input data $\boldsymbol{d}_b^{(\text{on})} \in \mathbb{C}^{|\mathcal{N}_{\text{on}}|\times 1}$ is mapped to $\boldsymbol{d}_b \in \mathbb{C}^{N\times 1}$ defined as $\boldsymbol{d}_b[k, m + K] = d_{k,m,b}$ such that $\boldsymbol{d}_b[\mathcal{N}_{\text{on}}] = \boldsymbol{d}_b^{(\text{on})}$.
2. *Core modulation:* This refers to generating the modulated block, which can be expressed in a vector $\boldsymbol{x}_b = \boldsymbol{A}\boldsymbol{d}_b \in \mathbb{C}^{N\times 1}$, where $\boldsymbol{A} \in \mathbb{C}^{N\times N}$ is the modulation matrix defined as $\boldsymbol{A}[nk + mK] = g_{k,m}[n]$.
3. *Additional processing:* This processing involves adding CP/CS, which corresponds to overhead of size N_{cp}. Thus, the CP-block is defined by $x_b^{(\text{cp})}[n] = \boldsymbol{x}_b[\langle n - N_{cp}\rangle_N]$, $n = 0. \dots, N + N_{cp} - 1$. Windowing and/or filtering can be applied on top. Thus, the actual transmitted block is $x_b^{(\text{tx})}[n] = f[n] * (w[n]x_b^{(\text{cp})}[n])$. The filter increases the block length by the filter tail N_f. Thereby, the total transmitted block length is

$$N_t = N + N_{cp} + N_f.$$

   The additional processing can be summarized by linear operation using the matrix $\boldsymbol{G}^{(\text{tx})} \in \mathbb{C}^{N_t\times N}$, such that $\boldsymbol{x}_b^{(\text{tx})} = \boldsymbol{G}^{(\text{tx})}\boldsymbol{x}_b$.
4. *Block multiplexing:* Block multiplexing is achieved by transmitting the signal with N_s samples spacing, such that $x[n] = \sum_b x_b^{(\text{tx})}[n - bN_s]$. The difference $N_o = N_t - N_s$ represents the overlapping between blocks. The case where $N_o < 0$ corresponds to zero padding.

The main complexity arises from the core block modulation, which in general requires N^2 complex multiplications. Nevertheless, this complexity can be significantly reduced by exploiting the modulation matrix structure.

4.1.3.2 Receiver

Synchronization techniques are used for the detection of transmitted frames at the receiver. For example, in preamble-based synchronization, a reference sequence is transmitted in front of a frame to enable the detection of the start of the frame. After the preamble, a header signal is used to provide information about the number of blocks in the frame and other modulation parameters. Afterwards, inverse operations take place as shown in Figure 4.5.

1. *Block demultiplexing:* Block demultiplexing extracts the modulated blocks. Depending on the block spacing and the number of discrete channel taps, a received block might be

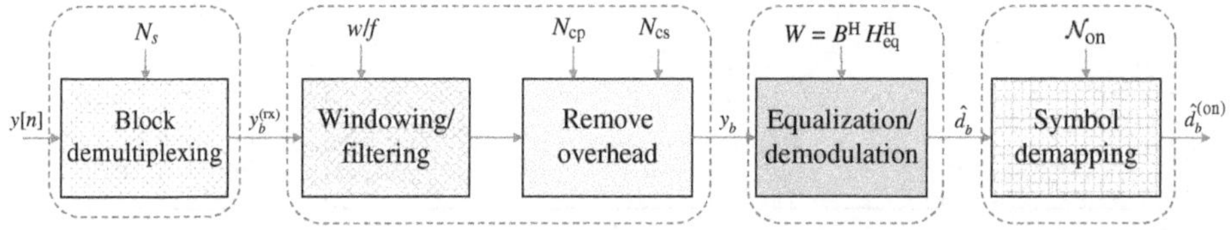

Figure 4.5 Stages of multicarrier waveforms receiver.

interfered with the previous and the next block. Let $\boldsymbol{H}_b$ be the discrete channel matrix representing the linear channel. The received block of N_t samples can be expressed in the vector

$$\boldsymbol{y}_b^{(rx)} = \boldsymbol{H}_b \boldsymbol{x}_b^{(tx)} + \boldsymbol{z}_b^{(rx)} + \boldsymbol{v}_b^{(rx)}$$

where $\boldsymbol{z}_b^{(rx)}$ is the inter-block-interference (IBI) and $\boldsymbol{v}_b^{(rx)}$ the additive noise.

2. *Further processing:* Receiver filtering and/or windowing can be used at the receiver, and the CP/CS are removed. This linear processing, which is expressed in a matrix $\boldsymbol{G}^{(rx)} \in \mathbb{C}^{N\times N_t}$, produces the core block, $\boldsymbol{y}_i = \boldsymbol{G}^{(rx)}\boldsymbol{y}_b^{(rx)}$, which can be expressed as

$$\boldsymbol{y}_b = \boldsymbol{H}_b^{(e)} \boldsymbol{x}_b + \boldsymbol{z}_b + \boldsymbol{v}_b, \tag{4.10}$$

where $\boldsymbol{H}^{(e)} = \boldsymbol{G}^{(rx)}\boldsymbol{H}\boldsymbol{G}^{(tx)} \in \mathbb{C}^{N\times N}$ is the effective channel matrix, $\boldsymbol{z}_b$ the IBI, and $\boldsymbol{v}_b$ the additive noise.

3. *Equalization and demodulation:* Several approaches can be used. In order to exploit the design architecture, channel equalization can be performed first using the equalizer matrix $\boldsymbol{H}_{eq}^{H}$, and the equalized block is fed to the demodulator with matrix $\boldsymbol{B}^{H}$. To simplify the channel equalization, it is convenient to design the waveforms to force the channel $\boldsymbol{H}^{(e)}$ to become circular, and $\boldsymbol{z}_b = \boldsymbol{0}$. This is feasible when the channel variation is negligible within the block and the CP is selected to be longer than the filter length plus the channel length. With that, single-tap FD channel equalization can be performed on the FD block similar to OFDM. Using the DFT matrix $\boldsymbol{F}_N$, $\boldsymbol{F}_N[m,n] = e^{-j2\pi\frac{nm}{N}}$, we get

$$\tilde{\boldsymbol{y}}_b = \Lambda^{(h)}\tilde{\boldsymbol{x}}_b + \tilde{\boldsymbol{v}}_b, \text{ where } \tilde{\boldsymbol{x}} = \boldsymbol{F}_N\boldsymbol{x}. \tag{4.11}$$

4. *Data demapping:* After the demodulation, the estimated symbols are extracted from the demodulated vector such that $\boldsymbol{d}_b^{(on)} = \boldsymbol{d}_b[\mathcal{N}_{on}]$.

4.2 GFDM As a Flexible Framework

In this section, we focus on the GFDM-based core modulator and demodulator. First, we provide several different GFDM representations. Then, we develop a unified time–frequency framework followed by a hardware implementation architecture. This architecture is enhanced with additional flexibility that extends GFDM beyond the scope of conventional waveforms.

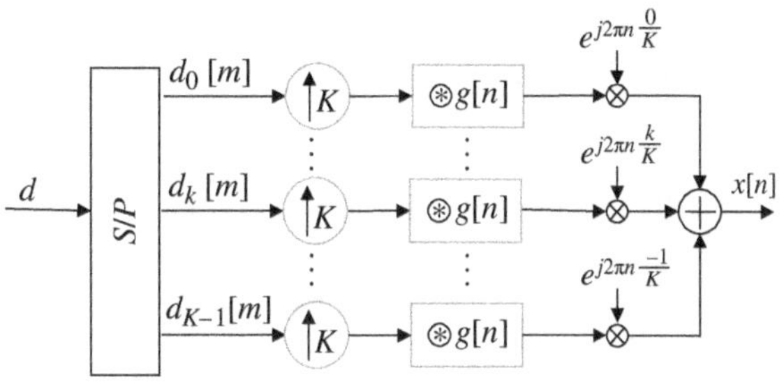

Figure 4.6 Circular convolution filter bank representation.

4.2.1 GFDM Representations

Recall that the pulse shapes in GFDM are generated using a prototype pulse $g[n]$, whose N-DFT is denoted as $\tilde{g}(s)$ such that

$$g_{k,m}[n] = g[\langle n - mK\rangle_N]e^{j2\pi\frac{nk}{K}},\ \tilde{g}_{k,m}[s] = \tilde{g}[\langle s - kM\rangle_N]e^{-j2\pi\frac{sm}{M}}. \tag{4.12}$$

The pulse shapes have a similar formulation in TD and FD, with the exchange of parameters. In the following, we use the data matrix $\boldsymbol{D} \in \mathbb{C}^{K\times M}$, where $\boldsymbol{D}[k,m] = d_{k,m}$. The TD-GFDM block can be expressed as

$$x[n] = \sum_{k=0}^{K-1}\sum_{m=0}^{M-1} \boldsymbol{D}[k,m]g[\langle n - mK\rangle_N]e^{j2\pi\frac{k}{K}n}. \tag{4.13}$$

4.2.1.1 Filter Bank Representation

The GFDM block can be seen as a result of circular convolution with a filter $g[n]$ followed by frequency upconversion, as seen in Figure 4.6:

$$x[n] = \sum_{k=0}^{K-1}\left[g[n] \circledast \sum_{m=0}^{M-1} \boldsymbol{D}[k,m]\delta[n - mK]\right]e^{j2\pi\frac{k}{K}n},$$

where $\circledast$ denotes the circular convolution. Circular filtering was introduced in the original GFDM publication [15] as an approach for the digital implementation of the classical filter bank. Circular convolution is equivalent to linear filtering with tail biting. It is used to reduce the overhead due to the filter tail and to allow inserting CP. This representation corresponds to the conventional GFDM, where the focus is on a specific type of prototype pulses that spans two subcarriers spacing [21].

4.2.1.2 Vector Representation

The TD block (4.13) can be presented in a vector form $\boldsymbol{x} = \boldsymbol{A}\boldsymbol{d}$, where $\boldsymbol{d} = \mathrm{vec}\{\boldsymbol{D}\}$ is the vectorization of $\boldsymbol{D}$, with $\boldsymbol{d}[k + mK] = \boldsymbol{D}[k,m]$. The modulation matrix $\boldsymbol{A}$ is defined by $\boldsymbol{A}[n, m + kM] = g_{k,m}[n]$. The FD block is accordingly $\tilde{\boldsymbol{x}} = \boldsymbol{F}_N\boldsymbol{A}\boldsymbol{d}$. Thus, $\tilde{\boldsymbol{A}} = \boldsymbol{F}_N\boldsymbol{A}$ represents the FD modulation matrix. Note that $\tilde{\boldsymbol{x}} = \tilde{\boldsymbol{A}}\boldsymbol{d}$ and $\boldsymbol{x} = \frac{1}{N}\boldsymbol{F}_N^{\mathrm{H}}\tilde{\boldsymbol{x}}$ can be seen as the core block corresponding to the precoded data vector $\tilde{\boldsymbol{x}}$ and the modulation matrix $\frac{1}{N}\boldsymbol{F}_N^{\mathrm{H}}$. Thereby, any linear modulation can be technically considered as precoded OFDM. Nevertheless, this precoding is required to achieve certain waveform properties, and the design and implementation of the precoding matrix $\tilde{\boldsymbol{A}}$ provides the waveform features, such as low out-of-band (OOB). More details on using GFDM for OFDM enhancement are introduced in Section 4.3.

4.2.1.3 2D-Block Representation

To clarify the time–frequency structure, the entries of the modulation block $\boldsymbol{x}$ can be rearranged in a matrix using the inverse of vectorization $\boldsymbol{X} = \text{unvec}_{K\times M}\{\boldsymbol{x}\} \in \mathbb{C}^{K\times M}$, where $\boldsymbol{X}[k,m] = \boldsymbol{x}[k+mK]$. The FD block is given by $\tilde{\boldsymbol{X}} = \text{unvec}_{M\times K}\{\tilde{\boldsymbol{x}}\} \in \mathbb{C}^{M\times K}$. The m-th column of $\boldsymbol{X}$ represents the samples transmitted within the subsymbols duration, and the k-th column of $\tilde{\boldsymbol{X}}$ corresponds to a subcarrier width of M samples. Using the notation $\boldsymbol{X}[q,p] = \boldsymbol{x}[q+pK]$, from (4.13) we get

$$\begin{aligned}\boldsymbol{X}[q,p] &= \sum_{m=0}^{M-1}\left[\sum_{k=0}^{K-1}\boldsymbol{D}[k,m]^{e^{j2\pi\frac{qk}{K}}}\right]\boldsymbol{G}[q,\langle p-m\rangle_M] \\ &= \sum_{m=0}^{M-1}(\boldsymbol{F}_K^{\mathrm{H}}\boldsymbol{D})[q,m]\boldsymbol{G}[q,\langle p-m\rangle_M],\end{aligned} \tag{4.14}$$

where $\boldsymbol{G} \in \mathbb{C}^{K\times M}$, $\boldsymbol{G}[q,p] = g[q+pK]$. Hence, the 2D-block results from the circular filtering of the rows of $\boldsymbol{F}_K^{\mathrm{H}}\boldsymbol{D}$ and the rows of $\boldsymbol{G}$. This is essentially the polyphase representation of the filter bank represented in Figure 4.6, where the q-th row of $\boldsymbol{G}$ is the q-th polyphase of $g[n]$. By computing the M-DFT of the rows and using the element-wise product notation $\odot$, we get

$$\boldsymbol{X}\boldsymbol{F}_M = [\boldsymbol{F}_K^{\mathrm{H}}\boldsymbol{D}\boldsymbol{F}_M] \odot [\boldsymbol{G}\boldsymbol{F}_M]. \tag{4.15}$$

Relation to Zak Transform

The matrix $\boldsymbol{Z}_{M,K}^{(\boldsymbol{x})} = \boldsymbol{F}_M\boldsymbol{X}^{\mathrm{T}}$ is known as the Zak transform of $\boldsymbol{x}$, which is invertible. The k-th row of $\boldsymbol{x}$, $\boldsymbol{x}_k^{\mathrm{T}} = \boldsymbol{X}[k,:]$ [3] results from the downsampling of $\boldsymbol{x}$ by factor K starting from the k-th sample. The matrix $\boldsymbol{X}^{\mathrm{T}}$ stacks the polyphase signals with respect to the sampling factor K, as illustrated in Figure 4.7. This transform couples the subsymbols and provides a

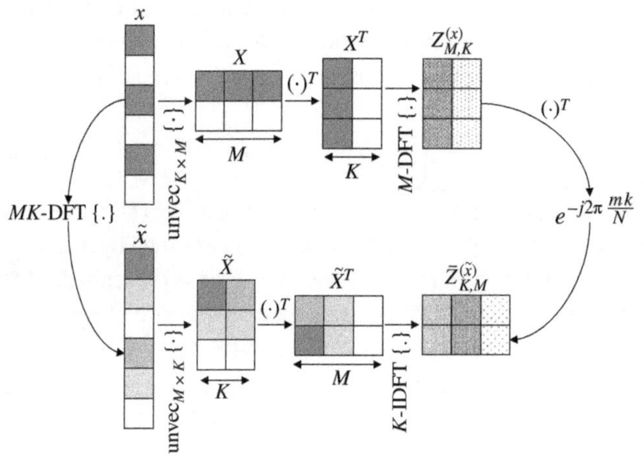

Figure 4.7 Zak transform.

3 Here, we use Matlab notations for indexing.

time–frequency relation. To determine the relation between the TD and FD block, we use the definition of N-DFT, and $\tilde{\boldsymbol{X}}[p,q] = \tilde{\boldsymbol{x}}[p+qM]$,

$$\tilde{\boldsymbol{X}}[p,q] = \sum_{k=0}^{K-1}\sum_{m=0}^{M-1} \boldsymbol{x}[k+mK]e^{-j2\pi\frac{(p+qM)(k+mK)}{N}}$$
$$= \sum_{k=0}^{K-1}(\boldsymbol{X}\boldsymbol{F}_M)[k,p]\boldsymbol{\Phi}[k,p]e^{-j2\pi\frac{qk}{K}},\ \boldsymbol{\Phi}[k,m] = e^{-j2\pi\frac{km}{N}}.$$

As a result, the time–frequency relation is given by

$$\frac{1}{K}\tilde{\boldsymbol{X}}\boldsymbol{F}_K^{\mathrm{H}} = \boldsymbol{\Phi}^{\mathrm{T}} \odot [\boldsymbol{F}_M\boldsymbol{X}^{\mathrm{T}}]. \tag{4.16}$$

Thus $\boldsymbol{Z}_{K,M}^{(\tilde{x})} = \frac{1}{K}\boldsymbol{F}_K^{\mathrm{H}}\tilde{\boldsymbol{X}}^{\mathrm{T}}$ is the dual FD Zak transform. Following similar derivation for the FD block $\tilde{\boldsymbol{X}}$, from (4.15), (4.16), and using $\boldsymbol{Z}_{K,M}^{(g)} = \boldsymbol{F}_M\boldsymbol{G}^{\mathrm{T}}$,

$$\frac{1}{K}\tilde{\boldsymbol{X}}\boldsymbol{F}_K^{\mathrm{H}} = (\boldsymbol{\Phi} \odot \{\boldsymbol{Z}_{M,K}^{(g)}\}^{\mathrm{T}} \odot [\boldsymbol{F}_K^{\mathrm{H}}\boldsymbol{D}\boldsymbol{F}_M])^{\mathrm{T}}$$
$$= [\boldsymbol{F}_M\boldsymbol{D}^{\mathrm{T}}\boldsymbol{F}_K^{\mathrm{H}}] \odot [\frac{1}{K}\tilde{\boldsymbol{G}}\boldsymbol{F}_K^{\mathrm{H}}], \tag{4.17}$$

where $\tilde{\boldsymbol{G}} \in \mathbb{C}^{M\times K}$, $\tilde{\boldsymbol{G}}[p,q] = \tilde{g}[p+qM]$. Therefore,

$$\tilde{\boldsymbol{X}}[p,q] = \sum_{k=0}^{K-1}(\boldsymbol{F}_M\boldsymbol{D}^{\mathrm{T}})[p,k]\tilde{\boldsymbol{G}}[p,\langle q-k\rangle_K]. \tag{4.18}$$

4.2.1.4 GFDM Matrix Structure

The structure of the GFDM modulation matrix can be derived by the vectorization of (4.15) and (4.17). First, we define the notations required to express the vectorization. For a matrix $\boldsymbol{Z} \in \mathbb{C}^{Q\times P}$,

$$\mathrm{vec}\{\boldsymbol{F}_Q\boldsymbol{Z}\} = \sqrt{Q}\boldsymbol{U}_{P,Q}\mathrm{vec}\{\boldsymbol{Z}\},\ \boldsymbol{U}_{P,Q} = \frac{1}{\sqrt{Q}}\boldsymbol{I}_P \otimes \boldsymbol{F}_Q$$
$$\boldsymbol{\Pi}_{Q,P} : \mathrm{vec}\{\boldsymbol{Z}^{\mathrm{T}}\} = \boldsymbol{\Pi}_{Q,P}\mathrm{vec}\{\boldsymbol{Z}\}. \tag{4.19}$$

$\boldsymbol{U}_{P,Q}$ is a unitary matrix and $\boldsymbol{\Pi}_{Q,P}$ is a permutation matrix known as the commutative matrix of size $QP \times QP$ defined by $\boldsymbol{\Pi}_{Q,P}[p+qPq+pQ] = 1$. Note that $\boldsymbol{\Pi}_{Q,P}^{-1} = \boldsymbol{\Pi}_{P,Q}$. Using the TD relation, $\boldsymbol{X}\boldsymbol{F}_M = [\boldsymbol{F}_K^{\mathrm{H}}\boldsymbol{D}\boldsymbol{F}_M] \odot [\boldsymbol{G}\boldsymbol{F}_M]$,

$$\mathrm{vec}\{\boldsymbol{X}\boldsymbol{F}_M\} = \boldsymbol{\Pi}_{M,K}\mathrm{vec}\{\boldsymbol{F}_M\boldsymbol{X}^{\mathrm{T}}\} = \sqrt{M}\boldsymbol{\Pi}_{M,K}\boldsymbol{U}_{K,M}\mathrm{vec}\{\boldsymbol{X}^{\mathrm{T}}\},$$
$$\mathrm{vec}\{\boldsymbol{F}_K^{\mathrm{H}}\boldsymbol{D}\boldsymbol{F}_M\} = \sqrt{K}\boldsymbol{U}_{M,K}^{\mathrm{H}}\mathrm{vec}\{\boldsymbol{D}\boldsymbol{F}_M\} = \sqrt{KM}\boldsymbol{U}_{M,K}^{\mathrm{H}}\boldsymbol{\Pi}_{M,K}\boldsymbol{U}_{K,M}\mathrm{vec}\{\boldsymbol{D}^{\mathrm{T}}\}.$$

Let $\boldsymbol{\Lambda}^{(g)} = \sqrt{K}\mathrm{diag}\{\mathrm{vec}\{\boldsymbol{G}\boldsymbol{F}_M\}\}$, and because $\boldsymbol{x} = \boldsymbol{A}\boldsymbol{d}$, then

$$\boldsymbol{A} = \underbrace{\boldsymbol{\Pi}_{M,K}\boldsymbol{U}_{K,M}^{\mathrm{H}}\boldsymbol{\Pi}_{K,M}}_{\boldsymbol{V}_t}\boldsymbol{\Lambda}^{(g)}\underbrace{\boldsymbol{U}_{K,M}^{\mathrm{H}}\boldsymbol{\Pi}_{M,K}\boldsymbol{U}_{K,M}\boldsymbol{\Pi}_{K,M}}_{\boldsymbol{U}_t}. \tag{4.20}$$

The matrices $\boldsymbol{V}_t$ and $\boldsymbol{U}_t$ are unitary matrices, and therefore, the condition of $\boldsymbol{A}$ is influenced by the diagonal elements of $\boldsymbol{\Lambda}^{(g)}$.[4] Similarly, the FD matrix , where $\tilde{\boldsymbol{x}} = \tilde{\boldsymbol{A}}\boldsymbol{d}$ can be derived

4 Different permutations of the elements of $\boldsymbol{\Lambda}^{(g)}$ lead to different $\boldsymbol{V}_t$ and $\boldsymbol{U}_t$. Thus, different formulas can be found in the literature.

from $\tilde{\boldsymbol{X}}\boldsymbol{F}_K^{\mathrm{H}} = [\boldsymbol{F}_M\boldsymbol{D}^{\mathrm{T}}\boldsymbol{F}_K^{\mathrm{H}}] \odot [\tilde{\boldsymbol{G}}\boldsymbol{F}_K^{\mathrm{H}}]$,

$$\tilde{\boldsymbol{A}} = \underbrace{\boldsymbol{\Pi}_{K,M}\boldsymbol{U}_{M,K}\boldsymbol{\Pi}_{M,K}}_{\boldsymbol{V}_t}\boldsymbol{\Lambda}^{(\mathrm{g})}\underbrace{\boldsymbol{U}_{K,M}\boldsymbol{\Pi}_{K,M}\boldsymbol{U}_{K,M}^{\mathrm{H}}}_{\boldsymbol{U}_t}, \tag{4.21}$$

where $\boldsymbol{\Lambda}^{(\mathrm{g})} = \sqrt{M}\mathrm{diag}\{\mathrm{vec}\{\tilde{\boldsymbol{G}}\}\boldsymbol{F}_K^{\mathrm{H}}\}$. Any matrix that fulfills one of those structures is called a GFDM matrix. As a result, the condition of the modulation matrix is determined from the TD or FD Zak transform of $g[n]$ given by $z_{m,k} = (\boldsymbol{G}\boldsymbol{F}_M)[k,m]$ and $\bar{z}_{k,m} = \frac{1}{K}(\tilde{\boldsymbol{G}}\boldsymbol{F}_K^{\mathrm{H}})[m,k]$, respectively. For instance, a real symmetric pulse produces a singular matrix when both K and M are even numbers [21].

Orthogonal Modulation

In order to achieve an orthogonal modulation matrix, it is necessary that $|\bar{z}_{k,m}| = |z_{m,k}| =$ const. Thus we get different orthogonal waveforms by designing the phases $\phi_{m,k} = \angle(z_{m,k})$ in the time domain or $\bar{\phi}_{k,m} = \angle(\bar{z}_{k,m})$ in the frequency domain. Two corner cases arise; the first is when $\phi_{m,k} = \phi_k$, which results in a frame of M consecutive OFDM symbols. In other words, $\boldsymbol{D}[:,m]$ can be seen as the frequency domain data symbol mapped to the m-th OFDM symbol of length K. In this case, there is clear separation of the signals corresponding to each subsymbol. In the other case, when $\bar{\phi}_{k,m} = \bar{\phi}_m$, we get the DFT-S-OFDM, and for the special case of $K = 1$, we get the single carrier. Therefore, $\boldsymbol{D}[k,:]$ can be seen as the time-domain symbols that use the k-th subchannel of bandwidth M. Thereby, the subcarriers can be separated. In all other cases, the subsymbols and subcarriers overlap, i.e. the data are spread over the time and frequency domains (Figure 4.8).

4.2.2 Architecture and Extended Flexibility

The TD and FD blocks in (4.15) and (4.17) can be rewritten in the form

$$\begin{aligned}\boldsymbol{X}^{\mathrm{T}} &= \frac{1}{M}\boldsymbol{F}_M^{\mathrm{H}}\left([\boldsymbol{G}\boldsymbol{F}_M]^{\mathrm{T}} \odot [\boldsymbol{F}_M[\boldsymbol{F}_K^{\mathrm{H}}\boldsymbol{D}]^{\mathrm{T}}]\right)\\ \tilde{\boldsymbol{X}}^{\mathrm{T}} &= \frac{1}{K}\boldsymbol{F}_K\left([\tilde{\boldsymbol{G}}\boldsymbol{F}_K^{\mathrm{H}}]^{\mathrm{T}} \odot [\boldsymbol{F}_K^{\mathrm{H}}[\boldsymbol{F}_M\boldsymbol{D}^{\mathrm{T}}]^{\mathrm{T}}]\right).\end{aligned} \tag{4.22}$$

Accordingly, the two cases can be presented in the form

$$\boldsymbol{X}_0^{\mathrm{T}} = \boldsymbol{U}_3(\boldsymbol{W}_{\mathrm{tx}} \odot [\boldsymbol{U}_2[\boldsymbol{U}_1\boldsymbol{D}_0]^{\mathrm{T}}]),\ \boldsymbol{x}_0 = \mathrm{vec}\{\boldsymbol{X}_0\}, \tag{4.23}$$

where $\boldsymbol{U}_x \in \mathbb{C}^{N_x \times N_x}$ is a normalized DFT or IDFT matrix of size N_x; for $x = 1,2,3$, $N_3 = N_2$, $\boldsymbol{W}_{\mathrm{tx}} \in \mathbb{C}^{N_2 \times N_1}$ corresponds to the Zak transform, $\boldsymbol{D}_0 \in \mathbb{C}^{N_1 \times N_2}$ represents the data matrix, $\boldsymbol{x}_0 \in \mathbb{C}^{N_1 N_2 \times 1}$ is the GFDM block, and $\boldsymbol{X}_0 \in \mathbb{C}^{N_1 \times N_2}$ is the GFDM block matrix. The concrete values corresponding to each domain are summarized in Table 4.2.

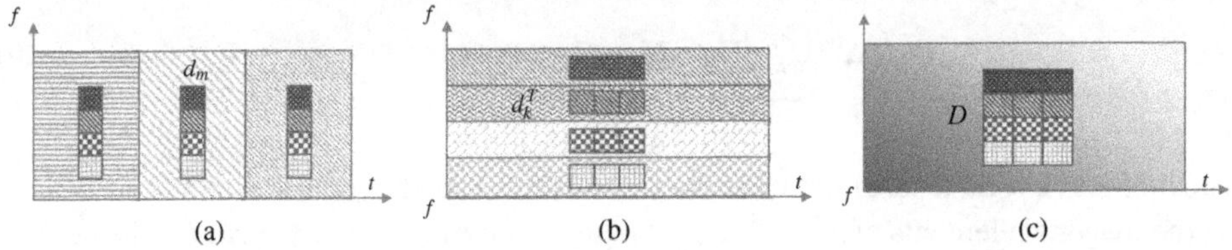

Figure 4.8 Orthogonal GFDM. (a) Separate subsymbols and overlapped subcarriers. (b) Separate subcarriers and overlapped subsymbols. (c) Overlapped subcarriers and subsymbols.

Table 4.2 Flexible GFDM parameters.

Parameter	TD	FD
$[\boldsymbol{X}_0, \boldsymbol{D}_0, \boldsymbol{W}_{\text{tx}}]$	$[\boldsymbol{X}, \boldsymbol{D}, \sqrt{K}\boldsymbol{F}_M\boldsymbol{G}^{\text{T}}]$	$[\tilde{\boldsymbol{X}}, \boldsymbol{D}^{\text{T}}, \sqrt{M}\boldsymbol{F}_K^{\text{H}}\tilde{\boldsymbol{G}}^{\text{T}}]$
$[\boldsymbol{U}_1, \boldsymbol{U}_2, \boldsymbol{U}_3]$	$[\frac{1}{\sqrt{K}}\boldsymbol{F}_K^{\text{H}}, \frac{1}{\sqrt{M}}\boldsymbol{F}_M, \frac{1}{\sqrt{M}}\boldsymbol{F}_M^{\text{H}}]$	$[\frac{1}{\sqrt{M}}\boldsymbol{F}_M, \frac{1}{\sqrt{K}}\boldsymbol{F}_K^{\text{H}}, \frac{1}{\sqrt{M}}\boldsymbol{F}_M^{\text{H}}]$

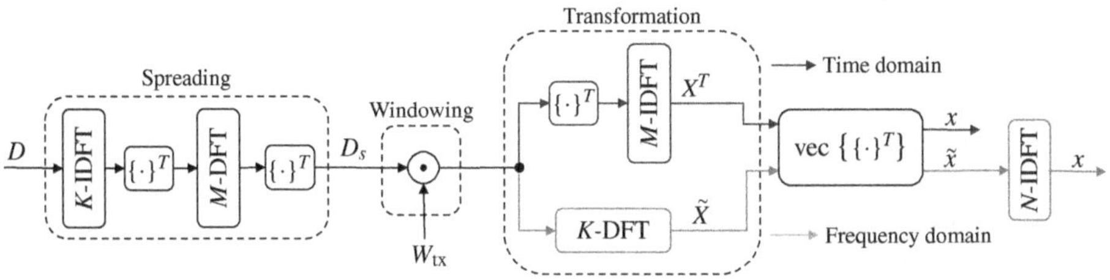

Figure 4.9 GFDM modulator in four steps.

4.2.2.1 Alternative Interpretation of GFDM

Based on the model (4.23), GFDM modulation can be achieved in four steps, as illustrated in Figure 4.9 for both TD and FD.

1. *Data spreading:* Spreading is achieved by applying DFT/IDFT on the columns and rows of $\boldsymbol{D}_0 \in \mathbb{C}^{N_1 \times N_2}$. Therefore, we get the spread data matrix $\boldsymbol{D}_s = \boldsymbol{U}_2\boldsymbol{D}_0^{\text{T}}\boldsymbol{U}_1 \in \mathbb{C}^{N_2 \times N_1}$.
2. *Windowing:* The spread data matrix $\boldsymbol{D}_s$ is element-wise multiplied with a transmitter windowing matrix $\boldsymbol{W}_{\text{tx}} \in \mathbb{C}^{N_2 \times N_1}$, which is generated based on the prototype pulse shape.
3. *Transformation:* The columns of the matrix $\boldsymbol{W}_{\text{tx}} \odot \boldsymbol{D}_s$ are transformed by means of IDFT or DFT, where the samples are expressed by matrix $\boldsymbol{X}_0^{\text{T}} = \boldsymbol{U}_3[\boldsymbol{W}_{\text{tx}} \odot \boldsymbol{D}_s]$.
4. *Sample allocation:* The final vector is achieved by allocating the generated samples to the corresponding indexes, specifically, $\boldsymbol{x}_0 = \text{vec}\{\boldsymbol{X}_0\}$. In the case of FD modulation, an additional N-IDFT is required to convert the signal to the time domain, i.e. $\boldsymbol{x} = \frac{1}{N}\boldsymbol{F}_N^{\text{H}}\tilde{\boldsymbol{x}}$.

The demodulator performs the inverse operations, as illustrated in Figure 4.10. Let $\boldsymbol{y}_0$ be the equalized block, and $\boldsymbol{W}_{\text{rx}}$ be the receiver window. First, the vector is reallocated to the matrix $\boldsymbol{Y}_0 = \text{unvec}_{N_2 \times N_1}\{\boldsymbol{y}_0\}$, then $\hat{\boldsymbol{D}}_0$ is computed by performing the inverses of transform, windowing, and spreading, as

$$\hat{\boldsymbol{D}}_0 = \boldsymbol{U}_1[\boldsymbol{U}_2^{\text{H}}(\boldsymbol{W}_{\text{rx}} \odot [\boldsymbol{U}_3^{\text{H}}\boldsymbol{Y}_0^{\text{T}}])]^{\text{T}}. \tag{4.24}$$

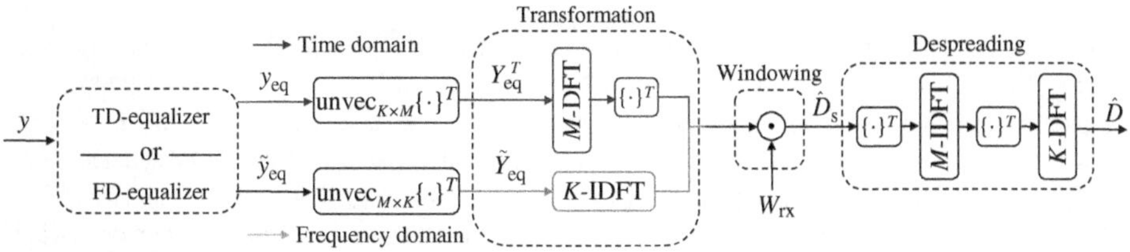

Figure 4.10 GFDM demodulator in four steps.

The receiver window is determined based on the transmitter window in the correct domain. For instance, at the transmitter, TD processing is performed, whereas FD processing is more appropriate at the receiver when FD channel equalization is used. Accordingly, the FD window needs to be determined from the equivalent FD transmitter window. Note that the linear demodulation is equivalent to the operation, $\hat{\boldsymbol{d}}_0 = \boldsymbol{B}_0^{\mathrm{H}}\boldsymbol{y}_0$, where $\boldsymbol{B}_0$ is the demodulation matrix, which also has the GFDM matrix structure, as discussed in Section 4.2.1.4.

4.2.2.2 Extended Flexibility

In addition to the original flexibility in selecting the prototype pulse (the transmitter window), the number of subcarriers and subsymbols, the flexibility can be extended without additional computational cost as follows:

1. *Flexible spreading:* It is achieved by switching one or both the spreading matrices. This is equivalent to replacing the spreading matrix by identity matrix. It can also be interpreted as GFDM precoding [22]. On the other hand, the DFT spreading can be replaced by Walsh–Hadamard transform, which requires a smaller complexity (only addition operation).
2. *Bypassing windowing and/or transform:* This flexibility can be exploited to reduce the latency when generating specific waveforms, such as OFDM and DFT-spread-OFDM.
3. *Flexible allocation:* Instead of transmitting $\boldsymbol{x}_0 = \mathrm{vec}\{\boldsymbol{X}_0^{\mathrm{T}}\}$, other types of mapping, such as $\boldsymbol{x}_0 = \mathrm{vec}\{\boldsymbol{X}_0\}$, which essentially produces orthogonal time frequency space modulation (OTFS) signal [23, 24] can be used.[5] Refer to Chapter 6 of this book for an in-depth discussion on OTFS. On the other hand, flexible allocation allows direct integration of FD-GFDM within an OFDM system. For that, the additional IDFT block used in the FD modulation needs to be flexible.

4.2.2.3 Flexible Hardware Architecture

To realize sufficient flexibility with reasonable resource cost, we consider radix-2 values of K and M, which enable the implementation of DFT with flexible fast Fourier transform (FFT) intellectual property (IP) cores.[6] This core allows the run-time reconfiguration of the size of DFT and setting the block either in DFT or IDFT mode. The architecture is illustrated in Figure 4.11. This design consists of four flexible FFT cores with the configuration parameter N_x for the DFT size and I_x to set the core in the direct (D) or inverse (I) mode. Additionally, each FFT block can be enabled (E) or disabled (D) with the parameter E_x, such that the disabled block forwards the samples to the next stage. The fourth FFT core is optional depending on the desired domain of implementation. Moreover, one memory block is used to store the result of the first transform and performs the transpose by configuring the indexing unit. The modulator window is stored in a memory. This memory is always

5 The OTFS modulation follows the TD GFDM modulation. The data spreading $\boldsymbol{D}_s = \frac{1}{\sqrt{KM}}\boldsymbol{F}_M\boldsymbol{D}^{\mathrm{T}}\boldsymbol{F}_K^{\mathrm{H}} \in \mathbb{C}^{M\times K}$ is denoted as the symplectic finite Fourier transform (SFFT). A window $\boldsymbol{W}_{\mathrm{tx}} \in \mathbb{C}^{M\times K}$ is applied. The resulting precoded data $\boldsymbol{W}_{\mathrm{tx}} \odot \boldsymbol{D}_s$ are transmitted over K OFDM symbols of M subcarriers, such that $\boldsymbol{x}_k = \frac{1}{\sqrt{M}}\boldsymbol{F}_M^{\mathrm{H}}(\boldsymbol{W}_{\mathrm{tx}} \odot \boldsymbol{D}_s)[:, k]$. A CP can be inserted as well per OFDM symbol.

6 Early studies on GFDM are constrained by even K and odd M, as a consequence of the singularity problem, which results from the assumption of real symmetric prototype pulse. This limitation was tackled by the design of Hermitian symmetric filter with similar spectrum properties [25].

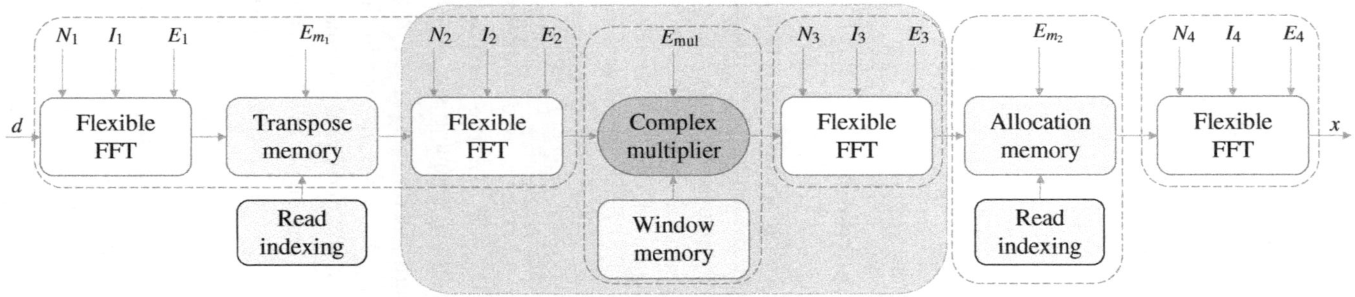

Figure 4.11 Unified architecture for TD and FD GFDM processing.

read incrementally and written with respect to the implementation domain. Furthermore, one high throughput complex multiplier is used to perform the element-wise multiplication. A third memory is required to store the samples prior to generating the final block by performing the transpose or customized allocation. The demodulator has a similar architecture in reversed order. The N_4-FFT block is enabled or disabled depending on the domain of the equalized signal and the configuration of the demodulator. For instance, if the input is an FD equalized signal and the demodulator is configured in TD, this block needs to be configured as N-IFFT. The allocation memory is used to store the received signal in columns and forward it in rows to the next block. The samples at the output of the demodulator correspond to the columns of $\hat{\boldsymbol{D}}$ in the TD and to the rows in the FD configuration. Usually, FD equalization is used. Thus, the modem can be configured as TD modulator and FD demodulator. In this way, the overall modulation and equalization processing requires only one N-FFT for the equalizer. On the other hand, including N_4-FFT preserves the symmetry of the architecture. Therefore, with fast reconfiguration, the modulator can be reconfigured as demodulator. This solution is typical for low-cost time division duplex transmission.

4.3 GFDM for OFDM Enhancement

Recalling from Section 4.2.1.2, the TD block can be expressed as $\boldsymbol{x} = \frac{1}{N}\boldsymbol{F}_N^{\mathrm{H}}\tilde{\boldsymbol{x}}$, where $\tilde{\boldsymbol{x}}$ is the FD generated with FD modulator. Consequently, the FD block can be seen as precoded data corresponding to the OFDM symbol of length N. However, the waveform is influenced by the FD modulation matrix $\tilde{\boldsymbol{A}}$. The idea can be extended to fit a small-size GFDM block within a subband that consists of a group of OFDM subcarriers.

4.3.1 Transmitter

Let N_u be the number of OFDM subcarriers used to allocate the GFDM signal, with the parameters M_u, K_u, g_u, and $N_u = K_uM_u$. The corresponding FD GFDM block is $\tilde{\boldsymbol{x}}_u = \tilde{\boldsymbol{A}}_u\boldsymbol{d}_u \in \mathbb{C}^{N_u\times 1}$. TD block with critical sampling is $\boldsymbol{x}_u = \frac{1}{N_u}\boldsymbol{F}N_u\tilde{\boldsymbol{x}}_u$. The upsampling and filtering preserve the spectral properties of the signal. Let $\tilde{\boldsymbol{x}}^{(\mathrm{up})} \in \mathbb{C}^{N\times 1}$, where $\tilde{\boldsymbol{x}}^{(\mathrm{up})}[\langle n\rangle_N] = \tilde{\boldsymbol{x}}_u[\langle n\rangle_{N_u}], -\frac{N_u}{n} \leq n < \frac{N_u}{n}$, i.e. $N - N_u$ zeros are inserted in the middle of $\tilde{\boldsymbol{x}}_u$. Therefore, $\boldsymbol{x}^{(\mathrm{up})} = \frac{1}{N}\boldsymbol{F}_N^{\mathrm{H}}\tilde{\boldsymbol{x}}^{(\mathrm{up})}$ is the upsampled and filtered signal corresponding to $\boldsymbol{x}_u$ with upsampling rate $\frac{N}{N_u}$. A frequency shift of s_u frequency samples that keeps the signal within the allocated band also preserves the spectral properties. Therefore, $\boldsymbol{x} \in \mathbb{C}^{N\times 1}$, where $\boldsymbol{x}[n] = \boldsymbol{x}^{(\mathrm{up})}e^{j2\Pi\frac{ns_u}{n}}$ has similar spectral properties of $\boldsymbol{x}_u$. All these linear steps can be expressed in the form

$$\boldsymbol{x} = \frac{1}{N}\boldsymbol{F}_N^{\mathrm{H}}\boldsymbol{P}_{\mathrm{u}}\tilde{\boldsymbol{A}}_{\mathrm{u}}\boldsymbol{d}_{\mathrm{u}}, \tag{4.25}$$

where $\boldsymbol{P}_u \in \mathbb{R}^{N\times N_u}$ is the allocation matrix, which maps the N_u FD samples of the FD block $\tilde{\boldsymbol{A}}_u\boldsymbol{d}_u$ to the N points seen as the input of the DFT matrix. Based on that, GFDM can be integrated in OFDM by inserting a flexible small-size FD-GFDM modulator to precode the input data of certain subcarriers. At the receiver side, channel equalization is required in addition to the FD-GFDM demodulator, as illustrated in Figure 4.12. One application of this model is to improve the spectral usage of OFDM. Instead of disabling large number of OFDM

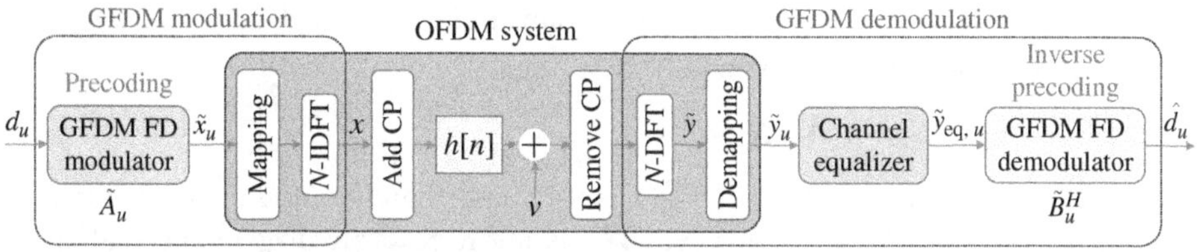

Figure 4.12 Precoded OFDM system model.

subcarriers to produce low OOB, GFDM can be used on the edge subcarriers with proper configurations to provide low side lobes. On the other hand, to preserve low complexity, the middle subcarriers can keep using plain OFDM via bypassing of the GFDM precoding. This feature can also be used to support asynchronous multiple access, which allows using fewer guard subcarriers between adjacent users. This model can also be used to introduce different type of orthogonal precoding, using the extended flexibility of GFDM. The precoding can be used, e.g. to spread the data over larger bandwidth to harvest frequency diversity.

4.3.2 Receiver

The receiver first performs the conventional OFDM processing, i.e. removing the CP followed by N-DFT, and then the deallocation. Therefore, we get the received FD-GFDM symbol

$$\tilde{\boldsymbol{y}}_u = \boldsymbol{\Lambda}^{(\tilde{h}_u)}\tilde{\boldsymbol{A}}_u\boldsymbol{d}_u + \tilde{\boldsymbol{v}}_u, \tag{4.26}$$

where $\boldsymbol{\Lambda}^{(\tilde{h}_u)}$ is the equivalent FD diagonal channel matrix and $\tilde{\boldsymbol{v}}_u$ is the additive white Gaussian noise (AWGN). In several theoretical works, the equivalent demodulation and channel equalization matrix $\boldsymbol{H}_u = \boldsymbol{\Lambda}^{(\tilde{h}_u)}\tilde{\boldsymbol{A}}_u$ is considered as a large-scale multiple-input multiple-output (MIMO) system. The linear receiver with matrix $\boldsymbol{W}_u$ is computed to get an estimate $\hat{\boldsymbol{d}}_u = \boldsymbol{W}_u^{\mathrm{H}}\tilde{\boldsymbol{y}}_u$. However, this approach may lead to an over-complicated matrix, which makes the hardware realization infeasible. In order to maintain low complexity, the receiver matrix needs to have the structure $\boldsymbol{W}_u = \boldsymbol{\Lambda}_u^{(\tilde{h}_{\mathrm{eq}})}\tilde{\boldsymbol{B}}_u$, where $\boldsymbol{\Lambda}_u^{(\tilde{h}_{\mathrm{eq}})}$ is a diagonal matrix for channel equalization, and $\tilde{\boldsymbol{B}}_u$ is a GFDM matrix. We denote this receiver as GFDM-based receiver. Accordingly, the OFDM channel equalizer function can be reused with slight modification in computing the equalizer coefficients. Moreover, small-size FD-GFDM demodulation needs to be performed after the equalization.

4.3.2.1 LMMSE GFDM-Based Receiver

For notation simplicity, we drop the index u and focus on the model of the received FD-GFDM signal: $\tilde{\boldsymbol{y}} = \boldsymbol{\Lambda}^{(\tilde{h})}\tilde{\boldsymbol{A}}\boldsymbol{d} + \tilde{\boldsymbol{v}}$. Assuming uncorrelated noise, $\boldsymbol{R}_v = E[\tilde{\boldsymbol{v}}\tilde{\boldsymbol{v}}^{\mathrm{H}}] = \sigma^2\boldsymbol{I}_N$, $\boldsymbol{R}_d = E[\boldsymbol{d}\boldsymbol{d}^{\mathrm{H}}]$ is the correlation matrix of the data symbols, and $\tilde{\boldsymbol{A}}$ is non-singular. The linear minimum mean squared error (LMMSE) receiver can be written as

$$\begin{aligned}\boldsymbol{W}^{\mathrm{H}} &= (\tilde{\boldsymbol{A}}^{\mathrm{H}}\boldsymbol{\Lambda}^{(\tilde{h})\mathrm{H}}\boldsymbol{R}_v^{-1}\boldsymbol{\Lambda}^{(\tilde{h})}\tilde{\boldsymbol{A}} + \boldsymbol{R}_d^{-1})^{-1}\tilde{\boldsymbol{A}}^{\mathrm{H}}\boldsymbol{\Lambda}^{(\tilde{h})\mathrm{H}}\boldsymbol{R}_v^{-1}\\ &= \tilde{\boldsymbol{A}}^{-1}(\boldsymbol{\Lambda}^{(\tilde{h})\mathrm{H}}\boldsymbol{\Lambda}^{(\tilde{h})} + \sigma^2\tilde{\boldsymbol{A}}^{-H}\boldsymbol{R}_d^{-1}\tilde{\boldsymbol{A}}^{-1})^{-1}\boldsymbol{\Lambda}^{(\tilde{h})\mathrm{H}}.\end{aligned} \tag{4.27}$$

Therefore, the LMMSE receiver can be decoupled into channel equalizer $\boldsymbol{H}_{\text{eq}}$ and zero-forcing (ZF) demodulation, where $\tilde{\boldsymbol{B}}^{\text{H}} = \tilde{\boldsymbol{A}}^{-1}$, and

$$\boldsymbol{H}_{\text{eq}} = (\boldsymbol{\Lambda}^{(\tilde{h})\text{H}}\boldsymbol{\Lambda}^{(\tilde{h})} + \sigma^2\tilde{\boldsymbol{A}}^{-H}\boldsymbol{R}_d^{-1}\tilde{\boldsymbol{A}}^{-1})^{-1}\boldsymbol{\Lambda}^{(\tilde{h})\text{H}}.$$

From (4.21), it is clear that $\tilde{\boldsymbol{B}}$ is a GFDM matrix, and therefore, the demodulator window $\boldsymbol{W}_{\text{rx}}[k,m] = [\boldsymbol{W}_{\text{tx}}[k,m]]^{-1}$. However, $\boldsymbol{H}_{\text{eq}}$ is not necessarily diagonal depending on $\boldsymbol{R}_x = \tilde{\boldsymbol{A}}\boldsymbol{R}_d\tilde{\boldsymbol{A}}^{\text{H}}$. The diagonal equalization corresponds to the element-wise equalization of the signal $\tilde{\boldsymbol{y}}[n] = \tilde{\boldsymbol{h}}[n]\tilde{\boldsymbol{x}}[n] + \tilde{\boldsymbol{v}}[n]$, where $E[\tilde{\boldsymbol{x}}[n]^2] = \boldsymbol{R}_x[n,n]$. Accordingly,

$$\tilde{\boldsymbol{h}}_{\text{eq}}[n] = \left(|\tilde{\boldsymbol{h}}[n]|^2 + \frac{\sigma^2}{\boldsymbol{R}_x[n,n]}\right)^{-1}\tilde{\boldsymbol{h}}^*[n].$$

In the special case when $\boldsymbol{R}_d = E_s\boldsymbol{I}_N$ and $\tilde{\boldsymbol{A}}$ is orthogonal, $\boldsymbol{R}_x = E_s\boldsymbol{I}_N$, which produces the exact LMMSE, because

$$\boldsymbol{H}_{\text{eq}} = \boldsymbol{\Lambda}^{(\tilde{h}_{\text{eq}})} = \left(\boldsymbol{\Lambda}^{(\tilde{h})\text{H}}\boldsymbol{\Lambda}^{(\tilde{h})} + \frac{\sigma^2}{E_s}\right)^{-1}\boldsymbol{\Lambda}^{(\tilde{h})\text{H}}. \tag{4.28}$$

This is the case where the extended GFDM modulation is used as orthogonal precoding. On the other hand, when conventional GFDM is used to reduce the OOB, either the correlation matrix $\boldsymbol{R}_d$ is not a scaled identity or the modulation matrix is non-orthogonal, the diagonal equalization with ZF demodulation is an approximation of LMMSE. The computation of $\boldsymbol{R}_x[n,n]$ is performed offline for different configurations and stored in lookup tables.

4.4 Conclusions

The extended GFDM framework combined with other waveform techniques provides an efficient platform for the processing and prototyping of waveforms. This is aligned with the design of modern communications systems architecture, which is based on virtualization and software-defined networking. To this extent, the flexible GFDM integrated with software-defined radio front end contributes to the virtualization of physical layer.

References

1 Bingham, J.A.C. (1990). Multicarrier modulation for data transmission: an idea whose time has come. *IEEE Communications Magazine* 28 (5): 5–14.

2 Schwartz, M. and Batchelor, C. (2008). The origins of carrier multiplexing: Major George Owen Squier and AT T. *IEEE Communications Magazine*. 46 (5): 20–24.

3 Cherubini, G., Eleftheriou, E., and Olcer, S. (1999). Filtered multitone modulation for VDSL. IEEE GLOBECOM, Volume 2, Rio de Janeireo, Brazil, December 1999, pp. 1139–1144.

4 Weinstein, S.B. (2009). The history of orthogonal frequency-division multiplexing [History of Communications]. *IEEE Communications Magazine* 47 (11): 26–35.

5 Saltzberg, B. (1967). Performance of an efficient parallel data transmission system. *IEEE Signal Processing Letters.* 15 (6): 805–811.

6 Farhang-Boroujeny, B. (2011). OFDM versus filter bank multicarrier. *IEEE Signal Processing Magazine* 28 (3): 92–112.

7 Weinstein, S. and Ebert, P. (1971). Data transmission by frequency-division multiplexing using the discrete Fourier transform. *IEEE Signal Processing Letters.* 19 (5): 628–634.

8 Chow, J.S., Tu, J.C., and Cioffi, J.M. (1991). A discrete multitone transceiver system for HDSL applications. *IEEE Journal on Selected Areas in Communications.* 9 (6): 895–908.

9 Peled, A. and Ruiz, A. (1980). Frequency domain data transmission using reduced computational complexity algorithms. IEE ICASSP, Denver, Colorado, USA, April 1980.

10 Walzman, T. and Schwartz, M. (1973). Automatic equalization using the discrete frequency domain. *IEEE Transactions on Information Theory* 19 (1): 59–68.

11 Sari, H., Karam, G., and Jeanclaude, I. (1995). Transmission techniques for digital terrestrial TV broadcasting. *IEEE Communications Magazine* 33 (2): 100–109.

12 Myung, H.G., Lim, J., and Goodman, D.J. (2006). Single carrier FDMA for uplink wireless transmission. *IEEE Vehicular Technology Magazine.* 1 (3): 30–38.

13 Dinis, R., Falconer, D., Lam, C.T., and Sabbaghian, M. (2004). A multiple access scheme for the uplink of broadband wireless systems. IEEE GLOBECOM, Dallas, TX, USA, November 2004.

14 Michailow, N., Matthé, M., Gaspar, I.S. et al. (2014). Generalized frequency division multiplexing for 5th generation cellular networks. *IEEE Transactions on Communications* 62 (9): 3045–3061.

15 Fettweis, G., Krondorf, M., and Bittner, S. (2009). GFDM - generalized frequency division multiplexing. IEEE VTC Spring, Barcelona, Spain, April 2009.

16 Gao, X., Wang, W., Xia, X. et al. (2011). Cyclic prefixed OQAM-OFDM and its application to single-carrier FDMA. *IEEE Transactions on Communications* 59 (5): 1467–1480.

17 Daubechies, I. , Landau, H.J., Landau, Z. (1994). Gabor time-frequency lattices and the Wexler-Raz identity. *Journal of Fourier Analysis and Applications.* 1 (4): 437–478.

18 Wexler, J. and Raz, S. (1990). Discrete Gabor expansions. *Signal Processing* 21 (3): 207–220.

19 Janssen, A.J.E.M. (1994). Signal analytic proofs of two basic results on lattice expansions. *Applied and Computational Harmonic Analysis.*1 (4): 350–354.

20 Gaspar, I., Matthé, M., Michailow, N., et al. (2015). Frequency-shift offset-QAM for GFDM. *IEEE Communications Letters* 19 (8): 1454–1457.

21 Nimr, A., Matthe, M., Zhang, D., and Fettweis, G.P. (2017). Optimal radix-2 FFT compatible filters for GFDM. *IEEE Communications Letters* 21 (7): 1497–1500.

22 Matthé, M., Mendes, L., Gaspar, I. et al. (2016). Precoded GFDM transceiver with low complexity time domain processing. *EURASIP Journal on Wireless Communications and Networking* 2016 (1): 138.

23 Hadani, R., Rakib, S., Tsatsanis, A., Monk, A., Goldsmith, A.J., Molisch, A.F., Calderbank, R. (2017). Orthogonal time frequency space modulation. IEEE WCNC, IEEE, pp. 1–6.

24 Nimr, A., Chafii, M., Matthe, M., and Fettweis, G.P. (2018). Extended GFDM framework: OTFS and GFDM comparison. GLOBECOM, IEEE, pp. 1–6.

25 Nimr, A., Zhang, D., Martinez, A.B., and Fettweis, G.P. (2017). A study on the physical layer performance of GFDM for high throughput wireless communication. EUSIPCO, IEEE, pp. 638–642.

5

Filter Bank Multicarrier Modulation

Behrouz Farhang-Boroujeny

Electrical and Computer Engineering Department, University of Utah, USA

5.1 Introduction

Despite the dominance of orthogonal frequency division multiplexing (OFDM) as the most popular broadband communications technology in both wired [1] and wireless [2] applications, there still exist use cases where other alternatives may suit better. For instance, OFDM may degrade significantly in channels with fast variation in time; see Chapter 6 for more explanation and a solution. Another situation where OFDM degradation may be noticeable is in channels with partial band interference. Poor response of the OFDM prototype filter makes suppression of such interference a difficult task. Although some filtering methods have been proposed to overcome this problem (see Chapter 2), these are somewhat limited in performance. The filter bank multicarrier (FBMC) techniques that are presented in this chapter provide a perfect solution to partial band interference, simply by deactivating the interfered subcarrier bands. Moreover, as discussed in Section 5.4, the FBMC prototype filter can be designed to balance its robustness to both time spreading (i.e. frequency selectivity) and frequency spreading (i.e. variation in time) of the channel.

This chapter is organized as follows. A summary of FBMC methods is discussed in Section 5.2. The fundamental theory behind the class of FBMC systems that achieve maximum compactness both in time and frequency is presented in Section 5.3. The subject of the prototype filter design is covered in Section 5.4. Synchronization and tracking methods are then discussed in Section 5.5. The topics of channel equalization and computational complexity are briefly discussed in Sections 5.6 and 5.7, respectively. Lastly, in Section 5.8, we discuss some potential applications of FBMC.

5.1.1 Notations:

The presentations in this chapter follow a mix of continuous-time and discrete-time formulations, as appropriate. While $x(t)$ refers to a continuous function of time t, $x[n]$ is used to refer to its discrete-time version, with n denoting the time index. The notation f is used as frequency variable in $X(f)$, the Fourier transform of the continuous-time signal $x(t)$, and also as normalized frequency in $X(e^{j2\pi f})$, the Fourier transform of the discrete-time signal $x[n]$. We use $\star$ to denote linear convolution, and the superscript $*$ to denote complex conjugate.

Radio Access Network Slicing and Virtualization for 5G Vertical Industries, First Edition.
Edited by Lei Zhang, Arman Farhang, Gang Feng, and Oluwakayode Onireti.

5.2 FBMC Methods

FBMC communication techniques were first developed in the mid-1960s. Chang [3] presented the conditions required for signaling a parallel set of pulse amplitude modulated (PAM) symbol sequences through a bank of overlapping filters within a minimum bandwidth. To transmit PAM symbols in a bandwidth-efficient manner, Chang proposed vestigial sideband (VSB) signaling for subcarrier sequences. Saltzberg [4] extended the idea and showed how Chang's method could be modified for transmission of quadrature amplitude modulation (QAM) symbols in a double-sideband (DSB)-modulated format. In order to keep the bandwidth efficiency of this method similar to that of Chang's signaling, Saltzberg noted that the in-phase and quadrature components of each QAM symbol should be time staggered by half a symbol interval. Another key development appeared in [5], where the authors noted that Chang's/Saltzberg's method could be adopted to match channel variations in doubly dispersive channels, and hence minimize intersymbol interference (ISI) and intercarrier interference (ICI).

Saltzberg's method has received broad attention in the literature and has been given different names. Most authors have used the name offset quadrature amplitude modulation (OQAM) to reflect the fact that the in-phase and quadrature components are transmitted with a time offset with respect to each other. Moreover, to emphasize the multicarrier feature of the method, the suffix OFDM has been added, and hence the name OQAM-OFDM. Others have chosen to call it staggered quadrature amplitude modulation (SQAM), equivalently, SQAM-OFDM. In [6] the shorter name, staggered multitone (SMT), was introduced. Moreover, in [6] Chang's method is named cosine-modulated multitone (CMT) and it is shown that SMT and CMT are equivalent and, thus, with a minor modification, an implementation for one can be applied to the other. Further details of SMT and CMT can be found in [7, 8]. We further note here that subcarrier bands in both SMT and CMT overlap to maximize the bandwidth efficiency.

Another multicarrier method that has received some attention is filtered multitone (FMT) [9]. FMT is based on non-overlapping subcarriers, and hence is less band efficient, but separation of subcarriers may simplify some signal processing tasks in certain applications. The discussion in this chapter is limited to SMT and CMT.

5.3 Theory

The theory of CMT and SMT has evolved over the past five decades through many researchers who have studied them from different angles. Early studies by Chang [3] and Saltzberg [4] have presented their finding in terms of continuous-time signals. The more recent studies have presented the formulations and conditions for ISI and ICI cancellation in discrete time, e.g. [10, 11]. On the other hand, a couple of recent works [6, 7], from the author of this chapter and his group, have revisited the more classical approach and presented the theory of CMT and SMT in continuous time. It is believed that this formulation greatly simplifies the essence of the theoretical concepts behind the theory of CMT and SMT and how these two waveforms are related. It also facilitates the design of prototype filters that are used for realization of CMT and SMT systems. Thus, here also we follow the continuous-time approach. We discuss the underlying theory mostly through the *time–frequency phase–space*, with minimum involvement in mathematical details. Interested readers who wish to see more mathematical details are referred to [6, 12].

5.3.1 CMT

In CMT, data symbols are from a PAM alphabet, and hence are real-valued. To establish a transmission with the maximum bandwidth efficiency, PAM symbols are distributed in a time–frequency phase–space lattice with a density of two symbols per unit area. This is equivalent to one complex symbol per unit area. Moreover, because of the reasons that are explained below a 90° phase shift is introduced to the respective carriers among the adjacent symbols. These concepts are presented in Figure 5.1. VSB modulation is applied to cope with the carrier spacing $F = 1/2T$. The pulse-shape used for this purpose at the transmitter as well as for matched filtering at the receiver is a square-root Nyquist (SR-Nyquist) waveform, $p(t)$, which has been designed such that $q(t) = p(t) \star p(-t)$ is a Nyquist pulse with regular zero crossings at $2T$ time intervals. Also, $p(t)$, by design, is a real-valued function of time and $p(t) = p(-t)$. These properties of $p(t)$, as demonstrated below, are instrumental for the correct functionality of CMT.

Figure 5.2 presents a set of magnitude responses of the modulated versions of the pulse-shape $p(t)$ for the data symbols transmitted at $t = 0$ and $t = T$. The colors used for the plots follow those in Figure 5.1, to reflect the respective phase shifts.

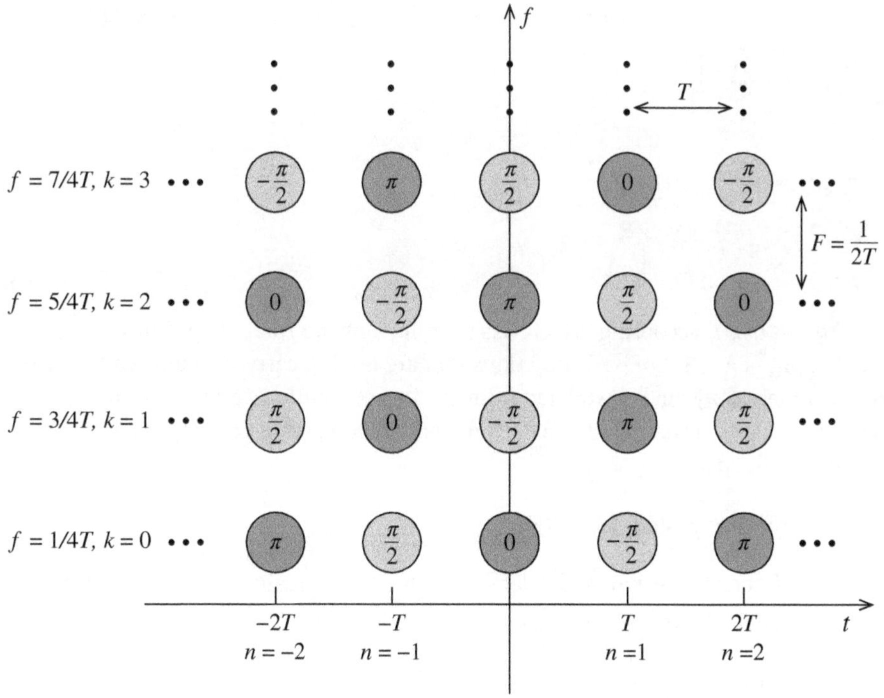

Figure 5.1 The CMT time–frequency phase–space lattice.

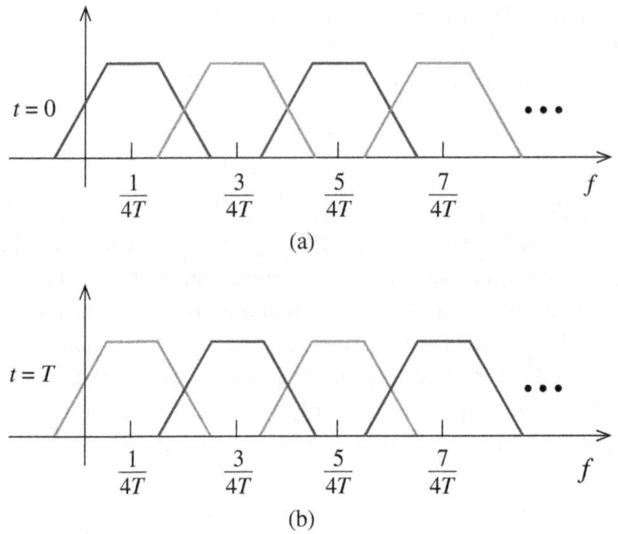

Figure 5.2 Magnitude responses of the CMT pulse-shaping filters at various subcarriers and time instants $t = 0$ and $t = T$.

Let $n = \dots, -2, -1, 0, 1, 2, \dots$ denote symbol time index, $k = 0, 1, 2, \dots$ denote symbol frequency index, $s_k[n]$ denote the (n, k) data symbol in the time–frequency lattice, and $\theta_k[n] = (k - n)\frac{\pi}{2}$ be the phase shift that is added to the carrier of $s_k[n]$. Accordingly, a CMT waveform that is constructed based on the pulse-shape/prototype filter $p(t)$ is expressed as

$$x(t) = \sum_n \sum_k s_k[n] a_{n,k}(t), \tag{5.1}$$

where

$$a_{n,k}(t) = e^{j\theta_k[n]} p_{n,k}(t), \tag{5.2}$$

and

$$p_{n,k}(t) = p(t - nT)e^{j\frac{(2k+1)\pi}{2T}t}. \tag{5.3}$$

The synthesis of $x(t)$ according to (5.1) has the following interpretations. The terms $a_{n,k}(t)$ may be thought of as a set of basis functions that are used to carry the data symbols $s_k[n]$. The data symbols $s_k[n]$ can be extracted from $x(t)$ in a straightforward manner if $a_{n,k}(t)$ are a set of orthogonal basis functions. The orthogonality for a pair of functions $v_1(t)$ and $v_2(t)$, in general, is defined as

$$\langle v_1(t), v_2(t) \rangle = \int_{-\infty}^{\infty} v_1(t) v_2^*(t) dt = 0. \tag{5.4}$$

For the case of interest here, where the data symbols $s_k[n]$ are real-valued, the orthogonality definition (5.4) can be replaced by the more relaxed definition

$$\langle v_1(t), v_2(t) \rangle_{\mathrm{R}} = \Re\left\{ \int_{-\infty}^{\infty} v_1(t) v_2^*(t) dt \right\} = 0 \tag{5.5}$$

where $\Re\{\cdot\}$ indicates the real part. The definition (5.5) is referred to as *real orthogonality*.

It is not difficult to show that

$$\langle a_{n,k}(t), a_{m,l}(t)\rangle_R = \begin{cases} 1, & n = m \text{ and } k = l \\ 0, & \text{otherwise} \end{cases} \tag{5.6}$$

and hence, for any pair of n and k,

$$s_k[n] = \langle x(t), a_{n,k}(t)\rangle_R. \tag{5.7}$$

To develop an in-depth understanding of the CMT signaling, it is instructive to explore a detailed derivation of (5.6). To this end, we begin with the definition

$$\langle a_{n,k}(t), a_{m,l}(t)\rangle_{\mathrm{R}} = \Re\left\{\int_{-\infty}^{\infty} e^{j\theta_k[n]}p_{n,k}(t)e^{-j\theta_l[m]}p^*_{m,l}(t)dt\right\}, \tag{5.8}$$

and note that this can be rearranged as

$$\langle a_{n,k}(t), a_{m,l}(t)\rangle_{\mathrm{R}} = \Re\left\{\int_{-\infty}^{\infty} e^{j(m-n+k-l)\frac{\pi}{2}}p(t-nT)p(t-mT)e^{j\frac{(k-l)\pi}{T}t}dt\right\}. \tag{5.9}$$

When $m = n$ and $k = l$, after a change of variable t to $t + nT$, (5.9) reduces to

$$\begin{aligned}\langle a_{n,k}(t), a_{n,k}(t)\rangle_R &= \int_{-\infty}^{\infty} p^2(t)dt \\ &= 1,\end{aligned} \tag{5.10}$$

where the second equality follows from the fact that $p(t)$ is a real-valued SR-Nyquist pulse and $p(t) = p(-t)$.

When $k = l$, $m \neq n$, and $m - n = 2r$, where r is an integer,

$$\begin{aligned}\langle a_{n,k}(t), a_{n,k}(t)\rangle_R &= (-1)^r \int_{-\infty}^{\infty} p(t-nT)p(t-mT)dt \\ &= 0,\end{aligned} \tag{5.11}$$

where the second equality follows since $p(t)$ is a SR-Nyquist pulse, designed for a symbol spacing $2T$. On the other hand, when $k = l$, but $m - n = 2r + 1$,

$$\begin{aligned}\langle a_{n,k}(t), a_{n,k}(t)\rangle_R &= \Re\left\{j(-1)^r \int_{-\infty}^{\infty} p(t-nT)p(t-mT)dt\right\} \\ &= 0.\end{aligned} \tag{5.12}$$

Next, consider the case where $k - l = 1$ and $m - n = 2r$. In that case, one finds that

$$\begin{aligned}\langle a_{n,k}(t), a_{m,l}(t)\rangle_R &= \Re\left\{j(-1)^r \int_{-\infty}^{\infty} p(t-nT)p(t-mT)e^{j\frac{\pi}{T}t}dt\right\} \\ &= -(-1)^r \int_{-\infty}^{\infty} p(t-nT)p(t-mT)\sin\left(\frac{\pi}{T}t\right)dt \\ &= 0\end{aligned} \tag{5.13}$$

where the last identity follows by applying the change of variable $t \to t + \frac{m+n}{2}T$ and noting that the expression under the integral will reduce to an odd function of t. Following similar procedures, it can be shown that the real orthogonality $\langle a_{n,k}(t), a_{m,l}(t)\rangle_R = 0$ also holds, when $k - l = 1$ and $m - n$ is an odd integer, and when $k - l = -1$ and $m - n$ is either an even or an odd integer. Finally, for the cases where $|k - l| > 1$, the real orthogonality $\langle a_{n,k}(t), a_{m,l}(t)\rangle_R = 0$ is trivially confirmed by noting that the underlying basis functions correspond to filters that have no overlapping bands. The stop-band quality of the frequency response of the prototype filter $p(t)$ determines the accuracy of the equality $\langle a_{n,k}(t), a_{m,l}(t)\rangle_R = 0$ when $|k - l| > 1$.

To summarize, the time–frequency phase–space lattice diagram of Figure 5.1 and the subsequent derivations revealed that the set of basis functions $a_{n,k}(t)$ that carry the PAM symbols $s_k[n]$ satisfy the real orthogonality condition (5.5) and as a result facilitate extraction of the PAM symbols at the receiver. Furthermore, making use of these results, one arrives at the CMT transmitter and receiver structures that are presented in Figure 5.3. As shown, a synthesis filter bank (SFB) is used to construct the transmit signal. This essentially modulates the basis functions $a_{n,k}(t)$ by the PAM symbols $s_k[n]$. At the receiver side, on the other hand, the received signal is passed through an analysis filter bank (AFB) that finds the (real) projections of the received signal over the basis functions $a_{n,k}(t)$. Furthermore, it is assumed that there are N subcarrier streams and the data stream of the kth subcarrier at the input to the SFB is represented by the impulse train

$$s_k(t) = \sum_n s_k[n]\delta(t - nT). \tag{5.14}$$

It is also worth noting that, in practice, the analyzed signals at the output of the AFB should be equalized. Here, to keep the presentation simple, we have not included the equalizers. Equalizer details are presented in Section 5.6.

5.3.2 SMT

SMT may be thought as an alternative to CMT. Its time–frequency phase–space lattice is obtained from that of CMT (Figure 5.1) through a frequency shift of the lattice points down by $1/4T$, scaling the time axis by a factor $1/2$ and hence the frequency axis by a factor 2. Moreover, to remain consistent with the past literature of SMT (equivalently, OQAM-OFDM), some adjustments to the carrier phases of the lattice points have been made. This leads to the time–frequency phase–space lattice that is presented in Figure 5.4. The magnitude responses of the SMT pulse-shaping filters at various subcarriers and time instants $t = 0$ and $t = T/2$ are presented in Figure 5.5. Note that here the PAM symbols are spaced by $T/2$ and subcarriers are spaced by $1/T$. In SMT, each pair of adjacent symbols along time in each subcarrier is treated as the real and imaginary parts of a QAM symbol. This leads to the transmitter and receiver structures that are presented in Figure 5.6. Here, each data symbol $s_k[n]$ belongs to a QAM constellation and, thus, may be written in terms of its in-phase and quadrature components as $s_k[n] = s_k^I[n] + js_k^Q[n]$. Accordingly, the inputs to the SFB in Figure 5.6 are

$$s_k^I(t) = \sum_n s_k^I[n]\delta(t - nT), \tag{5.15}$$

and

$$s_k^Q(t) = \sum_n s_k^Q[n]\delta(t - nT). \tag{5.16}$$

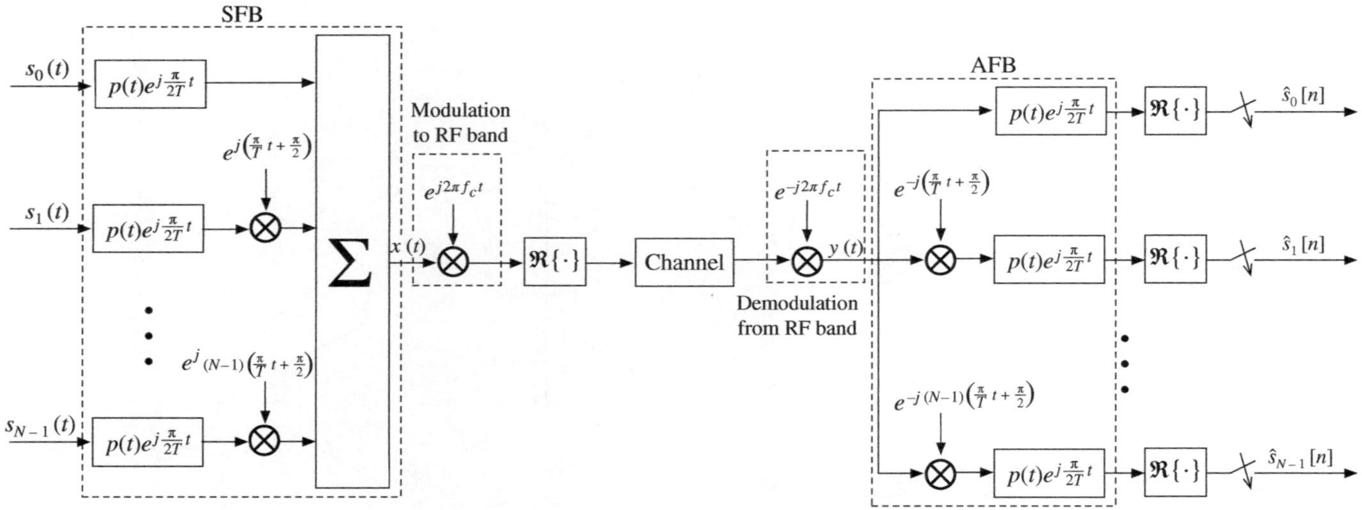

Figure 5.3 CMT transmitter and receiver blocks.

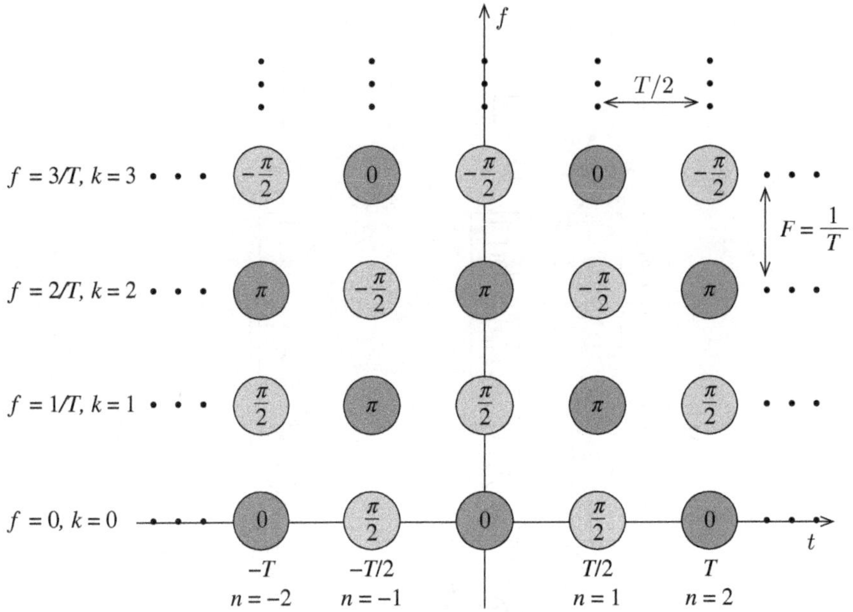

Figure 5.4 The SMT time–frequency phase–space lattice.

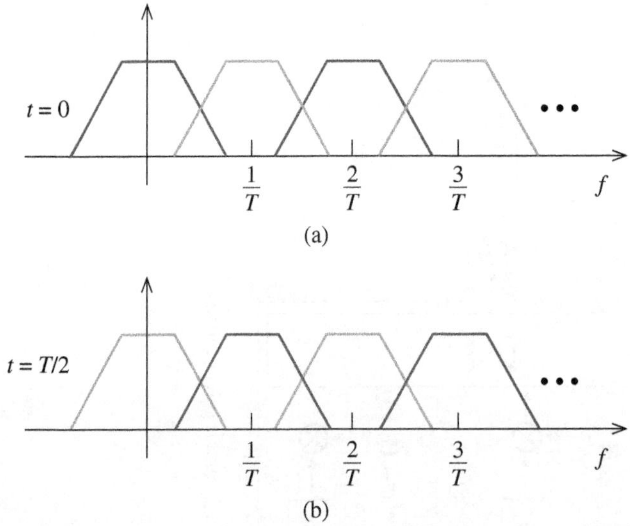

Figure 5.5 Magnitude responses of the SMT pulse-shaping filters at various subcarriers and time instants $t = 0$ and $t = T/2$.

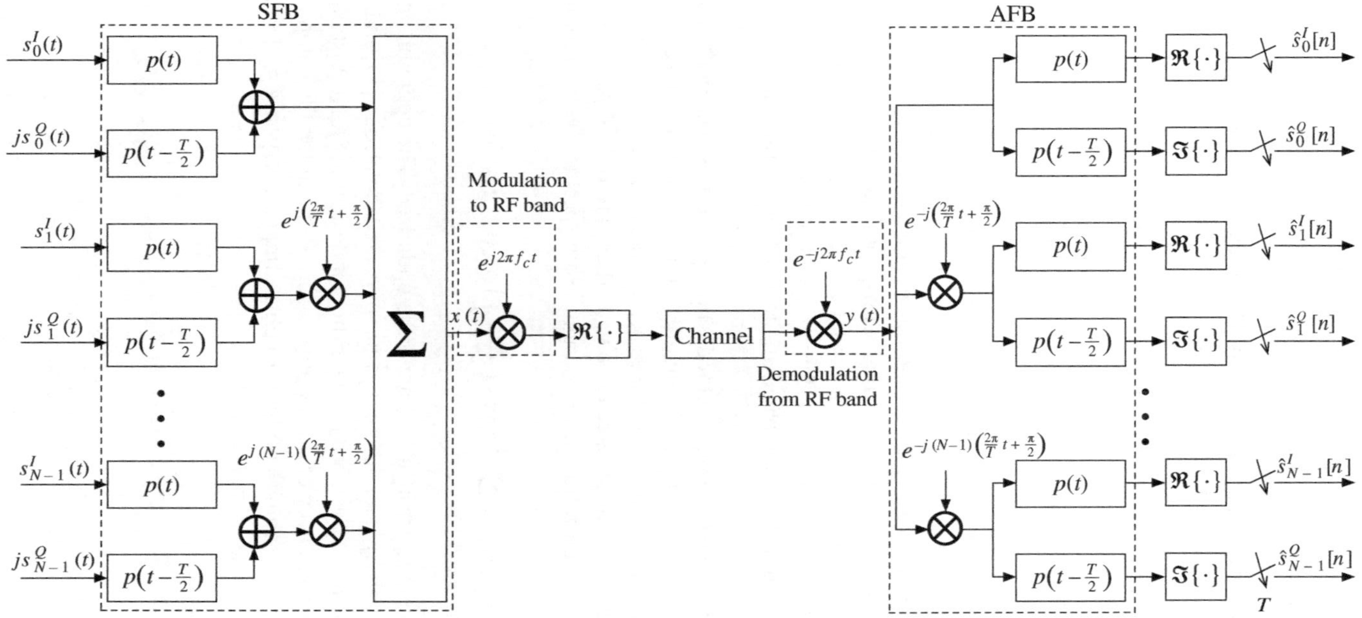

Figure 5.6 SMT transmitter and receiver blocks.

5.4 Prototype Filter Design

The results of the Section 5.3 suggest that linear phase SR-Nyquist filters that have been widely used for single carrier data transmission are the most trivial choice for prototype filter in FBMC systems. This indeed remains an accurate statement as long as the underlying channel is time invariant or varies slowly. However, in cases where the channel is fast varying, a more general class of SR-Nyquist filters that satisfy the Nyquist condition along both the time and frequency axis should be adopted. Taking note of this, here we discuss prototype filter designs for time-invariant and time-varying channels separately.

5.4.1 Prototype Filters for Time-Invariant Channels

The classical prototype filter for FBMC systems that was suggested in [3, 4] was the square-root raised-cosine (SRRC) filter. More specifically, both [3, 4] suggested using an SRRC filter with the roll-off factor $\alpha = 1$. In practice, where FBMC systems are implemented in discrete time, SRRC response should be sampled and truncated. The truncation of the SRRC response may result in a filter with poor frequency response, thus making SRRC a poor choice. Advancements in digital filters design have led to very effective methods for designing SR-Nyquist filters for any specified finite length. These designs, two of which are presented here, aim at balancing the stop-band attenuation and the Nyquist property of the designs.

Martin [13] has proposed a design method whose goal is to satisfy the Nyquist criterion approximately, while achieving good attenuation in the stopband. A nice property of this method is that when the number of samples per symbol interval is N and the filter length is $\beta N + 1$, where β is an integer greater than 1, there are only β parameters that need to be found for the optimization of the design. More specifically, the prototype filter is constructed as

$$p[n] = \begin{cases} \frac{1}{\beta N+1}\left(k_0 + 2\sum_{l=1}^{\beta-1} k_l \cos\left(\frac{2\pi l n}{\beta N+1}\right)\right), & 0 \leq n \leq \beta N \\ 0, & \text{otherwise} \end{cases} \tag{5.17}$$

where the coefficients k_0 through $k_{\beta-1}$ are to be optimized. The optimum choices of the coefficients k_0 through $k_{\beta-1}$, for different values of β, are tabulated in [13] and, also in [14].

More recently, we have developed an algorithm for designing SR-Nyquist filters that can balance between the accuracy of the Nyquist constraints and the filter stop-band attenuation [15]. To select the samples of the zero-phase impulse response $p[n] = p[-n]$ of the SR-Nyquist filter, this algorithm defines

$$q[n] = p[n] \star p[-n] \tag{5.18}$$

and minimizes the cost function

$$\xi = (q[0] - 1)^2 + \sum_{m=1}^{K/2} q^2[mN] + \gamma \int_{\frac{1+\alpha}{2N}}^{1-\frac{1+\alpha}{2N}} |P(e^{j2\pi f})|^2 df \tag{5.19}$$

with respect to the elements $p[n]$. The first two terms on the right-hand side of (5.19) are to enforce Nyquist property on $q[n]$, and the third term is included to minimize the stop-band response of $p[n]$. Also, the length of $p[n]$ is assumed to be equal to $\beta N + 1$. One may note

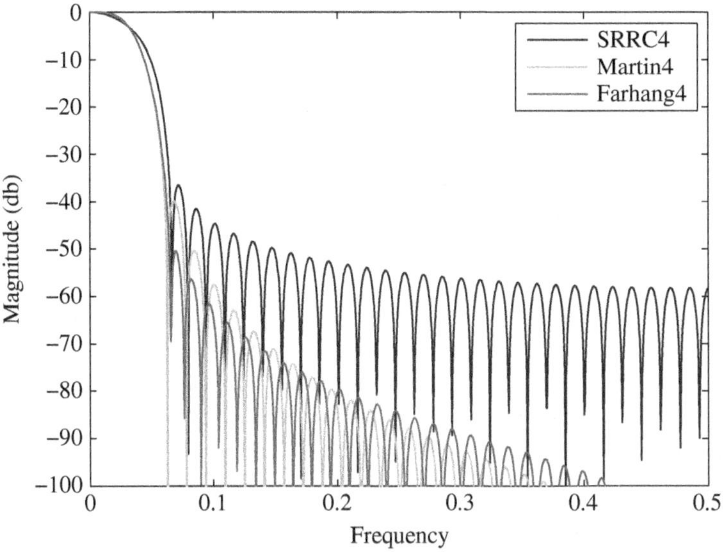

Figure 5.7 Magnitude responses of a sampled and truncated SRRC filter, and two discrete-time designed SR-Nyquist filters.

that since ξ is a fourth order function of the parameters $p[n]$, it is a multimodal function and hence its minimization may not be straightforward. However, fortunately, if $p[n]$ could be initialized to near its optimum choice, an iterative solution may be applied to search for the local minimum of ξ.

Figure 5.7 presents the magnitude responses of three designs based on (i) a sampled and truncated version of the impulse response of a SRRC filter: SRRC4; (ii) an SR-Nyquist design obtained following [13]: Martin4; and (iii) an SR-Nyquist design obtained following [15]: Farhang4. All designs are based on the roll-off factor $\alpha = 1$, are for an FBMC system with $N = 16$ subcarriers, and the suffix 4 on the names indicates that for all designs, $\beta = 4$, i.e. the designed filters have a length $\beta N + 1 = 4 \times 16 + 1 = 65$. In addition, for Farhang4, the parameter γ is set equal to 0.1.

5.4.2 Prototype Filters for Time-Varying Channels

When a channel is subject to dispersion both in time (due to multipath effects) and in frequency (due to variation of the channel in time), we say the channel is *doubly dispersive*. We also note that while the dominant effect of time dispersion in a channel is ISI, the dominant effect of frequency dispersion is ICI. Hence, in order to design prototype filters that address both time and frequency dispersions, it has been argued [5] that one should choose a pulse-shape $p(t)$ with similar behavior along time and frequency. In particular, if $p(t)$ is selected to be a SR-Nyquist along the time axis, it should also be ensured that $P(f)$ is an SR-Nyquist along the frequency axis. To this end, the pulse-shape, $p(t)$, with the following property may be adopted:

$$P(f) = p(\eta f), \quad \text{for a constant scaling factor } \eta, \tag{5.20}$$

i.e. a function that has the same form in both the time and frequency domain.

The parameter η in (5.20) is related to the symbol spacing in time, T, and frequency, F, and is given by

$$\eta = \frac{T}{F}. \tag{5.21}$$

Also, as one may understand intuitively, T and F are, respectively, chosen proportional to time dispersion, $\Delta\tau$, and frequency dispersion, $\Delta\nu$, of the channel, i.e. $T/F = \Delta\tau/\Delta\nu$. Hence, the following identity also holds.

$$\eta = \frac{\Delta\tau}{\Delta\nu}. \tag{5.22}$$

The definitions for $\Delta\tau$ and $\Delta\nu$ are usually vague. The time dispersion $\Delta\tau$ may be thought of as a coarse estimate of the duration of the channel impulse response, equivalently, the span of the multipaths of the channel. Similarly, the frequency dispersion $\Delta\nu$ may be thought of as a coarse estimate of the span of Doppler spread of the channel.

The design of the prototype filter for time-varying channels is closely tied to the ambiguity function; e.g. see [16]:

$$A_p(\tau, \nu) = \int_{-\infty}^{\infty} p(t + \tau/2)p^*(t - \tau/2)e^{-j2\pi\nu t}dt, \tag{5.23}$$

where τ is a time delay and ν is a frequency shift.

Let $p(t)$ be a prototype filter and $p_{n,k}(t)$ be a time frequency translated version of the same, which is defined as

$$p_{n,k}(t) = p(t - nT)e^{j2\pi kFt}. \tag{5.24}$$

We note that

$$\begin{aligned}\langle p_{m,k}(t), p_{n,l}(t)\rangle &= \int_{-\infty}^{\infty} p(t - mT)e^{j2\pi kFt}p(t - nT)e^{-j2\pi lFt}dt \\ &\propto A_p((n - m)T, (l - k)F),\end{aligned} \tag{5.25}$$

where $\propto$ denotes proportionality. The proportionality factor is a phase shift due to the delays of mT and nT whose value is irrelevant to our discussion here. Using (5.25), one may note that $p_{n,k}(t)$ of (5.24), for all choices of n and k, will form a set of complex-valued orthogonal basis functions, if $p(t)$ is chosen such that

$$A_p(nT, lF) = \begin{cases} 1, & n = l = 0 \\ 0, & \text{otherwise.} \end{cases} \tag{5.26}$$

Filter design methods that satisfy the orthogonality conditions (5.26) have been proposed. A more widely known design is presented in [5] and is called *isotropic* orthogonal transform algorithm (IOTA) filter. An alternative design proposed in [17], on the other hand, constructs the filter $p(t)$ based on a linear combination of a set of Hermite functions. This design gives a great degree of flexibilities as demonstrated in [18]. Discussion on IOTA design can be found in [5, 7]. More on both IOTA design and Hermite design can be found in [8].

5.5 Synchronization and Tracking Methods

Synchronization methods are necessary in any receiver to compensate for carrier and clock mismatch between the transmitter and receiver. In majority of standards, including those of OFDM and FBMC, pilot aided approaches have been adopted. For instance, all the OFDM packet formats that have been proposed for WiFi and WiMAX as well as those suggested in Long-Term Evolution (LTE), LTE-Advanced, and 5G new radio (NR) start with a short-training periodic preamble, for coarse carrier acquisition. This is followed by a long preamble, consisting of two or more similar OFDM symbols that are used for fine-tuning of the carrier, timing acquisition, and channel estimation; e.g. see [19]. In addition, to be able to track channel variations during the payload of each packet, it is often proposed to insert distributed pilots in the time–frequency space. Interpolation methods are then used to obtain the channel gain, and accordingly equalizer gains, at the data points.

A few researchers who have looked at the packet design for FBMC have also suggested a similar packet format to those of the OFDM [20–22]. In particular, a periodic training consisting of a few repetitions of a pilot FBMC symbol is suggested to be used along with a least squares or a maximum likelihood estimator to estimate carrier frequency offset (CFO) and symbol timing. This is equivalent to long preamble in OFDM. Moreover, in [20], it is noted that the use of a short preamble, similar to that of OFDM, may be useful for coarse estimation of carrier, so that fine-tuning of carrier and timing phase based on the long preamble can be established more robustly. Distributed pilots to track channel variations in payload part of FBMC packets have also been proposed [23, 24]. In the rest of this section, we review these methods in some detail.

5.5.1 Preamble Design

Figure 5.8 presents the packet format that has been proposed in [20–22, 25, 26]. The short preamble consists of a few tones with wider separation than the subcarrier spacing, to allow detection of large CFO values. It is used to obtain a coarse estimate of the CFO, so that the residual CFO (after removal of the estimated CFO) will be within the range of half symbol spacing. The long preamble, then, may be used for fine-tuning the carrier. The works presented in [21, 25, 26] do not mention the short preamble, but they make assumption that CFO is within the range of half symbol spacing. Naturally, to satisfy this assumption a mechanism similar to short preamble may be necessary.

There is also some difference in the long preamble proposed in [21, 25, 26] and the one proposed in [20, 22]. In the context of SMT waveform, the long preamble proposed in [21, 25, 26] is constructed based on a set of complex-valued pilot symbols that occupy all subcarriers. The long preamble proposed in [20, 22], on the other hand, is real-valued and only occupies the alternate subcarriers; the other subcarriers are left empty. This modification/simplification of the long preamble, as demonstrated through numerical results in [22], improves the accuracy of the CFO estimates significantly.

Here, to make suggestions to further improve the preamble design, we first note that the preamble is a packet component that constitutes a loss in the bandwidth efficiency of

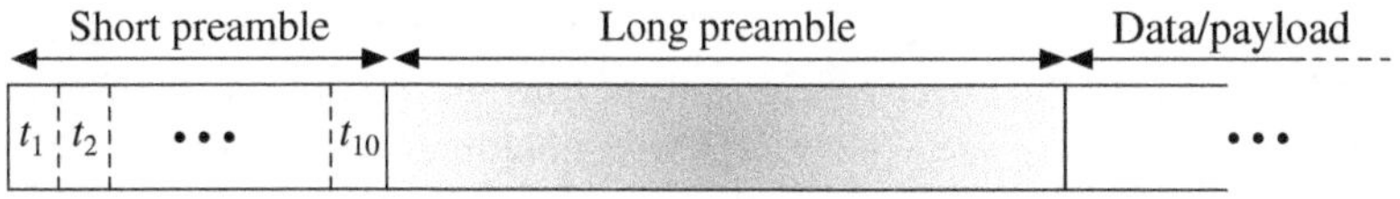

Figure 5.8 An FBMC data packet with short and long preambles.

transmission. Clearly, an attempt to reduce the length of the preamble improves the bandwidth efficiency. Research in the above cited works has led to the following observations. (i) A preamble with a set of pilot symbols that are well separated in frequency is needed to assure detection and correction of large CFO values. (ii) A longer preamble that allows a more accurate estimate of CFO may be also necessary. On the other hand, one may realize that to obtain the CFO and symbol timing estimates, it is not necessary to send pilots on all the subcarriers. The number of parameters that characterize the channel response (determined by the length of the channel impulse response) is usually much smaller than the number of subcarriers. This means, a fraction of subcarriers, whose number is greater than or equal to the length of the channel impulse response, would suffice for pilot subcarriers in the preamble. This, in turn, implies that one may propose to combine the short and long preambles into one preamble, and by taking such an approach, obviously, a shorter preamble will result.

More recent research into channel estimation methods for FBMC has noted that the above mentioned works may be classified as frequency-domain approaches. They assume frequency flat gain over each subcarrier band, which may not be a good approximation in some cases. It has been noted that this problem may be resolved by adopting a time domain approach, e.g. see [27]. In these works, the channel is estimated by comparing the transmitted signal and received signal in the time domain. Extension of these methods to multiuser cases where multiple users transmit pilots simultaneously and base station (BS) estimates their channels simultaneously has also been made; see [28] and the references therein. This method obviously leads to high spectral efficiency since multiple users share the same time slots for sending their pilot symbols.

A clever method of shortening an FBMC packet, and hence improving on its spectral efficiency, has been proposed in [29]. In this work, the authors have developed a method of adding a few (virtual) symbols at the beginning and the end of each FBMC packet to suppress signal samples at both its starting and ending tail.

5.5.2 Channel Tracking

Tracking of channel variation (including variations due to CFO and clock timing drift), during the payload transmission, is usually made through distributed pilots that are transmitted along with the data symbols. This topic, as noted above, has been well studied for OFDM. However, it has received very limited attention in the FBMC literature. Stitz et al. [24] have studied the problem of pilot-aided synchronization and tracking methods in CMT/SMT-type systems. They have shown that by applying specific pilots designed for CMT/SMT, the CFO, fractional time delay (FTD), and channel response can be accurately estimated. The work presented in [30] has taken a different approach. It takes note of the fact that each signal component at the output of the AFB, before performing equalization and taking its real (or imaginary) part, has the form of $h_k[n](s_k[n] + j\psi_k[n])$ (or $h_k[n](js_k[n] + \psi_k[n])$), where $\psi_k[n]$ arises from ICI and ISI. It is thus argued that if the same (known) symbol is transmitted over two instants of time at the same subcarrier, or allocated to the real and imaginary parts of an OQAM symbol in an SMT waveform, the analyzed signal samples give a pair of equations that can be solved to obtain an estimate of $h_k[n]$. In a more recent work [31], a coding method that removes $\psi_k[n]$ from the pilot symbols and thus facilitates the channel estimation has been proposed.

In [32], it has been noted that the use of PAM symbols in CMT (and similarly in SMT, if the real and imaginary part of each OQAM symbol is viewed as a pair of PAM symbols) allows development of a blind equalization algorithm. This algorithm may also be used for channel tracking. Hence, in CMT and SMT channel tracking is possible without any use of pilot symbols.

5.5.3 Timing Tracking

As noted above, any difference between the clock frequency at the transmitter and its counterpart at the receiver introduces a timing drift. The common method of compensating for this timing drift is to add a timing recovery loop to the system. In a single carrier receiver, the addition of a timing recovery loop as part of the receiver processing is a relatively straightforward task and is well understood; e.g. see [19]. In an OFDM system, with a not very long payload, the timing drift may be absorbed by the cyclic prefix (CP), provided that not the full length of the CP has been consumed by the channel transient response. Drift of timing phase introduces a linear phase across frequency and that can be easily compensated by the single-tap equalizers at the fast Fourier transform (FFT) outputs. In FBMC, on the other hand, the absence of CP/guard interval makes an FBMC receiver more sensitive to timing drift. However, fortunately, this problem can be resolved trivially through the use of a set of multi-tap fractionally spaced equalizers, one per subcarrier. For more details, the interested reader may refer to [8].

5.6 Equalization

In FBMC receivers, equalization is performed at the output of the AFBs. It is often assumed that each subcarrier has a small bandwidth, and hence the channel may be assumed flat over each subcarrier band. In that case, a single-tap equalizer per subcarrier would suffice. In cases where the flat gain approximation may be insufficient, a multi-tap equalizer per subcarrier band may be necessary. For CMT and SMT systems, it is necessary to use a fractionally spaced equalizer. A tap-spacing of half symbol interval is the most convenient choice, and hence has been suggested in the literature, e.g. [33].

Although, in many scenarios, it may appear that a single-tap equalizer per subcarrier is sufficient, in practice, where carrier and clock mismatch between the transmitter and receiver is inevitable, a multi-tap equalizer can be instrumental. In particular, the difference between the clock used at the transmitter to pass the signal samples to a digital-to-analog converter (DAC) and the one used at the receiver to control the rate of samples taken by an analog-to-digital converter (ADC) at the receiver introduces a constant drift in the timing phase of the sampled signal. This drift, as explained in Section 5.5, can deteriorate the receiver performance. The use of a multi-tap equalizer can resolve this problem to a great extent. Numerical results that demonstrate this can be found in [8].

5.7 Computational Complexity

Polyphase structures are commonly used to implement FBMC transmitter and receiver [19]. Direct mimicking of the continuous-time structures of Figures 5.3 and 5.6 to implement CMT and SMT, respectively, may lead to structures whose complexity are not optimized. It leads to one polyphase structure per set of real symbols. However, taking advantage of some multirate techniques, there exist clever methods that combine the polyphase structures for each pair of real symbols, equivalent to an OQAM symbol, to cut the complexity to about one half at the transmitter side of FBMC systems. At the receiver side, special attention has to be paid so that proper equalizers can be applied to the analyzed signal components. A detailed study along this line has been presented in [8]. The conclusion made in this study is that SMT and CMT have a complexity that is roughly two to three times that of OFDM.

5.8 Applications

In [7], OFDM and FBMC are contrasted with respect to their suitability for different applications. It is noted that while for many applications OFDM is a perfect choice, FBMC may be a superior choice for other applications. The applications in which FBMC may be preferred are (i) the uplink of multiuser multicarrier systems; (ii) cognitive radios; (iii) doubly dispersive channels; and (iv) digital subscriber lines and power line communications. Detailed discussion on why FBMC is a better choice in these applications is provided in [7]. Here, we emphasize on the class of massive (multiuser) multiple-input multiple-output (MIMO) communication systems.

Massive MIMO, in essence, is a multiuser technique, somewhat similar to code division multiple access (CDMA). In its simplest form, each mobile terminal (MT) has a single antenna, but the BS has a large number of antennas. The spreading code for each user is then the vector of channel gains between the respective MT antenna and the multiple antennas at the BS. Accordingly, by increasing the number of antennas at the BS, the processing gain of each user can be increased to an arbitrarily large value. In the pioneering work of Marzetta [34] it is noted that, in the limit, as the number of BS antennas tends to infinity, the processing gain of the system tends to infinity and, as a result, the effects of both noise and multiuser interference (MUI) are completely removed. Therefore, the network capacity (in theory) can be increased unboundedly by increasing the number of antennas at the BS [34].

Motivated by Marzetta's observations [34], multiple research groups in the recent past have studied a variety of implementation issues related to massive MIMO systems. An assumption made by Marzetta [34] and followed by other researchers is that OFDM may be used to convert the frequency-selective channels between each MT and the multiple antennas at the BS to a set of flat fading channels. Accordingly, the flat gains associated with the set of channels within each subcarrier constitute the spreading gain vector that is used for dispreading of the respective data in upstream, as well as for precoding of the respective downstream data.

In a recent work [35] embarking on the above concepts, we have introduced the application of FBMC to the area of massive MIMO communications. The interesting finding in this work is that unlike in conventional MIMO where applicability of FBMC (CMT and SMT, in particular) is found limited, FBMC offers a number of appealing properties in the application of massive MIMO. In particular, the linear combining of the signal components from different antennas at BS (equivalent to dispreading in CDMA) smooths channel distortion across each subcarrier band. An additional equalization step that further flattens the channel across each subcarrier band has been recently proposed in [36]. This property of FBMC allows one to reduce the number of subcarriers in an assigned bandwidth. Reducing the number of subcarriers has the following impacts, which all position FBMC as a strong candidate in the application of massive MIMO.

1. The complexity of both transmitter and receiver will be reduced, since the underlying polyphase structures will be based on smaller size inverse fast Fourier transform (IFFT)/FFT blocks.
2. The system latency will be reduced. This is because the length and, thus, the corresponding group delay of the underlying prototype filter, are reduced.
3. The system sensitivity to CFO will be reduced.
4. The peak to average power ratio (PAPR) of the transmit signal will be reduced.

References

1 Chen, W.Y. (1998). *DSL Simulation Techniques and Standards Development for Digital Subscriber Line Systems*. Indianapolis: Macmillan Publishing.

2 Van Nee, R. and Prasad, R. (2000). *OFDM for Wireless Multimedia Communications*. Boston, MA: Arthec House.

3 Chang, R.W. (1966). High-speed multichannel data transmission with bandlimited orthogonal signals. *Bell System Technical Journal* 45: 1775–1796.

4 Saltzberg, B. (1967). Performance of an efficient parallel data transmission system. *IEEE Transactions on Communication Technology* 15 (6): 805–811.

5 Le Floch, B., Alard, M., and Berrou, C. (1995). Coded orthogonal frequency division multiplex. *Proceedings of the IEEE* 83 (6): 982–996. doi: https://doi.org/10.1109/5.387096.

6 Farhang-Boroujeny, B. and (George) Yuen, C. (2010). Cosine modulated and offset QAM filter bank multicarrier techniques: a continuous-time prospect. *EURASIP Journal on Applied Signal Processing* 1–16. doi: https://doi.org/10.1155/2010/165654.

7 Farhang-Boroujeny, B. (2011). OFDM versus filter bank multicarrier. *IEEE Signal Processing Magazine* 28 (3): 92–112.

8 Farhang-Boroujeny, B. (2014). Filter bank multicarrier modulation: a waveform candidate for 5G and beyond. *Advances in Electrical Engineering, Hindawi* 2014: 1–25.

9 Sjoberg, F., Nilsson, R., Isaksson, M. et al. (1999). Asynchronous zipper. Proceedings of the IEEE ICC '99, Vancouver, Canada, Volume 1 (6–10 June), pp. 231–235.

10 Fliege, N.J. (1994). *Multirate Digital Signal Processing*. New York, NY: Wiley.

11 Siohan, P. and Roche, C. (2000). Cosine-modulated filterbanks based on extended Gaussian functions. *IEEE Transactions on Signal Processing* 48 (11): 3052–3061. doi: https://doi.org/10.1109/78.875463.

12 Sahin, A., Guvenc, I., and Arslan, H. (2014). A survey on multicarrier communications: prototype filters, lattice structures, and implementation aspects. *IEEE Communications Surveys & Tutorials* 16 (3): 1312–1338. doi: https://doi.org/10.1109/SURV.2013.121213.00263.

13 Martin, K.W. (1998). Small side-lobe filter design for multitone data-communication applications. *IEEE Transactions on Circuits and Systems II: Analog and Digital Signal Processing* 45 (8): 1155–1161. doi: https://doi.org/10.1109/82.718830.

14 Mirabbasi, S. and Martin, Ken. (2003). Overlapped complex-modulated transmultiplexer filters with simplified design and superior stopbands. *IEEE Transactions on Circuits and Systems II: Analog and Digital Signal Processing* 50 (8): 456–469. doi: https://doi.org/10.1109/TCSII.2003.813592.

15 Farhang-Boroujeny, B. (2008). A square-root Nyquist (M) filter design for digital communication systems. *IEEE Transactions on Signal Processing* 56 (5): 2127–2132. doi: https://doi.org/10.1109/TSP.2007.912892.

16 Kozek, W. and Molisch, A.F. (1998). Nonorthogonal pulseshapes for multicarrier communications in doubly dispersive channels. *IEEE Journal on Selected Areas in Communications* 16 (8): 1579–1589. doi: https://doi.org/10.1109/49.730463.

17 Haas, R. and Belfiore, J.-C. (1997). A time-frequency well-localized pulse for multiple carrier transmission. *Wireless Personal Communications* 5 (1): 1–18.

18 Amini, P., Chen, R.-R., and Farhang-Boroujeny, B. (2014). Filterbank multicarrier communications for underwater acoustic channels. *IEEE Journal of Oceanic Engineering* doi: https://doi.org/10.1109/Allerton.2011.6120228.

19 Farhang-Boroujeny, B. (2009). *Signal Processing Techniques for Software Radios*, 2e. Lulu Publication House.

20 Amini, P. and Farhang-Boroujeny, B. (2010). Packet format design and decision directed tracking methods for filter bank multicarrier systems. *EURASIP Journal on Advances in Signal Processing* 2010. (6): 1–13. doi: https://doi.org/10.1155/2010/307983

21 Fusco, T., Petrella, A., and Tanda, M. (2010). Joint symbol timing and CFO estimation for OFDM/OQAM systems in multipath channels. *EURASIP Journal on Advances in Signal Processing* 2010 (5): 1–11. doi: https://doi.org/10.1155/2010/897607.

22 Saeedi-Sourck, H., Sadri, S., Wu, Y., and Farhang-Boroujeny, B. (2013). Near maximum likelihood synchronization for filter bank multicarrier systems. *IEEE Wireless Communications Letters* 2 (2): 235–238. doi: https://doi.org/10.1109/WCL.2013.012513.120713.

23 Stitz, T.H., Viholainen, A., Ihalainen, T., and Renfors, M. (2009). CFO estimation and correction in a WiMAX-like FBMC system. IEEE 10th Workshop on Signal Processing Advances in Wireless Communications, 2009. SPAWC '09, pp. 633–637. doi: https://doi.org/10.1109/SPAWC.2009.5161862.

24 Stitz, T.H., Ihalainen, T., Viholainen, A., and Renfors, M. (2010). Pilot-based synchronization and equalization in filter bank multicarrier communications. *EURASIP Journal on Advances in Signal Processing* 2010. http://dblp.uni-trier.de/db/journals/ejasp/ejasp2010.

25 Fusco, T., Petrella, A., and Tanda, M. (2009). Joint symbol timing and CFO estimation in multiuser OFDM/OQAM systems. IEEE 10th Workshop on Signal Processing Advances in Wireless Communications, 2009. SPAWC '09, pp. 613–617. doi: https://doi.org/10.1109/SPAWC.2009.5161858.

26 Fusco, T., Petrella, A., and Tanda, M. (2009). Data-aided symbol timing and CFO synchronization for filter bank multicarrier systems. *IEEE Transactions on Wireless Communications* 8 (5): 2705–2715. doi: https://doi.org/10.1109/TWC.2009.080860.

27 Kong, D., Qu, D., and Jiang, T. (2014). Time domain channel estimation for OQAM-OFDM systems: algorithms and performance bounds. *IEEE Transactions on Signal Processing* 62 (2): 322–330.

28 Hosseini, H., Farhang, A., and Farhang-Boroujeny, B. (2011). Spectrally efficient pilot structure and channel estimation for multiuser FBMC systems. IEEE International Conference on Communications (ICC), pp. 1–7. doi: https://doi.org/10.1109/GLOCOM.2011.6133591.

29 Qu, D., Wang, F., Wang, Y. et al. (2017). Improving spectral efficiency of FBMC-OQAM through virtual symbols. *IEEE Transactions on Wireless Communications* 16 (7): 4204–4215.

30 Bellanger, M. and Dao, T. (2006). Multicarrier digital transmission system using an OQAM transmultiplexer. Granted Patent, US7072412.

31 Cui, W., Qu, D., Jiang, T., and Farhang-Boroujeny, B. (2016). Coded auxiliary pilots for channel estimation in FBMC-OQAM systems. *IEEE Transactions on Vehicular Technology* 65 (5): 2936–2946.

32 Farhang-Boroujeny, B. (2003). Multicarrier modulation with blind detection capability using cosine modulated filter banks. *IEEE Transactions on Communications* 51 (12): 2057–2070.

33 Hirosaki, B. (1981). An orthogonally multiplexed QAM system using the discrete Fourier transform. *IEEE Transactions on Communications* 29 (7): 982–989.

34 Marzetta, T.L. (2010). Noncooperative cellular wireless with unlimited numbers of base station antennas. *IEEE Transactions on Wireless Communications* 9 (11): 3590–3600.

35 Farhang, A., Marchetti, N., Doyle, L., and Farhang-Boroujeny, B. (2014). Filter bank multicarrier for massive MIMO. 2014 IEEE 80th Vehicular Technology Conference (VTC2014-Fall), Vancouver, BC (June 2014), 1–7.

36 Aminjavaheri, A., Farhang, A., and Farhang-Boroujeny, B. (2018). Filter bank multicarrier in massive MIMO: analysis and channel equalization. *IEEE Transactions on Signal Processing* 66 (15): 3987–4000.

6

Orthogonal Time–Frequency Space Modulation: Principles and Implementation

Arman Farhang[1] and Behrouz Farhang-Boroujeny[2]

[1] *Department of Electronic Engineering, Maynooth University, Maynooth, Co., Ireland*
[2] *Electrical and Computer Engineering Department, University of Utah, USA*

6.1 Introduction

While orthogonal frequency division multiplexing modulation (OFDM) scheme achieves a performance near the capacity limits in linear time-invariant channels, it leads to a poor performance in doubly dispersive channels. This is due to the large amount of interference that is imposed by the channel Doppler spread. The common approach to cope with this issue in the existing wireless standards such as IEEE 802.11a and digital video broadcasting systems is to shorten OFDM symbol duration in time so that the channel variations over each OFDM symbol are negligible. However, this reduces the spectral efficiency of transmission since the cyclic prefix (CP) length should remain constant. A thorough analysis of OFDM in such channels is conducted in [1]. Another classical approach for handling time-varying channels is to utilize filtered multicarrier systems that are optimized for a balanced performance in doubly dispersive channels [2].

Recently, new signaling techniques have emerged in the literature to tackle the TV channels [3, 4]. In [3], the authors introduce a new waveform called frequency division multiplexing with a frequency-domain cyclic prefix (FDM-FDCP). This waveform outperforms OFDM in channels with high Doppler and low delay spread. However, its performance in channels with high Doppler and medium to high delay spread is yet to be understood [3].

Orthogonal time–frequency space (OTFS) modulation has recently been proposed as another emerging signaling technique [4]. OTFS is an effective waveform for signaling over TV channels that takes advantage of the time diversity (that is, variation of channel with time) to improve the reliability of wireless links [4]. This waveform was first introduced in the pioneering work of Hadani et al. [4] where the two-dimensional (2D) delay-Doppler domain was proposed for multiplexing the transmit data. OTFS modulation is a generalized signaling framework where precoding and post-processing units are added to the modulator and demodulator of a multicarrier waveform allowing for taking advantage of full time and frequency diversity gain of doubly dispersive channels. This process also converts the time-varying channel to a time-invariant one in the delay-Doppler domain.

In OTFS, transmit data symbols are treated as the values of the regularly spaced grid points in delay-Doppler plane. A transformation step takes each data symbol and spreads

Radio Access Network Slicing and Virtualization for 5G Vertical Industries, First Edition.
Edited by Lei Zhang, Arman Farhang, Gang Feng, and Oluwakayode Onireti.

it over the entire time–frequency plane. The result of this transformation is then passed to a multicarrier system for modulation and transmission. In this way, all data symbols are equally affected by the channel frequency selectivity and time diversity and, as a result, the time-varying channel, with a good approximation, converts to a unified time-invariant impulse response for all the data symbols in delay-Doppler domain. To recover the transmitted data symbols at the receiver, the respective multicarrier demodulation followed by an inverse transformation step takes the received signal back to the delay-Doppler domain.

The authors in [4] deployed an inverse symplectic finite Fourier transform (SFFT^{-1}) operation to translate data symbols from the delay-Doppler to the time–frequency domain, at the transmitter. Consequently, a symplectic finite Fourier transform (SFFT) operation, i.e. the reverse operation to the one at the transmitter, is performed at the receiver to estimate the transmit data symbols. Transformation of data symbols from the delay-Doppler to time–frequency domain at the transmitter and the corresponding inverse transformation at the receiver, clearly leads to full diversity gain across both time and frequency. The whole OTFS modulation and demodulation process based on OFDM for time–frequency transmission is shown in Figure 6.1.

In the above setup, the equivalent channel between the transmitted and the received data symbols can be modeled by a two-dimensional time-invariant impulse response in the delay-Doppler domain. Soft detectors/equalizers, like some extensions to those presented in [5], may thus be used for near-optimal recovery of the transmitted information. In order to simplify such detectors, [4] has proposed that proper windows should be applied to the time–frequency signals at both the transmitter and receiver sides to sparsify the OTFS channel. However, very little has been said about how such windows could be selected.

The goal of this chapter is to introduce the main principles behind OTFS and provide a discrete time formulation that not only brings deep insights into OTFS but also leads to a simplified modem structure for such systems. Therefore, Section 6.2 explains the basics of OTFS where a generic multicarrier waveform is deployed for transmission of OTFS time–frequency samples. To simplify the derivations, OFDM is considered for time–frequency transmission of OTFS signal. Consequently, Section 6.3 covers

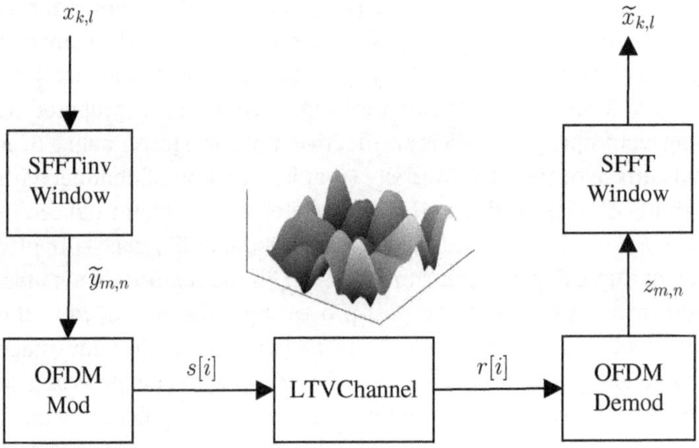

Figure 6.1 OTFS transmitter and receiver structure.

OFDM-based OTFS systems. In Section 6.4, mathematical formulation of Section 6.3 is further expanded to provide a concise representation of the effect of each signal processing unit on the input–output relationship of OFDM-based OTFS systems in delay-Doppler domain. This leads to an interesting finding that leads to a substantial amount of saving in transceiver complexity, which will be discussed in Section 6.5. Based on our findings in Sections 6.4 and 6.5, we note that window functions (at the transmitter and receiver) that may be effective in converting the channel to a sparse one in the delay-Doppler space require knowledge of channel variations on a given data packet at the transmitter, which obviously is unknown at the time of transmission. Hence, it is hard to say whether any effective window in this sense could be applied at the transmitter. On this basis, we will ignore the windowing step of the OTFS at the transmitter side. At the receiver side, on the other hand, an iterative channel estimation and equalization may allow the use of an effective window that leads to a sparse channel, e.g. [6, 7]. This is an interesting topic that falls out of the scope of this chapter and may be left for future studies. Thus, in Section 6.5, we include time–frequency signal windowing at the receiver side, but limit our discussion to the case where this is a separable window along the time and frequency dimensions and emphasize on channel-independent window functions with some reasonable impact. From system implementation point of view, computational complexity is a very important aspect that needs to be analyzed. Therefore, Section 6.6 analyzes and compares the computational load of the simplified modem structure that is presented in Section 6.5 with that of the conventional OTFS implementation and also OFDM. Finally, Section 7 covers recent results on OTFS from the literature while discussing some potential research directions in this area.

Notations

Throughout this chapter, matrices, vectors, and scalar quantities are denoted by boldface uppercase, boldface lowercase, and normal letters, respectively. $[\mathbf{A}]_{mn}$, $\mathbf{a}(n)$, and $\mathbf{A}^{-1}$ represent the mn^{th} element, the column n, and the inverse of $\mathbf{A}$, respectively. vec$\{\mathbf{A}\}$ vectorizes $\mathbf{A}$ by stacking its columns on the top of one another in a column vector. $\mathbf{I}_m$ is the identity matrix of size m. The superscripts $(\cdot)^{\text{T}}$, $(\cdot)^{\text{H}}$, and $(\cdot)^*$ indicate transpose, conjugate transpose, and conjugate operations, respectively. $< f[n], g[n] >$ denotes the inner product of the functions $f[n]$ and $g[n]$. Finally, $\odot$, $\otimes$, $\overset{2\text{D}}{\circledast}$, and $((\cdot))_N$ represent elementwise multiplication, Kronecker product, 2D circular convolution, and modulo-N operations, respectively.

6.2 OTFS Principles

OTFS modulation has two stages. First, it maps a set of data symbols with quadrature amplitude modulation (QAM) from delay-Doppler domain to time–frequency domain through a set of two-dimensional basis functions. The first stage also includes a time–frequency windowing operation. This can be implemented through an SFFT^{-1} operation together with a windowing procedure to be applied at the SFFT^{-1} output. In the second stage, the resulting time–frequency samples are fed into a modulator to form the time-domain transmit signal using a set of orthogonal time–frequency basis functions. After the signal is passed through the channel, the received signal needs to be mapped back to the delay-Doppler domain before detection. To this end, OTFS demodulator performs the reverse operations to those at the modulator. This procedure is shown in Figure 6.1. In the following, we cast the above explanation into mathematical formulation, paving the way for our further derivations in this chapter.

We consider an OTFS system transmitting a block of $M \times N$ delay-Doppler QAM data symbols $x_{k,l}$, where $k = 0, \dots, M-1$ and $l = 0, \dots, N-1$. The data symbols are from a zero mean independent and identically distributed (i.i.d) random process with the variance of unity. In OTFS modulation, the data symbols, $x_{k,l}$, are first modulated by a set of 2D basis functions, $b_{k,l}[m,n] = \frac{1}{\sqrt{MN}} e^{-j2\pi(\frac{mk}{M} - \frac{nl}{N})}$, in the time–frequency domain as

$$y_{m,n} = \sum_{k=0}^{M-1} \sum_{l=0}^{N-1} x_{k,l} b_{k,l}[m,n], \tag{6.1}$$

where the time–frequency samples $y_{m,n}$ for $m = 0, \dots, M-1, n = 0, \dots, N-1$ appear at the output of the SFFT^{-1} block. Then, the time–frequency samples $y_{m,n}$ are windowed through a 2D transmit window $W_{\text{Tx}}[m,n]$ and are fed into the time–frequency modulator to form the OTFS transmit signal as

$$s[i] = \sum_{k=0}^{M-1} \sum_{l=0}^{N-1} \widetilde{y}_{m,n}\, g_{m,n}[i], \tag{6.2}$$

where $\widetilde{y}_{m,n} = W_{\text{Tx}}[m,n] y_{m,n}$ are the windowed time–frequency samples and $g_{m,n}[i]$ are the time–frequency basis functions of the transmitter.

The received signal samples after transmission over a linear time-varying (LTV) channel can be obtained as

$$r[i] = \sum_{l=0}^{L-1} h[i,l] s[i-l] + v[i], \tag{6.3}$$

where $h[i,l]$ is the channel impulse response with length L at time instant i, and $v[i] \sim \mathcal{CN}(0, \sigma_v^2)$ is the circularly symmetric complex additive white Gaussian noise (AWGN).

To obtain data symbols that are affected by the channel in the delay-Doppler domain, the received signal, $r[i]$, needs to be first projected onto the receiver's basis functions, $f_{m,n}[i]$, corresponding to time–frequency points m, k.

$$z_{m,n} = < f_{m,n}[i], r[i] > . \tag{6.4}$$

For inter-symbol interference-free communication, transmit and receive basis functions are designed to satisfy the perfect reconstruction criterion, i.e.

$$< f_{m,n}[i], g_{m',n'}[i] > = \delta(m-m')\delta(n-n'). \tag{6.5}$$

It is worth noting that if the underlying time–frequency modulation scheme is orthogonal, e.g. OFDM, $f_{m,n}[i] = g_{m,n}[i]$. However, for a non-orthogonal waveform, such as generalized frequency division multiplexing (GFDM) [8], the transmitter and receiver basis functions may not be the same.

Recalling Figure 6.1, at the second stage of the OTFS demodulation, we form an $M \times N$ matrix, $\mathbf{Z} = [\mathbf{z}(0), \dots, \mathbf{z}(N-1)]$ where $\mathbf{z}(n) = [z_{0,n}, \dots, z_{M-1,n}]^{\text{T}}$, window its elements using a receive window $W_{\text{Rx}}[m,n]$, and then perform a SFFT operation. This process can be represented as

$$\tilde{x}_{k,l} = \sum_{m=0}^{M-1}\sum_{n=0}^{N-1} W_{\text{Rx}}[m,n] z_{m,n} b_{k,l}^{*}[m,n], \tag{6.6}$$

where $k = 0, \dots, M-1$ and $l = 0, \dots, N-1$.

6.3 OFDM-Based OTFS

In this section, we consider OFDM for transmission of the generated time–frequency signals after SFFT^{-1} and windowing operations. Therefore, the signal samples $\tilde{y}_{m,n}$ are fed into the OFDM transmitter to form the OTFS transmit signal as

$$\mathbf{S} = \mathbf{A}_{\text{CP}}\mathbf{F}_M^{\text{H}}\tilde{\mathbf{Y}}, \tag{6.7}$$

where $\tilde{\mathbf{Y}} = [\tilde{\mathbf{y}}(0), \dots, \tilde{\mathbf{y}}(N-1)]$ is an $M \times N$ matrix containing the time–frequency windowed samples $\tilde{y}_{m,n} = W_{\text{Tx}}[m,n] y_{m,n}$ on its mn^{th} elements. The nth column of $\tilde{\mathbf{Y}}$, i.e. $\tilde{\mathbf{y}}(n) = [\tilde{y}_{0,n}, \dots, \tilde{y}_{M-1,n}]^{\text{T}}$, includes the frequency domain samples to be transmitted within the nth OFDM symbol. $\mathbf{F}_M$ is the normalized M-point discrete Fourier transform (DFT) matrix with the elements $[\mathbf{F}_M]_{pq} = \frac{1}{\sqrt{M}} e^{\frac{-j2\pi pq}{M}}$ for $p, q = 0, \dots, M-1$. $\mathbf{A}_{\text{CP}} = [\mathbf{G}_{\text{CP}}^{\text{T}}, \mathbf{I}_M^{\text{T}}]^{\text{T}}$ is the CP addition matrix where the $M_{\text{CP}} \times M$ matrix $\mathbf{G}_{\text{CP}}$ includes the last M_{CP} rows of the identity matrix $\mathbf{I}_M$ and M_{CP} is the CP length. Finally, the $(M + M_{\text{CP}}) \times N$ matrix $\mathbf{S}$ includes the OFDM time domain transmit signals on its columns. After parallel to serial conversion of $\mathbf{S}$, the OTFS transmit signal at the baseband can be formed as $\mathbf{s} = \text{vec}\{\mathbf{S}\}$.

The received signal samples after transmission over an LTV channel with length L can be obtained as was shown in Eq. (6.3). Assuming $M_{\text{CP}} \geq L$, the received OFDM symbols are free of inter-symbol interference. Thus, the received OFDM symbol n at the output of the OFDM demodulator can be written as

$$\begin{aligned}\mathbf{z}(n) &= \mathbf{F}_M \mathbf{R}_{\text{CP}} \mathbf{H}_n \mathbf{A}_{\text{CP}} \mathbf{F}_M^{\text{H}} \tilde{\mathbf{y}}(n) + \bar{\mathbf{v}}(n)\\ &= \mathbf{F}_M \bar{\mathbf{H}}_n \mathbf{F}_M^{\text{H}} \tilde{\mathbf{y}}(n) + \bar{\mathbf{v}}(n)\\ &= \tilde{\mathbf{H}}_n \tilde{\mathbf{y}}(n) + \bar{\mathbf{v}}(n),\end{aligned} \tag{6.8}$$

where $\bar{\mathbf{H}}_n = \mathbf{R}_{\text{CP}}\mathbf{H}_n\mathbf{A}_{\text{CP}}$, $\mathbf{R}_{\text{CP}} = [\mathbf{0}_{M\times M_{\text{CP}}}, \mathbf{I}_M]$ is the CP removal matrix, $\tilde{\mathbf{H}}_n = \mathbf{F}_M\bar{\mathbf{H}}_n\mathbf{F}_M^{\text{H}}$, and $\mathbf{H}_n$ is the LTV channel matrix realizing the convolution operation on the samples of the nth OFDM symbol. The vector $\bar{\mathbf{v}}(n)$ contains the noise samples at the OFDM demodulator output and $\mathbf{z}(n) = [z_{0,n}, \dots, z_{M-1,n}]^{\text{T}}$.

As mentioned in Section 6.2, stacking blocks of N OFDM symbols at the output of the demodulator, we form an $M \times N$ matrix $\mathbf{Z}$, window its elements, and perform a SFFT operation to obtain the received data symbols at the delay-Doppler domain. This procedure is mathematically shown in Eq. (6.6). It is shown in [4] that the signal samples, $\tilde{x}_{k,l}$, at the output of OTFS receiver can be obtained as 2D circular convolution of the QAM data symbols $x_{k,l}$ and the channel impulse response in the delay-Doppler domain. More details and derivations on the baseband channel response in the delay-Doppler domain will be presented in Section 6.4.

6.4 Channel Impact

To study the channel impact, in this section, we expand further the formulation that was presented in Section 6.3. This leads to deeper insights into OTFS while paving the way for the derivation of a simplified modem structure in Section 6.5. As the first step, let us form an $M \times N$ matrix $\mathbf{Y}$ that includes the time–frequency samples $y_{m,n}$ at the output of the SFFT^{-1} block as its elements. In this chapter, we consider the transmit and receive windows to be separable functions of time and frequency indices. This makes our derivations simple while leading to an in-depth understanding of the windowing effect on the channel, in either delay or Doppler dimension. Consequently, $W_{\text{Tx}}[m,n] = W_{\text{Tx,c}}[m]W_{\text{Tx,r}}[n]$ where $W_{\text{Tx,c}}[m]$ and $W_{\text{Tx,r}}[n]$ perform windowing operation on the columns and rows of $\mathbf{Y}$, respectively. It is worth noting that the time–frequency windowing operation is equivalent to filtering in the delay-Doppler domain. Thus, the transmit windowing operation is applied through the multiplication of two diagonal matrices to $\mathbf{Y}$ from left and right, respectively, i.e. $\mathbf{W}_{\text{Tx,c}} = \text{diag}\{[W_{\text{Tx,c}}[0], \dots, W_{\text{Tx,c}}[M-1]]\}$ and $\mathbf{W}_{\text{Tx,r}} = \text{diag}\{[W_{\text{Tx,r}}[0], \dots, W_{\text{Tx,r}}[N-1]]\}$. This procedure can be represented as

$$\widetilde{\mathbf{Y}} = \mathbf{W}_{\text{Tx,c}}\mathbf{Y}\mathbf{W}_{\text{Tx,r}}. \tag{6.9}$$

From (6.9) the nth column of the matrix $\widetilde{\mathbf{Y}}$ can be obtained as

$$\widetilde{\mathbf{y}}(n) = \mathbf{W}_{\text{Tx,c}}\mathbf{Y}\mathbf{w}_{\text{Tx,r}}(n), \tag{6.10}$$

where $\mathbf{w}_{\text{Tx,r}}(n)$ is the nth column of $\mathbf{W}_{\text{Tx,r}}$.

The SFFT^{-1} operation that was shown in (6.1) can be represented as

$$\mathbf{Y} = \mathbf{F}_M\mathbf{X}\mathbf{F}_N^{\text{H}}, \tag{6.11}$$

where $\mathbf{X}$ is an $M \times N$ matrix containing the transmit QAM data symbols $x_{k,l}$ on its kl^{th} elements.

Substituting Eqs. (6.10) and (6.11) in (6.8), the received OFDM symbol n at the OFDM demodulator output can be expanded as

$$\begin{aligned}\mathbf{z}(n) &= \widetilde{\mathbf{H}}_n\mathbf{W}_{\text{Tx,c}}\mathbf{Y}\mathbf{w}_{\text{Tx,r}}(n) + \bar{\mathbf{v}}(n)\\ &= \mathbf{F}_M\bar{\mathbf{H}}_n\mathbf{F}_M^{\text{H}}\mathbf{W}_{\text{Tx,c}}\mathbf{F}_M\mathbf{X}\mathbf{F}_N^{\text{H}}\mathbf{w}_{\text{Tx,r}}(n) + \bar{\mathbf{v}}(n)\\ &= \mathbf{F}_M\bar{\mathbf{H}}_n\bar{\mathbf{W}}_{\text{Tx,c}}\mathbf{X}\mathbf{f}_N^*(n)W_{\text{Tx,r}}[n] + \bar{\mathbf{v}}(n),\end{aligned} \tag{6.12}$$

where $\bar{\mathbf{W}}_{\text{Tx,c}} = \mathbf{F}_M^{\text{H}}\mathbf{W}_{\text{Tx,c}}\mathbf{F}_M$ and $\mathbf{f}_N(n)$ is the nth column of the N-point DFT matrix. After packing the vectors $\mathbf{z}(n)$ for $n = 0, \dots, N-1$ on the columns of the matrix $\mathbf{Z} = [\mathbf{z}(0), \dots, \mathbf{z}(N-1)]$ and performing windowing and SFFT operations based on Eq. (6.6), the received signal samples at the output of OTFS receiver can be packed in an $M \times N$ matrix $\widetilde{\mathbf{X}}$ as

$$\widetilde{\mathbf{X}} = \mathbf{F}_M^{\text{H}}(\mathbf{W}_{\text{Rx,c}}\mathbf{Z}\mathbf{W}_{\text{Rx,r}})\mathbf{F}_N, \tag{6.13}$$

with the elements $\tilde{x}_{k,l}$. Similar to the transmitter, we consider a separable window at the receiver, i.e. $W_{\text{Rx}}[m,n] = W_{\text{Rx,c}}[m]W_{\text{Rx,r}}[n]$ where $\mathbf{W}_{\text{Rx,c}} = \text{diag}\{[W_{\text{Rx,c}}[0], \dots, W_{\text{Rx,c}}[M-1]]\}$ and $\mathbf{W}_{\text{Rx,r}} = \text{diag}\{[W_{\text{Rx,r}}[0], \dots, W_{\text{Rx,r}}[N-1]]\}$ window the columns and rows of $\mathbf{Z}$, respectively. To derive a clear-cut representation of the channel effect on the transmit data symbols, $x_{k,l}$, at the output of the receiver, we expand the nth column of the matrix $\widetilde{\mathbf{X}}$. This leads to

$$\begin{aligned}\widetilde{\mathbf{x}}(n) &= \mathbf{F}_M^{\mathrm{H}}(\mathbf{W}_{\mathrm{Rx,c}}\mathbf{Z}\mathbf{W}_{\mathrm{Rx,r}})\mathbf{f}_N(n)\\ &= \frac{1}{\sqrt{N}}\mathbf{F}_M^{\mathrm{H}}\mathbf{W}_{\mathrm{Rx,c}}\sum_{i=0}^{N-1}\mathbf{z}(i)W_{\mathrm{Rx,r}}[i]e^{\frac{j2\pi ni}{N}}.\end{aligned} \tag{6.14}$$

Substituting (6.12) into (6.14), we get

$$\begin{aligned}\widetilde{\mathbf{x}}(n) &= \frac{1}{\sqrt{N}}\bar{\mathbf{W}}_{\mathrm{Rx,c}}\sum_{i=0}^{N-1}\left(\bar{\mathbf{H}}_i\bar{\mathbf{W}}_{\mathrm{Tx,c}}\mathbf{X}\mathbf{f}_N^*(i)W_{\mathrm{Tx,r}}[i]W_{\mathrm{Rx,r}}[i]e^{\frac{j2\pi ni}{N}}\right)\\ &\quad+\frac{1}{\sqrt{N}}\mathbf{F}_M^{\mathrm{H}}\mathbf{W}_{\mathrm{Rx,c}}\sum_{i=0}^{N-1}\bar{\mathbf{v}}(n)W_{\mathrm{Rx,r}}[i]e^{\frac{j2\pi ni}{N}}\\ &= \frac{1}{N}\bar{\mathbf{W}}_{\mathrm{Rx,c}}\sum_{i=0}^{N-1}\sum_{k=0}^{N-1}\left(\bar{\mathbf{H}}_i\bar{\mathbf{W}}_{\mathrm{Tx,c}}\mathbf{x}(k)W_{\mathrm{Tx,r}}[i]W_{\mathrm{Rx,r}}[i]e^{\frac{j2\pi(n-k)i}{N}}\right)\\ &\quad+\frac{1}{\sqrt{N}}\mathbf{F}_M^{\mathrm{H}}\mathbf{W}_{\mathrm{Rx,c}}\sum_{i=0}^{N-1}\bar{\mathbf{v}}(n)W_{\mathrm{Rx,r}}[i]e^{\frac{j2\pi ni}{N}}\\ &= \bar{\mathbf{W}}_{\mathrm{Rx,c}}\sum_{k=0}^{N-1}\mathbf{H}_{((n-k))_N}\bar{\mathbf{W}}_{\mathrm{Tx,c}}\mathbf{x}(k)+\widetilde{\mathbf{v}}(n),\end{aligned} \tag{6.15}$$

where the $M \times 1$ vector $\widetilde{\mathbf{v}}(n)$ contains the noise samples at the OTFS demodulator output, $\bar{\mathbf{W}}_{\mathrm{Rx,c}} = \mathbf{F}_M^{\mathrm{H}}\mathbf{W}_{\mathrm{Rx,c}}\mathbf{F}_M$, and

$$\mathbf{H}_{((k))_N} = \frac{1}{N}\sum_{i=0}^{N-1}\bar{\mathbf{H}}_i W_{\mathrm{Tx,r}}[i]W_{\mathrm{Rx,r}}[i]e^{\frac{j2\pi ki}{N}}. \tag{6.16}$$

Using (6.15), defining an $MN \times MN$ block circulant matrix

$$\mathbf{H}_{\mathrm{BC}} = \begin{bmatrix}\mathbf{H}_0 & \mathbf{H}_1 & \cdots & \mathbf{H}_{N-1}\\ \mathbf{H}_{N-1} & \mathbf{H}_0 & \cdots & \mathbf{H}_{N-2}\\ \vdots & \vdots & \ddots & \vdots\\ \mathbf{H}_1 & \mathbf{H}_2 & \cdots & \mathbf{H}_0\end{bmatrix}, \tag{6.17}$$

the $MN \times MN$ block diagonal windowing matrices $\bar{\mathbf{W}}_{\mathrm{Rx,c}} = \bar{\mathbf{W}}_{\mathrm{Rx,c}} \otimes \mathbf{I}_N$, and $\bar{\mathbf{W}}_{\mathrm{Tx,c}} = \bar{\mathbf{W}}_{\mathrm{Tx,c}} \otimes \mathbf{I}_N$, the data stream at the output of the OTFS receiver, $\widetilde{d}_\kappa$ for $\kappa = 0, \dots, MN-1$, can be obtained in a compact form as

$$\widetilde{\mathbf{d}} = \bar{\mathbf{W}}_{\mathrm{Rx,c}}\mathbf{H}_{\mathrm{BC}}\bar{\mathbf{W}}_{\mathrm{Tx,c}}\mathbf{d} + \mathrm{vec}\{\widetilde{\mathbf{V}}\}, \tag{6.18}$$

where $\widetilde{\mathbf{d}} = [\widetilde{d}_0, \dots, \widetilde{d}_{MN-1}]^{\mathrm{T}} = \mathrm{vec}\{\widetilde{\mathbf{X}}\}$, $\mathbf{d} = [d_0, \dots, d_{MN-1}]^{\mathrm{T}} = \mathrm{vec}\{\mathbf{X}\}$, and $\widetilde{\mathbf{V}} = [\widetilde{\mathbf{v}}(0), \dots, \widetilde{\mathbf{v}}(N-1)]$. From (6.18), one may realize that if the submatrices $\mathbf{H}_n$ in $\mathbf{H}_{\mathrm{BC}}$ are circulant, i.e. the channel is time-invariant within each OFDM symbol interval, multiplication of $\mathbf{H}_{\mathrm{BC,w}} = \bar{\mathbf{W}}_{\mathrm{Rx,c}}\mathbf{H}_{\mathrm{BC}}\bar{\mathbf{W}}_{\mathrm{Tx,c}}$ to $\mathbf{d}$ realizes the 2D circular convolution of the windowed delay-Doppler channel impulse response $\mathbf{H}_{\mathrm{DD,w}}$ with the data matrix $\mathbf{X}$. This is due to the fact that all the matrices $\bar{\mathbf{W}}_{\mathrm{Rx,c}}$, $\mathbf{H}_{\mathrm{BC}}$, and $\bar{\mathbf{W}}_{\mathrm{Tx,c}}$ are block circulant with circulant submatrices and their multiplication result in a matrix preserving their circulant properties.

Therefore, $\widetilde{\mathbf{X}}$ can be written as

$$\widetilde{\mathbf{X}} = \mathbf{H}_{\mathrm{DD,w}} \overset{\mathrm{2D}}{\circledast} \mathbf{X} + \widetilde{\mathbf{V}}, \tag{6.19}$$

where the columns of the $M \times N$ matrix $\mathbf{H}_{\mathrm{DD,\,w}}$ include the first columns of the matrices $\mathbf{H}_n$ that are windowed by the window functions $W_{\mathrm{Tx,c}}[m]$ and $W_{\mathrm{Rx,c}}[m]$ in the delay domain. In particular, the elements kl of $\mathbf{H}_{\mathrm{DD,w}}$ for $k = 0, \ldots, M-1$ and $l = 0, \ldots, N-1$ are equal to $[\bar{\mathbf{W}}_{\mathrm{Rx,c}}\mathbf{H}_l\bar{\mathbf{W}}_{\mathrm{Tx,c}}]_{k0}$. When the channel is time-invariant within each OFDM symbol interval, we can reorder the matrices $\bar{\mathbf{W}}_{\mathrm{Rx,c}}$, $\mathbf{H}_{\mathrm{BC}}$, and $\bar{\mathbf{W}}_{\mathrm{Tx,c}}$ in (6.18) as $(\bar{\mathbf{W}}_{\mathrm{Rx,c}}\bar{\mathbf{W}}_{\mathrm{Tx,c}})\mathbf{H}_{\mathrm{BC}}$. This implies that given the transmit window in the delay domain is channel independent, it can be combined with the receive filter to simplify the transmitter.

Based on Eq. (6.18), one may realize that channel effect in the delay-Doppler domain can be equalized using linear equalizers such as zero forcing (ZF) or minimum mean square error (MMSE). Since the ZF equalizer may suffer from the noise enhancement problem, MMSE equalization seems to be a more appropriate choice. Hence, through MMSE criterion, the transmit data stream $\mathbf{d}$ can be estimated as

$$\hat{\mathbf{d}}_{\mathrm{MMSE}} = (\mathbf{H}^{\mathrm{H}}_{\mathrm{BC,w}}\mathbf{H}_{\mathrm{BC,w}} + \sigma_v^2\mathbf{I}_{MN})^{-1}\mathbf{H}^{\mathrm{H}}_{\mathrm{BC,w}}\widetilde{\mathbf{d}}. \tag{6.20}$$

It is worth noting that the matrix $(\mathbf{H}^{\mathrm{H}}_{\mathrm{BC,w}}\mathbf{H}_{\mathrm{BC,w}} + \sigma_v^2\mathbf{I}_{MN})$ is block circulant and thus can be block diagonalized using block DFT and IDFT matrices and efficiently inverted. Alternatively, soft equalization techniques such as turbo equalization that was used in [4] can be utilized for data detection.

6.5 Simplified Modem Structure

The SFFT^{-1}/SFFT and 2D windowing operations in OTFS modem make it more complex than a simple OFDM system to implement. To resolve this issue, this section presents derivations that lead to simple OTFS transmitter and receiver structures. It is worth noting that through these derivations, OTFS modem becomes even simpler than OFDM, when using similar parameters to 3GPP LTE systems. An in-depth computational complexity analysis of the modem structure that is developed in this section is presented in Section 6.6.

The class of separable window functions that are channel independent is considered in this chapter. This provides control over the Doppler and delay spread of the LTV channel separately. To understand the effect of windowing, the discussion here starts from using 2D rectangular window functions at both transmitter and receiver sides. Usually, the channel length L is much smaller than the number of frequency domain subcarriers M, i.e. $L \ll M$. Hence, the 2D delay-Doppler channel impulse response is sparse along the delay dimension and has the length of L while being zero elsewhere in delay dimension. In contrast, the channel is not very sparse in the Doppler dimension, especially, as mobility and thus, the channel variations increase. Windowing along the frequency bins with a channel-independent window function other than the rectangular window translates into circular convolution in the delay dimension. This slightly increases the delay spread of the channel that is due to the filter transient response. Consequently, channel independent filtering in the delay dimension at both transmitter and/or receiver does not lead to channel shortening effect.

This degrades channel sparsity in the delay dimension and hence can be harmful. However, Doppler spread can lead to a great amount of leakage all along the Doppler dimension. Therefore, windowing/filtering with a well-localized filter in the Doppler dimension is very effective in reducing the leakage and making the channel sparse.

On the basis of the above discussion, utilization of a rectangular window along the frequency domain, while using a window with smooth corners along the consecutive OFDM symbols in time leads to a sparse channel response. Hence, the 2D delay-Doppler channel impulse response keeps its minimum delay spread, i.e. L, while becoming quite sparse in the Doppler dimension. Furthermore, from (6.15) and (6.16), one may realize that the transmit and receive windows applied to N consecutive OFDM time symbols can be combined and utilized only at the receiver side. Accordingly, the modem structure that was discussed in Section 6.3 is reduced to an OTFS modulator without windowing, i.e. $\widetilde{\mathbf{Y}} = \mathbf{Y}$ in (6.7), and an OTFS demodulator with windowing/filtering only along the Doppler dimension.

As mentioned earlier, the SFFT^{-1} operation in (6.11) can be represented as

$$\mathbf{Y} = \mathbf{F}_M \mathbf{X} \mathbf{F}_N^{\mathrm{H}}. \tag{6.21}$$

Inserting $\mathbf{Y}$ from (6.21) into (6.7), one may realize that the IDFT operation at OFDM modulator and the M-point DFT operation in SFFT^{-1} block cancel out. Consequently, (6.7) boils down to

$$\mathbf{S} = \mathbf{A}_{\mathrm{CP}} \mathbf{X} \mathbf{F}_N^{\mathrm{H}}. \tag{6.22}$$

Thus, OTFS modulator can be implemented through a similar structure to OFDM including M number of N-point IDFT operations and a CP addition block. The IDFT operations are performed on the rows of the data block $\mathbf{X}$ and the CP addition block appends a CP to the beginning of every M samples. This structure is shown in Figure 6.2 where d_κ is the high

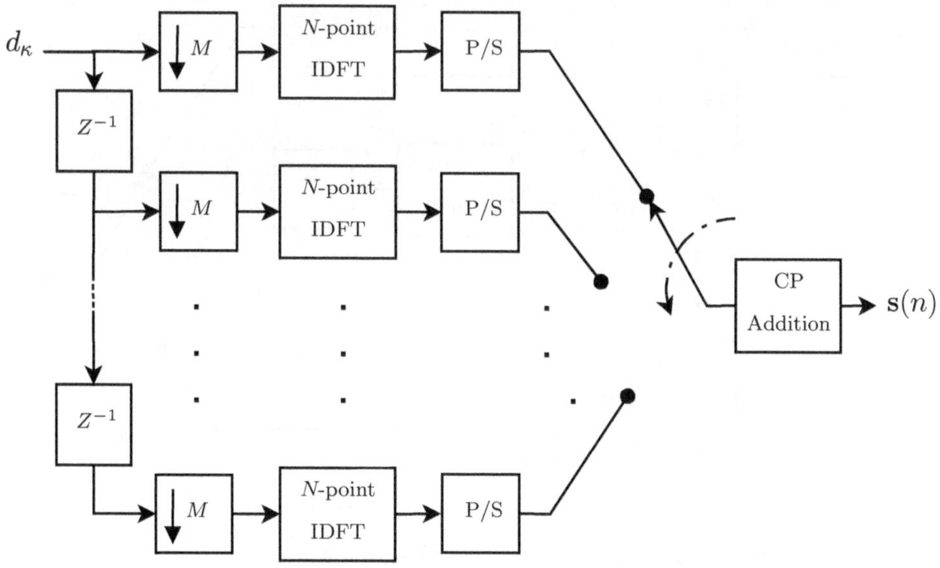

Figure 6.2 Simplified OFDM-based OTFS modulator structure.

rate QAM data stream before getting split into blocks of the size $M \times N$, i.e. the data blocks $\mathbf{X}$, located on the grid points of the delay-Doppler plane.

To arrive at a simplified OTFS receiver structure, we replace $\bar{\mathbf{W}}_{\text{Tx,c}}$ and $W_{\text{Tx,r}}[n]$ in (6.12) with $\mathbf{I}_M$ and 1, respectively. Thus, $\mathbf{Z}$, i.e. the OFDM receiver output signal, can be expanded as

$$\begin{aligned}\mathbf{Z} &= [\mathbf{F}_M\bar{\mathbf{H}}_0\mathbf{X}\mathbf{f}_N^*(0), \dots, \mathbf{F}_M\bar{\mathbf{H}}_{N-1}\mathbf{X}\mathbf{f}_N^*(N-1)] + \bar{\mathbf{V}} \\ &= \mathbf{F}_M[\bar{\mathbf{H}}_0\mathbf{X}\mathbf{f}_N^*(0), \dots, \bar{\mathbf{H}}_{N-1}\mathbf{X}\mathbf{f}_N^*(N-1)] + \bar{\mathbf{V}},\end{aligned} \tag{6.23}$$

where the $M \times N$ matrix $\bar{\mathbf{V}} = [\bar{\mathbf{v}}(0), \dots, \bar{\mathbf{v}}(N-1)]$ includes the noise samples at the output of the OFDM receiver. Replacing $\bar{\mathbf{W}}_{\text{Rx,c}}$ with $\mathbf{I}_M$ in (6.13), while using (6.23), one realizes that similar to the transmitter, the DFT operation at the OFDM demodulator and the IDFT in the SFFT block cancel out. As a result, (6.13) can be rearranged as

$$\widetilde{\mathbf{X}} = [\bar{\mathbf{H}}_0\mathbf{X}\mathbf{f}_N^*(0), \dots, \bar{\mathbf{H}}_{N-1}\mathbf{X}\mathbf{f}_N^*(N-1)]\mathbf{W}_{\text{Rx,r}}\mathbf{F}_N + \widetilde{\mathbf{V}}, \tag{6.24}$$

where the $M \times 1$ vectors $\bar{\mathbf{H}}_n\mathbf{X}\mathbf{f}_N^*(n)$ are the OFDM received symbols after CP removal and $\widetilde{\mathbf{V}} = \mathbf{F}_M^{\text{H}}\bar{\mathbf{V}}\mathbf{F}_N$. Equation (6.24) shows that after discarding the CP, OTFS receiver only requires scaling the OFDM symbols n with the scalar $W_{\text{Rx,r}}[n]$ and performs M number of N-point DFT operations. Therefore, the OTFS demodulator that is shown in Figure 6.1 boils down to the structure that is presented in Figure 6.3 where the $(M + M_{\text{CP}}) \times 1$ vector $\mathbf{r}(n)$ includes the samples of the nth time domain OTFS received symbol and $\tilde{d}_\kappa$ is the high rate data stream at the output of the OTFS demodulator. As was pointed out earlier, the OTFS demodulator output can be fed into a channel equalizer for data detection while taking advantage of full channel diversity.

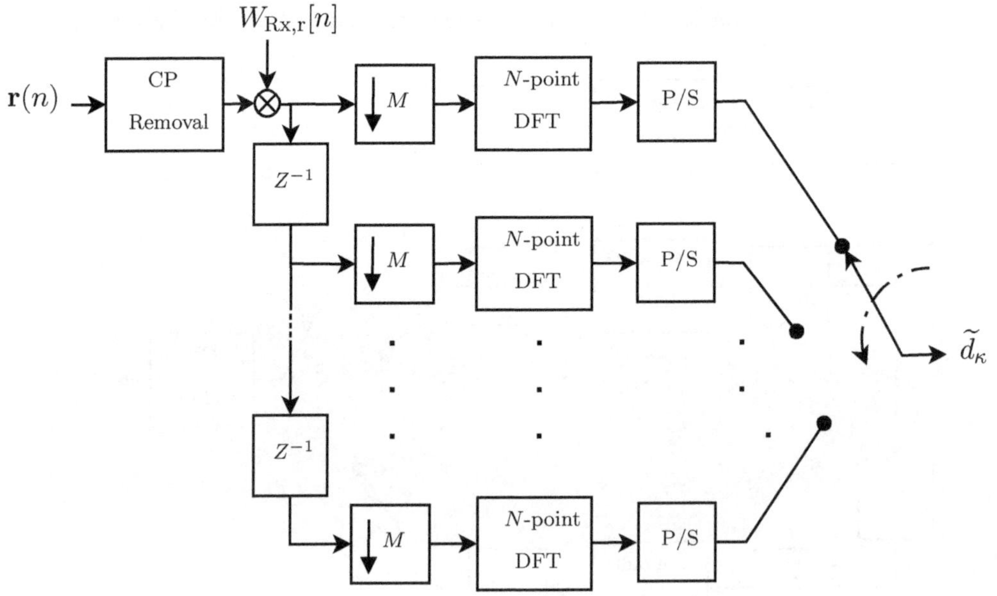

Figure 6.3 Simplified OFDM-based OTFS demodulator structure.

Table 6.1 Computational complexity of different modem structures.

Structure	Number of modulator CMs	Number of demodulator CMs
OTFS mod./demod. in Section 6.5	$\frac{MN}{2}\log_2 N$	$\frac{MN}{2}(1+\log_2 N)$
OTFS mod./demod. in [4]	$MN\log_2 M + \frac{MN}{2}\log_2 N$	$MN\log_2 M + \frac{MN}{2}(1+\log_2 N)$
OFDM mod./demod.	$\frac{MN}{2}\log_2 M$	$\frac{MN}{2}\log_2 M$

6.6 Complexity Analysis

This section focuses on the computational complexity of the presented modem structure in Section 6.5 and compares it with the conventional modem structure in [4] as well as that of OFDM. Let us consider transmission of a data block with the total number of M subcarriers and N time symbols. We assume M and N to be powers of two so that fast Fourier transform (FFT) and its inverse (IFFT) can be deployed for the DFT and IDFT operations, respectively. Table 6.1 summarizes the computational complexity of different modulator/demodulator structures in terms of the number of complex multiplications (CMs).

Conventional OFDM-based OTFS modulator/demodulator in [4] includes the following operations; SFFT^{-1}/SFFT to take the data from delay-Doppler domain to the time–frequency domain and vice versa, 2D time–frequency windowing, and M-point IFFT/FFT operations for OFDM modulation/demodulation. Based on discussions in Section 6.5, transmit windowing is not considered. Thus, for a fair comparison, only receiver windowing is considered in the complexity analysis here. Implementation of SFFT^{-1}/SFFT block requires M number of N-point IFFT and N number of M-point FFT operations. OFDM modulation/demodulation demands N number of M-point IFFT/FFT operations. Usually, channel-independent window functions are real valued and thus windowing demands $\frac{MN}{2}$ number of CMs.

Based on the results of Section 6.5, OTFS modulator and demodulator can be efficiently implemented by combining SFFT^{-1} and SFFT operations with OFDM modulator and demodulator blocks, respectively. Following this approach, Figure 6.2 shows that implementation of OTFS modulator requires only M number of N-point IFFT operations. Being the dual of the modulator, implementation of OTFS demodulator demands M number of N-point FFT operations with an additional windowing operation over the received OFDM time symbols; see Figure 6.3.

To give a more quantitative impression about our complexity analysis and comparisons above, let $M = 2048$ and $N = 8$, which are roughly in line with the parameters in the 3GPP LTE systems. In this case, the modem structure that was presented in Section 6.5 is around eight times simpler than the conventional one in [4] while being around 3.5 times simpler than OFDM. Compared with the conventional demodulator in [4] and its OFDM counterpart, the demodulator structure in Figure 6.3 is 6.5 and 2.75 times simpler, respectively. To further emphasize, we note that in a practical system, the number of subcarriers M is much larger than the number of time symbols N. Hence, lower computational complexity of OTFS modem in Section 6.5 compared with that of OFDM relies on the fact that OTFS requires N-point IFFT/FFT operations while OFDM demands M-point IFFT/FFT operations.

6.7 Recent Results and Potential Research Directions

The emergence of the diverse set of services and applications in the forthcoming 5G and beyond 5G mobile networks such as enhanced mobile broadband (eMBB), massive machine-type communications (mMTC), ultra-reliable low-latency communications (URLLC), high-speed and/or high frequency applications calls for development of a flexible air interface in different levels and the associated modulation scheme [9]. This air interface technology has to overcome the fading effects of frequency- and/or time-selective channels in many diverse scenarios such as high or low Doppler, delay spread, and carrier frequency. To deal with these scenarios, different sets of numerologies for subcarrier spacing/symbol duration of OFDM are designated for the air interface of 5G New Radio (5G NR). However, such an approach to tackle the time variations of wireless channels in high Doppler scenarios or high frequency wireless systems requires advanced channel tracking and estimation often with large training and signal overheads. For instance, to deal with the channel delay spread when the symbol duration is shortened, a CP length of 25% instead of 7% of the symbol duration, as a guard interval, is introduced in the recent 3GPP 5G NR standard [10]. This leads to bandwidth efficiency loss. To tackle the inherent wireless channel-fading problems, i.e. multipath and Doppler effects, this chapter focused on a new generation of modulation technologies, namely, OTFS modulation, that simultaneously utilizes physical resources in multiple dimensions such as time, frequency, and space to achieve the maximum diversity gains that are inherent to the wireless channels.

As explained earlier, OTFS is a novel and fresh approach in waveform design that takes advantage of full channel diversity in multiple dimensions by deploying the delay-Doppler instead of the time–frequency plane to multiplex the transmit data where the time variations of the channel are integrated over time. Hence, the equivalent channel relating the input and output of the system boils down to a sparse and time-invariant one. This is a very important result leading to longer coherence times for the wireless channel. Consequently, a much slower channel tracking and much lower reference-signal/training overheads are required for OTFS compared with its OFDM counterpart with adaptive subcarrier spacing/symbol duration. Additionally, OTFS can be implemented on the top of any given multicarrier waveform that utilizes the time–frequency grid for data transmission, only with the addition of precoding and post-processing units to the modulator and demodulator, respectively. This makes OTFS attractive especially to industry, as it does not demand fundamental changes to the baseband processing units of the currently existing 3GPP LTE standard and 5G NR. Additionally, as it was shown in the Sections 6.5 and 6.6, OTFS transceiver can be efficiently implemented with significantly lower computational complexity than the existing transceiver structures. Owing to all these benefits, OTFS has gained significant momentum in both industry and academia as a candidate waveform to deal with TV wireless channel environments that appear in the future wireless networks.

OTFS was first proposed in the landmark paper [4], where a continuous-time formulation of a single-antenna system was presented. In [11–13], OTFS was studied and compared with OFDM in mmWave systems, 28 GHz band, where the authors illustrated the resilience of OTFS against a high amount of frequency dispersion due to oscillator phase noise and Doppler effect and its superior performance to that of OFDM. In [14], a matrix-form discrete-time formulation of a single-antenna OTFS transceiver with rectangular windows at both transmitter and receiver was presented. The authors in [15] derived a more accurate matrix formulation for OTFS where separable receive window functions were considered. This brought deep insights into understanding of OTFS and the channel impact on the transmitted data symbols in delay-Doppler domain in light of the derived matrix formulation. It is worth mentioning that most of the mathematical derivations that are presented in this chapter are based on the results of [15] with more generalizations on the channel impact and the choice of transmit and receive windows. In [16], an end-to-end formulation for OTFS in multiple-input multiple-output (MIMO) channels was presented, which led to derivation of ergodic capacity for OTFS and its comparison with that of OFDM. In [17], an in-depth diversity analysis of OTFS for both SISO and MIMO OTFS systems is presented. Based on their analysis, the authors in [17] propose a phase rotation method for OTFS that achieves the full diversity gain in delay-Doppler domain in both cases of SISO and MIMO channels. More recently, effective diversity for OTFS is analyzed in [18], which provides a more in-depth and accurate diversity analysis than [17].

The authors in [6] presented a discrete-time formulation of OTFS by sampling the continuous-time channel for rectangular transmit and receive window functions. They also proposed a detection technique based on message passing algorithm. In a more recent work [19], the authors leverage the vectorized formulation for OTFS and propose a detection method based on Markov chain Monte Carlo (MCMC) sampling technique. Moreover, a channel estimation method by deploying a pseudo-random noise (PN) sequence in delay-Doppler domain was proposed in [19]. In [20], a channel estimation method was proposed where the entire OTFS block was utilized for pilot transmission. However, this method may not be effective if the channel estimates become outdated. In [21], a pilot-aided channel estimation method was presented for OTFS with the assumption of a channel causing integer Doppler shifts with respect to the delay-Doppler grid. This is while in reality, multipath channels not only cause integer but also fractional Doppler shifts. To address this issue, an embedded pilot-based channel estimation method was recently proposed in [22] where a single isolated pilot symbol in each OTFS block is deployed. This pilot symbol is surrounded by guard symbols in delay-Doppler grid to avoid interference due to the delay and Doppler spread of the channel between the pilot and data symbols. Starting from channel estimation in point-to-point single-input single-output (SISO) systems, the authors extend their solution to MIMO and multiuser uplink/downlink scenarios. However, a drawback of this approach is an increased training overhead due to the large number of guard symbols as the number of users increases. This is not favorable in vehicular channels with high density of users. In a massive MIMO setup, the aforementioned solutions are not effective in finding the channel estimates, since for a large number of antennas it requires a large number of orthogonal pilot sequences. This leads to a large amount of pilot overhead. To deal with this issue, in a recent work [23], the authors showed that the OTFS MIMO channel is sparse along the delay, Doppler, and angle dimensions. This three-dimensional sparsity allows for formulating the channel estimation problem as a sparse signal recovery problem and channel estimation with a low pilot overhead.

Apart from the abovementioned detection methods for OTFS, the authors in [24] propose an inter-symbol interference (ISI) cancellation method deploying soft-symbol feedback that was initially proposed for multicarrier code division multiple access systems in [25]. This method is based on the maximum likelihood criterion and two implementations for this method in time–frequency and delay-Doppler domain are presented in [24]. To reduce the computational burden of the available OTFS detectors, the authors in [26–28] propose low complexity linear detectors based on ZF and MMSE criteria.

Another interesting application of OTFS is for URLLC in vehicular scenarios where coherence between the uplink and downlink cannot be guaranteed. This leads to channel aging issues as the channel impulse response varies with time for the mobile users. Thus, the estimated channel responses in the uplink become outdated and cannot be used for downlink precoding in massive MIMO systems. This leads to the breakdown of channel hardening effect in massive MIMO and thus, performance degradation [29]. In contrast, when an orthogonal precoding (OP) such as the one in OTFS is deployed, both delay and Doppler spread can be used to recover the channel hardening effect [30]. As a result, massive MIMO with OP, when the channel estimates are outdated, brings two orders of magnitude bit error rate (BER) performance improvement as compared with the case without OP [30].

Multiple access is another important aspect of vehicular channels. There are different multiple access techniques being discussed in OTFS literature so far. Some are based on the traditional orthogonal multiple access (OMA) in delay-Doppler domain, e.g. [22], while some are considering asymptotic orthogonal multiple access in massive MIMO systems, e.g. [30]. As an alternative to these approaches, a new non-orthogonal multiple-access (NOMA) transmission protocol for OTFS has been recently proposed in [31, 32] that enables provision of service to users with heterogeneous mobility profiles, i.e. users that are moving at high speeds and users that are static or quasi-static. Based on the results of [31], both high mobility and low mobility users benefit from this multiple access scheme. This is due to the fact that NOMA enables the users with high mobility to spread their transmit data symbols in time–frequency plane while the static users directly deploy time–frequency grid for data transmission. Hence, low mobility or static users can simultaneously access the time–frequency resources that would be exclusive to the high mobility users in OMA systems. This brings improvements in terms of spectral efficiency and latency [31].

Similar to any other multicarrier waveform, OTFS also suffers from large peak-to-average power ratio (PAPR). Thus, the authors in [33] study PAPR performance of OTFS and show its superiority to OFDM and GFDM. In [34], a two-dimensional impulse as an isolated pilot was considered for channel estimation similar to the one that was proposed in [22]. In this work, a trade-off between the power allocated to data symbols and the pilot symbol was reported. The authors also investigated the effect of pilot power on PAPR of the transmitted OTFS signal. Moreover, influence of power amplifier (PA) nonlinearity on OTFS was studied and it was observed that PA nonlinearities limit the BER performance of OTFS.

OTFS is also analyzed in other communication media than the radio frequency bands, e.g. underwater acoustic channels [35] and optical-wireless communication channels [36]. In

[35], performance of OTFS in underwater acoustic channels in a multiuser massive MIMO setup is analyzed and compared against OFDM. In this study, it is shown that OTFS outperforms the conventional OFDM systems in terms of BER performance. Inspired by OTFS, the authors in [36] design a 2D modulation scheme with Hermitian symmetry that is similar to OTFS and outperforms OFDM in terms of BER performance.

Even though a reasonably large body of literature is formed around the topic of OTFS since its emergence, there is still space for pursuing new lines of research on this topic. Therefore, we close this chapter by itemizing a number of different research directions through which the available literature on OTFS can be extended.

- In all the available literature that was explained above, rectangular transmit and receive windows are considered for OTFS. This is while the transmit and receive windows can be designed to improve, for instance, PAPR, channel estimation, and signal detection performance of OTFS.
- All the available literature on OTFS considers perfect synchronization in both time and frequency domain while this is not the case in practical systems. Hence, timing and frequency offset estimation and compensation in both single and multiuser scenarios need to be studied for OTFS.
- Most of the literature on OTFS is based on OFDM for time–frequency transmission and the performance of OTFS when waveforms other than OFDM are utilized is unknown to date. This is another important line of research that requires attention in the near future.
- The study of OTFS in massive MIMO channels is still in its infancy and some issues such as pilot contamination problem of massive MIMO when OTFS is deployed for data transmission is unknown.
- Even though there is some literature available on channel equalization and detection of OTFS, channel equalization and multiuser detection in the uplink of OTFS in both OMA and NOMA modes need further study and analysis.
- Dynamic resource allocation plays a pivotal role in interference reduction and optimization of resource utilization in multiuser vehicular communication systems. Thus, resource allocation is another aspect of OTFS that needs to be studied in the future.

References

1 Wang, T., Proakis, J.G., Masry, E., and Zeidler, J.R. (2006). Performance degradation of OFDM systems due to Doppler spreading. *IEEE Transactions on Wireless Communications* 5 (6): 1422–1432. doi: https://doi.org/10.1109/TWC.2006.1638663.

2 Farhang-Boroujeny, B. (2014). Filter bank multicarrier modulation: a waveform candidate for 5G and beyond. *Advances in Electrical Engineering* 2014. 25p.

3 Dean, T., Chowdhury, M., and Goldsmith, A. (2017). A new modulation technique for Doppler compensation in frequency-dispersive channels. 2017 IEEE 28th Annual International Symposium on Personal, Indoor, and Mobile Radio Communications (PIMRC), October 2017, pp. 1–7. doi: https://doi.org/10.1109/PIMRC.2017.8292240.

4 Hadani, R., Rakib, S., Tsatsanis, M. et al. (2017). Orthogonal time frequency space modulation. IEEE Wireless Communications and Networking Conference (WCNC), March 2017, pp. 1–6. doi: https://doi.org/10.1109/WCNC.2017.7925924.

5 Tuchler, M., Koetter, R., and Singer, A.C. (2002). Turbo equalization: principles and new results. *IEEE Transactions on Communications* 50 (5): 754–767. doi: https://doi.org/10.1109/TCOMM.2002.1006557.

6 Raviteja, P., Phan, K.T., Jin, Q. et al. (2018). Low-complexity iterative detection for orthogonal time frequency space modulation. IEEE Wireless Communications and Networking Conference (WCNC), Barcelona, Spain. pp. 1–6.

7 (Alex) Li, L., Wei, H., Huang, Y. et al. (2017). A simple two-stage equalizer with simplified orthogonal time frequency space modulation over rapidly time-varying channels, arXiv preprint arXiv:1709.02505.

8 Michailow, N., Matthé, M., Gaspar, I.S. et al. (2014). Generalized frequency division multiplexing for 5th generation cellular networks. *IEEE Transactions on Communications* 62 (9): 3045–3061. doi: https://doi.org/10.1109/TCOMM.2014.2345566.

9 3GPP TR 38.913 (2017). Study On Scenarios and Requirements for Next Generation Access Technologies. Technical Report, Version 14.2.0 Release 14. ETSI.

10 3GPP TS 38.211 (2018). 5G NR; Physical Channels and Modulation. Technical specification.

11 Hadani, R., Rakib, S., Molisch, A.F. et al. (2017). Orthogonal time frequency space (OTFS) modulation for millimeter-wave communications systems. IEEE MTT-S International Microwave Symposium (IMS), Honolulu, Hawaii, USA. pp. 681–683.

12 Wiffen, F., Sayer, L., Bocus, M.Z. et al. (2018). Comparison of OTFS and OFDM in ray launched sub-6 GHz and mmWave line-of-sight mobility channels. IEEE Annual International Symposium on Personal, Indoor and Mobile Radio Communications (PIMRC), Bologna, Italy, pp. 73–79.

13 Surabhi, G.D., Ramachandran, M.K., and Chockalingam, A. (2019). OTFS modulation with phase noise in mmWave communications. IEEE Vehicular Technology Conference (VTC2019-Spring), Kuala Lumpur, Malaysia. pp. 1–5.

14 Li, L., Wei, H., Huang, Y. et al. (2017). A simple two-stage equalizer with simplified orthogonal time frequency space modulation over rapidly time-varying channels, arXiv preprint arXiv:1709.02505.

15 Farhang, A., RezazadehReyhani, A., Doyle, L.E., and Farhang-Boroujeny, B. (2017). Low complexity modem structure for OFDM-based orthogonal time frequency space modulation. *IEEE Wireless Communications Letters* 1–1. doi: https://doi.org/10.1109/LWC.2017.2776942.

16 RezazadehReyhani, A., Farhang, A., Ji, M. et al. (2018). Analysis of discrete-time MIMO OFDM-based orthogonal time frequency space modulation. IEEE International Conference on Communications (ICC), Kansas City, Kansas, USA. pp. 1–6.

17 Surabhi, G.D., Augustine, R.M., and Chockalingam, A. (2019). On the diversity of uncoded OTFS modulation in doubly-dispersive channels. *IEEE Transactions on Wireless Communications* 18 (6): 3049–3063.

18 Raviteja, P., Hong, Y., Viterbo, E., and Biglieri, E. (2020). Effective diversity of OTFS modulation. *IEEE Wireless Communications Letters* 9 (2): 249–253.

19 Murali, K.R. and Chockalingam, A. (2018). On OTFS modulation for high-Doppler fading channels. Information Theory and Applications Workshop (ITA), Pacific Beach, San Diego, USA. pp. 1–10.

20 Ramachandran, M.K. and Chockalingam, A. (2018). MIMO-OTFS in high-Doppler fading channels: signal detection and channel estimation. IEEE Global Communications Conference (GLOBECOM), Abu Dhabi, United Arab Emirates, pp. 206–212.

21 Hadani, R. and Rakib, S.S. (2016). OTFS methods of data channel characterization and uses thereof. US Patent 9, 444, 514 B2, September 2016.

22 Raviteja, P., Phan, K.T., and Hong, Y. (2019). Embedded pilot-aided channel estimation for OTFS in delay–Doppler channels. *IEEE Transactions on Vehicular Technology* 68 (5): 4906–4917.

23 Shen, W., Dai, L., An, J. et al. (2019). Channel estimation for orthogonal time frequency space (OTFS) massive MIMO. *IEEE Transactions on Signal Processing* 67 (16): 4204–4217.

24 Zemen, T., Hofer, M., and Loeschenbrand, D. (2017). Low-complexity equalization for orthogonal time and frequency signaling (OTFS), arXiv preprint aarXiv:1710.09916.

25 Zemen, T., Mecklenbrauker, C.F., Wehinger, J., and Muller, R.R. (2006). Iterative joint time-variant channel estimation and multi-user detection for MC-CDMA. *IEEE Transactions on Wireless Communications* 5 (6): 1469–1478.

26 Surabhi, G.D. and Chockalingam, A. (2020). Low-complexity linear equalization for OTFS modulation. *IEEE Communications Letters* 24 (2): 330–334.

27 Long, F., Niu, K., Dong, C., and Lin, J. (2019). Low complexity iterative LMMSE-PIC equalizer for OTFS. IEEE International Conference on Communications (ICC), Shanghai, China, pp. 1–6.

28 Tiwari, S., Das, S.S., and Rangamgari, V. (2019). Low complexity LMMSE receiver for OTFS. *IEEE Communications Letters* 23 (12): 2205–2209.

29 Truong, K.T. and Heath, R.W. (2013). Effects of channel aging in massive MIMO systems. *Journal of Communications and Networks* 15 (4): 338–351.

30 Zemen, T. and Löschenbrand, D. (2019). Combating massive MIMO channel aging by orthogonal precoding. IEEE 2nd 5G World Forum (5GWF), Dresden, Germany, pp. 97–101.

31 Ding, Z., Schober, R., Fan, P., and Vincent Poor, H. (2019). OTFS-NOMA: an efficient approach for exploiting heterogenous user mobility profiles. *IEEE Transactions on Communications* 67 (11): 7950–7965.

32 Ding, Z. (2020). Robust beamforming design for OTFS-NOMA. *IEEE Open Journal of the Communications Society* 1: 33–40.

33 Surabhi, G.D., Augustine, R.M., and Chockalingam, A. (2019). Peak-to-average power ratio of OTFS modulation. *IEEE Communications Letters* 23 (6): 999–1002.

34 Marsalek, R., Blumenstein, J., Prokes, A., and Gotthans, T. (2019). Orthogonal time frequency space modulation: pilot power allocation and nonlinear power amplifiers. IEEE International Symposium on Signal Processing and Information Technology (ISSPIT), Ajman, United Arab Emirates, pp. 1–4.

35 Bocus, M.J., Doufexi, A., and Agrafiotis, D. (2020). Performance of OFDM-based massive MIMO OTFS systems for underwater acoustic communication. *IET Communications* 14 (4): 588–593.

36 Zhong, J., Zhou, J., Liu, W., and Qin, J. (2020). Orthogonal time-frequency multiplexing with 2D Hermitian symmetry for optical-wireless communications. *IEEE Photonics Journal* 12 (2): 1–10.

Part II

RAN Slicing and 5G Vertical Industries

7

Multi-Numerology Waveform Parameter Assignment in 5G

Ahmet Yazar[1,*] *and Hüseyin Arslan*[1,2]

[1] *Electrical-Electronics Engineering, Istanbul Medipol University, Turkey*
[2] *Electrical Engineering, University of South Florida, USA*

7.1 Introduction

Third Generation Partnership Project (3GPP) Release 15 for 5G new radio (NR) is developed to meet the requirement diversity of different users and services [1]. Moreover, Release 16 and 17 are coming in two years respectively to meet further requirements of ultra-reliable low-latency communications (URLLC) and massive machine-type communications (mMTC) services [2]. Flexibility is increasing to be able to meet different requirements. The concept of multi-numerology is defined in 5G NR standardization to increase flexibility from the waveform perspective. Besides, scheduling and resource allocation become more related with waveform design thanks to the multi-numerology concept. However, frame flexibility brings more load at a scheduler [3]. There are more waveform parameter options that can be configured by the scheduler in NR compared to long-term evolution (LTE) [4].

Waveform parameter options can be analyzed with respect to the waveform components that include lattice, pulse shape, and frame [5]. As the first component, lattice is multidimensional resource mapping and each point shows a location of one resource element. The possible distances between lattice points give numerology structures for a waveform [6, 7]. For example, the distances between LTE lattice points are fixed as LTE employs a single-numerology waveform. However, 5G NR uses a flexible lattice structure that indicates that there are multiple numerologies. Additionally, both LTE and 5G NR use a multidimensional lattice on time–frequency planes. The other waveform component, pulse shape, gives the main characteristic to a waveform by deciding how to transmit the symbols on lattice points. LTE generally employs rectangular pulses for orthogonal frequency division multiplexing (OFDM) symbols but it is also possible to implement windowing. 5G NR can have different amounts of windowing for a pulse shape of OFDM symbols. As a last main waveform component, frame structure can change in addition to lattice and pulse shape. For example, multiple numerologies are multiplexed in frequency domain in 5G NR and the amounts of inter-numerology guard bands can be configured adaptively as a part

* Corresponding Author: Ahmet Yazar; ayazar@medipol.edu.tr

Radio Access Network Slicing and Virtualization for 5G Vertical Industries, First Edition.
Edited by Lei Zhang, Arman Farhang, Gang Feng, and Oluwakayode Onireti.

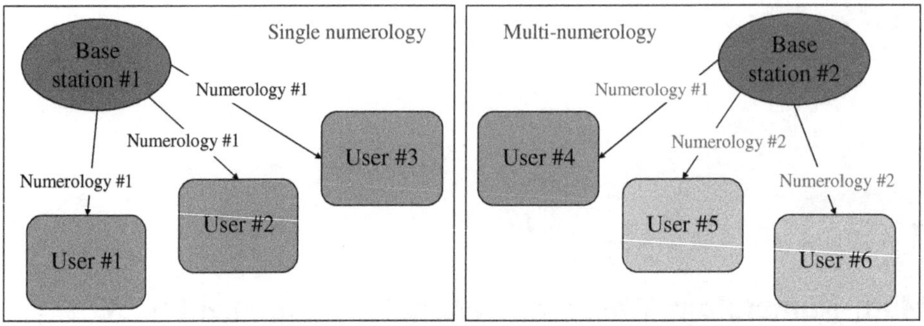

Figure 7.1 An example demonstration for communications with single numerology and multi-numerology waveforms.

of the frame structure [8]. From these definitions, waveform parameter options contain multiple numerology structures, windowing and filtering parameters, and the amount of inter-numerology guard bands in frequency domain. Besides, all of these parameters can change for different users and services in the same frame [7].

Figure 7.1 shows an example demonstration for single-numerology and multi-numerology communications. The users do not communicate with a base station (BS) with different numerologies in a single-numerology system. Forcing to use only one numerology does not provide high flexibility to meet several requirements of different users and services [3]. Diversifying the number of parameters increases flexibility [9]. Hence, multi-numerology waveforms enhance flexibility. However, there is a non-orthogonality problem between multiple numerologies because of subcarrier spacing variations [10]. If windowing, filtering, or inter-numerology guard bands is not employed, orthogonality cannot be ensured for multi-numerology OFDM systems [6, 11]. Therefore, there is an important trade-off situation regarding the flexibility and inter-numerology interference (INI) in 5G NR.

5G NR comes with more waveform parameters than the previous generations. Moreover, there are new problem sources such as INI. The importance of controlling mechanisms in medium access control (MAC) layer (e.g. parameter optimization, scheduling, radio resource control, evaluating feedback signals and measurements) increases. Through multi-numerology based 5G systems, waveform studies become directly related with scheduling, resource allocation, and radio access network (RAN) slicing [3, 12–27]. Waveform parametrization and waveform parameter assignment are important topics from this perspective. Details of the relationships for different studies are provided in Sections 7.2 and 7.3.

This chapter mainly aims to present a vision on waveform parameter assignment considering the current waveform design in 5G NR. The objective of the chapter is to draw attention to waveform parameter assignment by more efficiently employing the available waveform without designing a new one. At that point, multi-numerology concept plays a crucial role in waveform parametrization. Hence, waveform parameter assignment literature is reviewed from the multi-numerology perspective. After that, case studies for the optimizations in numerology assignment, waveform processing, and joint methods are discussed to reveal the relationships between multi-numerology concept and waveform parameter assignment in 5G NR.

7.1.1 Problem Definitions

Two main problems can be analyzed related with the waveform parameter assignment [25]: (i) A necessity of optimal parameter assignment strategy for each single user while meeting user requirements one by one; (ii) a necessity of optimal parameter assignment strategy for multiple users while meeting their individual requirements together. Meeting several requirements of one user can be difficult in some cases. Moreover, multiple users can have more challenging requirements to be met in the same frame of a waveform. These two main problems can be solved separately [3, 14, 15, 17, 18, 22]. However, they are not independent problems so it is also possible to solve them together [20, 21, 25].

The subproblems of the two main problems given above can be listed as (i) numerology selection or assignment for each user, (ii) waveform processing based on the user–numerology association, (iii) resource allocation between all users, and (iv) optimization(s) for numerology assignment, waveform processing, and resource allocation. Figure 7.2 illustrates these problems. These subproblems can be solved independently or jointly. For example, waveform processing techniques such as adaptive inter-numerology guard bands and subframe windowing with different roll-off factors can be designed based on the assumption of perfect user–numerology association [3, 14, 15, 17, 18, 22]. However, it is more realistic to develop this type of waveform processing techniques together with adaptive user–numerology association methods because there can be a different waveform processing technique or parameter necessity while numerologies of users are changing [20, 25].

More requirements need more flexibility. Flexibility brings in the necessity for more parametrization. Then the high number of parameters makes the given optimization problems more difficult. Under the scope of these problems, the topic of waveform parameter

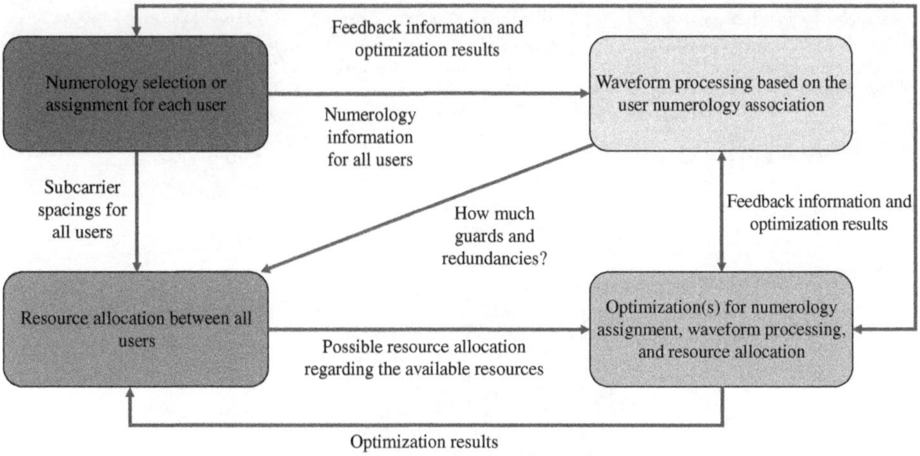

Figure 7.2 Illustrations of the given subproblems and their example relationships.

assignment has relationships with the other multi-numerology subjects and some of the relationships are provided in Section 7.1.2 from a wider perspective.

7.1.2 Literature Review

We can relate different multi-numerology studies [6] with the waveform parameter assignment. Multi-numerology studies can be grouped as (i) general multi-numerology concept, (ii) INI and non-orthogonality issues, (iii) time-domain analysis of multi-numerology signals, and (iv) scheduling and resource allocation. Figure 7.3 shows the relationships between different topics for multi-numerology waveforms. Waveform parameter assignment can be done regarding the information on requirements and frame design considerations for multiple numerologies (Group-1). Therefore, relationships between the multi-numerology structures defined in standards and different requirements with frame design considerations should be established properly. Meanwhile, INI analysis and the related non-orthogonality issues (Group-2) are important feedbacks for waveform parameter assignment. INI is a strong drawback considering the resulting non-orthogonality. Time-domain analysis of multi-numerology signals (Group-3) has also effects on the waveform parameter assignment algorithms. Scheduling and resource allocation units (Group-4) include all of these feedback mechanisms and waveform parameter assignment algorithms to solve the problems that are defined in Section 7.1.1.

For Group-1 studies, the general concept on multi-numerology structures is discussed in [7, 28–30]. Research opportunities for multi-numerology waveform are given in [6]. In [3] and [31], requirements and frame design considerations for multiple numerologies are explained. Requirement analysis is very important for the feedback of solution mechanisms as regards the optimization problems. If the user and service requirements are mapped into the waveform parameters correctly, all of these requirements can be met in a better way. Requirements should be matched with suitable waveform parameters as in [3]. Otherwise, employing the multi-numerology concept is not meaningful.

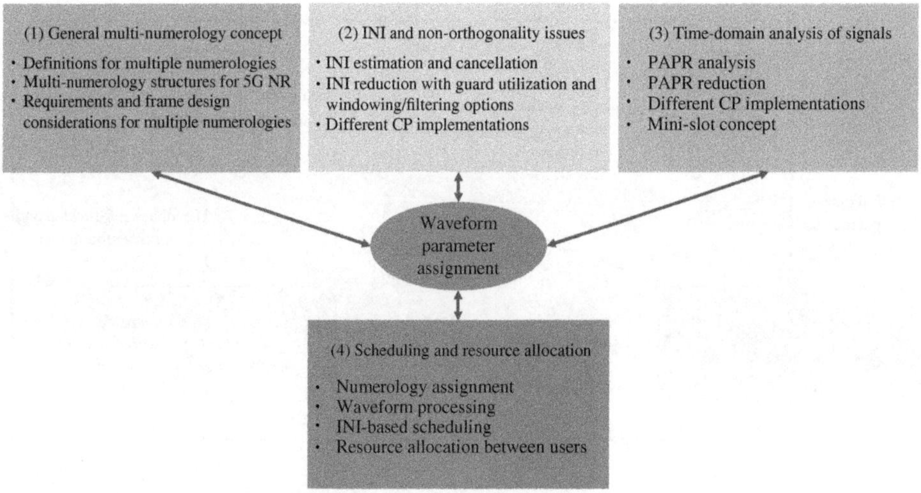

Figure 7.3 All of the multi-numerology based research topics are directly associated with waveform parameter assignment subject because the other topics cannot be investigated without waveform parameter decisions.

For Group-2 studies, INI analysis and the related non-orthogonality issues are discussed in [10, 11, 32–36]. INI cancellation methods are provided in [11, 32, 35, 36]. Multi-numerology waveforms provide high flexibility; however, their main disadvantage is the INI effects. Reduction or cancellation of INI enhances reliability and then the capacity. Besides, INI analysis is one of the main feedbacks for the waveform parameter assignment algorithms.

For Group-3 studies, peak to average power ratio (PAPR) reduction methods for multi-numerology OFDM signals are developed in [37] and [38]. Moreover, PAPR analysis is provided in [39] based on multiple numerologies. Different cyclic prefix (CP) implementations are discussed in [6] and [10]. Using only one CP for the composite signal that is formed by subframes of different numerologies provides several advantages such as higher spectral efficiency for the shorter CP cases rather than the fixed individual CPs of multiple numerologies. CP implementation variety has direct effects on the waveform parameter assignment.

For Group-4 studies, multi-numerology based scheduling and resource allocation papers are investigated under five subgroups that are listed below. In Section 7.3, some of these studies are revisited to give a more detailed perspective of waveform parameter assignment. As a short summary, Table 7.1 shows the differences among the related studies in multi-numerology literature for scheduling and resource allocation.

Table 7.1 Comparison of multi-numerology based scheduling and resource allocation studies.

Study	Numerology assignment Topic-1	Waveform processing Topic-2	INI-based scheduling Topic-3	Resource allocation Topic-4	Joint optimization Topic-5
Bag et al. [12]				✓	
Chang et al. [13]				✓	
Demir and Arslan [14]		✓			
Demmer et al. [15]		✓			
Gonzalez et al. [16]				✓	
Lagen et al. [17]	✓				
Marijanovic et al. [18]	✓				
Marijanovic et al. [19]			✓	✓	
Marijanovic et al. [20]	✓		✓	✓	✓
Marijanovic et al. [21]			✓	✓	✓
Soni et al. [22]	✓				
Sui et al. [23]				✓	
Yazar and Arslan [3]	✓				
Yazar and Arslan [24]			✓		
Yazar and Arslan [25]	✓	✓			✓
You et al. [26]				✓	
Zhang et al. [27]				✓	

Topic-1: numerology assignment, Topic-2: waveform processing, Topic-3: INI-based scheduling, Topic-4: resource allocation, Topic-5: joint optimization.

1) Numerology assignment
2) Waveform processing
3) INI-based scheduling
4) Resource allocation between users
5) Joint optimization

Numerology assignment: In most of the studies under multi-numerology waveform literature, numerology assignment is given as an assumption to make the other optimizations. Some of the studies present joint optimization with the numerology assignment as in [17, 20, 25]. Yazar and Arslan [25] use machine learning (ML) methods to optimize the numerology assignment and guard band selection between different numerologies jointly. In [17], the authors assume that there is a direct mapping between numerologies and service types. The paper optimizes the numerology configuration and the downlink-uplink (DL-UL) duplexing ratio in a time division duplexing (TDD). The focus of [20] is the joint optimization of bandwidth allocations and numerology assignments for four users. There are also pure numerology assignment optimization studies in the literature [3, 17, 18, 22]. Yazar and Arslan [3] make optimal numerology assignment regarding the requirements and frame design considerations. The authors try to find the effective number of multiple numerologies. In [18], the authors find the best subcarrier spacing for all users. Adaptive numerology selection method is developed for V2X service in [22].

Waveform processing: Adaptive guard band concept is analyzed in [14] and [15] for multiple numerologies. Putting guard band between different numerologies has an important effect on INI but using large guard bands decreases spectral efficiency. Moreover, it is possible to use different amounts of guard bands between different numerology pairs. Yazar and Arslan [25] investigate the ML usage for the guard band decisions. Additionally, INI analysis studies also include guard band usage in the INI equations.

INI-based scheduling: INI is an important feedback information for scheduling decisions because it is the main disadvantage of multi-numerology systems. In different studies, INI is used as an input for scheduling mechanisms [19–21, 24]. INI can be called as inter-band interference (IBI) in some of these studies.

Resource allocation between users: Most of the multi-numerology scheduling studies in the literature focus on resource allocation [12, 13, 16, 19–21, 23, 26, 27]. The main aim of these studies is the optimization of bandwidth allocations. They are not directly related with the waveform parameter assignment but resource allocation and waveform parameter assignment should be handled jointly.

Joint optimization: Numerology assignment, waveform processing, the amount of INI, and resource allocation can be jointly optimized. Some of the studies make optimizations for some of them together [17, 20, 21, 25]. A summary of the literature is presented in Table 7.1.

7.2 Waveform Parameter Options

In the cellular communications systems, there is a trend that shows an increase in the number of configurable parameter options with each generation [4]. There is a linear increment until now. The number of configurable parameters for a waveform design is also increased with 5G. This section presents a brief discussion on the waveform parameter options that are defined in NR [40].

There are two key points from the flexibility perspective. The first one is the number of configurable parameters in the system. Availability of more configurable parameters generally means high flexibility. The second key point is to have the capability to provide different parametrization options in the same frame for the users in one coverage area. 5G NR has availability of different configurable parameters and it also provides the capability to assign different parameters to users at a time [40].

In future, all the different parameters may be present for each user in a frame but this is not the case for 5G NR. There are drawbacks while employing multiple waveform parameters in one frame and they limit the number of parameter options that are available for the assignment [3]. Therefore, the current cellular systems have limited parameter sets that are defined in the standards.

Figure 7.4 illustrates the waveform parameter options in 5G NR [25]. Numerologies change mainly with the subcarrier spacing parameter in 5G NR. CP ratio is fixed and CP durations depend on the subcarrier spacings. However, there are normal and extended CP versions as an exception for 60 kHz subcarrier spacing. Extended CP usage is not available for the other subcarrier spacings. Subcarrier spacings of 15, 30, and 60 kHz are used up to 6 GHz spectrum. The 60 and 120 kHz subcarrier spacings are preferred for above 6 GHz. For 52.6 GHz and beyond, there may be larger subcarrier spacings [41].

In [5], different pulse shapes are compared to each other. 5G NR provides options for rectangular and raised-cosine windowing. There is a flexibility to apply windowing with different or same roll-off factors on the subframes for each numerology or composite signal of multiple numerologies at the transmitter. Receiver windowing is the other option for 5G NR. Filtering is also possible in addition to windowing.

Inter-numerology guard band utilization aims to decrease the INI effects rather than changing the pulse shape as in windowing and filtering. All of them can be called as waveform processing techniques but inter-numerology guard bands are utilized while designing the frame structure. As discussed in Section 7.1, multiple numerologies are related with the lattice structure, applying window and filter are related with the pulse shape, and inter-numerology guard band is related with the frame structure. Guard band utilization can be in a fixed or adaptive manner. For example, some of the subcarriers that are not affected by INI in the guard bands are used for data communications in [42]. It is possible to develop different types of adaptive guard band utilization methods. Additionally, the

5G numerology structures

Subcarrier spacing (kHz)	CP Duration (μs)	Slot duration (ms)	Number of symbols in one slot
15	4.76	1	14
30	2.38	0.5	14
60	1.19 | 4.17	0.25	12 | 14
120	0.60	0.125	14

Optional waveform processing techniques

- Different or same guard bands between multiple numerologies.
- Windowing with different or same roll-off factors for each user or numerology.
- Filtering with different or same filter coefficients for each user or numerology.

Figure 7.4 Waveform parameter options in 5G NR.

amount of inter-numerology guard bands is left flexible in 3GPP standards, and there are no limitations.

7.3 Waveform Parameter Assignment

3GPP standards give the BS and user equipment (UE) manufacturers the freedom to implement any additional algorithm they desire as long as it is transparent to the receiver [43]. Scheduling algorithms are also left implementation-dependent. In this context, waveform parameter assignment can be done flexibly with different optimization steps. In this section, details of various example case studies for waveform parameter assignment are provided.

Numerology assignments will be done via bandwidth part (BWP) selections in 5G NR. Each BWP has one specific numerology, and BWP operations in frequency domain are implementation-dependent. They are controlled by BS. One BWP can belong to a single user or multiple users. It is also possible to make a BWP for one service. In this subsection, the focus is on numerology assignment rather than on BWP operations. Therefore, BWP issues are not discussed in the remaining parts.

An example case study for numerology assignment for users and services is given in [3]. They first map the user and service requirements to numerology parameters in a specific parameter set for 5G NR. It is assumed that there is a feedback mechanism from users to BSs regarding the requirements, as shown in Figure 7.5. Then, they use the scheduler to find the best suitable numerologies from a parameter set for each user and services. This solution makes all users satisfied but it is not realistic. Increasing the number of effective numerologies in a system brings in the necessity for more redundancies. Hence, the capacity is affected negatively.

To make a more realistic system, the authors of [3] remove the least selected numerology from a parameter set and repeat the numerology assignment step without the eliminated numerology. They try to decrease redundancy in the system while maintaining the

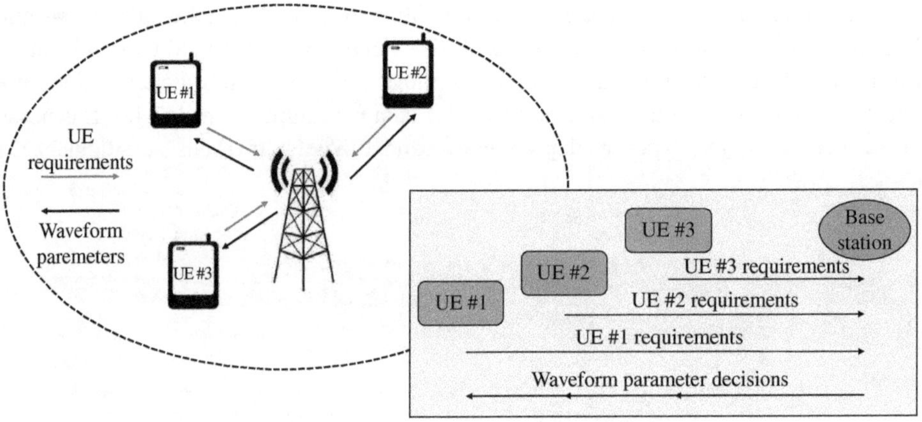

Figure 7.5 An example demonstration for the numerology assignment to UEs. Source: Reprinted and adapted with permission from Yazar and Arslan [3]. Copyright 2018, IEEE.

flexibility as much as possible. In the subsequent steps, they check the amount of redundancy, and then stop or continue to eliminate numerologies that are selected less frequently. This method finds an optimum point considering the redundancies and system flexibility. However, it does not include waveform processing techniques while optimizing the numerology assignments.

5G NR optional waveform processing methods include windowing and filtering for each numerology subframe or composite signal subframe. Windowing can be at the transmitter or receiver. Additionally, inter-numerology guard band is optional and it can be in a fixed or an adaptive manner. For example, Demir and Arslan [14] present one of the adaptive guard band algorithm examples. The authors adaptively change the guard bands between multiple numerologies considering the average desired signal-to-noise-plus-interference ratio (SINR) and power levels of each user. On the contrary, employing fixed amount of guard bands rather than adaptive ones is not an optimum method but it decreases the scheduling complexity.

In [14], it is assumed that optimal numerology assignment is done before applying adaptive guard band algorithms. However, it is not a realistic scenario because different amounts of guard bands can have effects on the numerology assignment recursively. Therefore, there is a need for joint optimization methods.

Optimization methods for waveform parameter assignment generally affect each other recursively. These optimizations can be done in multiple steps or in one joint step. Joint optimization research is exemplified in [25] using ML approach. Figure 7.6 provides example (a) feature set and (b) class labels for this ML-based joint optimization method. The number of numerologies and the amount of inter-numerology guard bands are decided using channel- and service-based features. They do not make direct user–numerology association because it requires a high number of classes in the system. Also, windowing parameter is not included because of the same reason. It is not an easy task but all of these numerologies and parameters need to be optimized together jointly in an ideal scheduling mechanism for 5G NR.

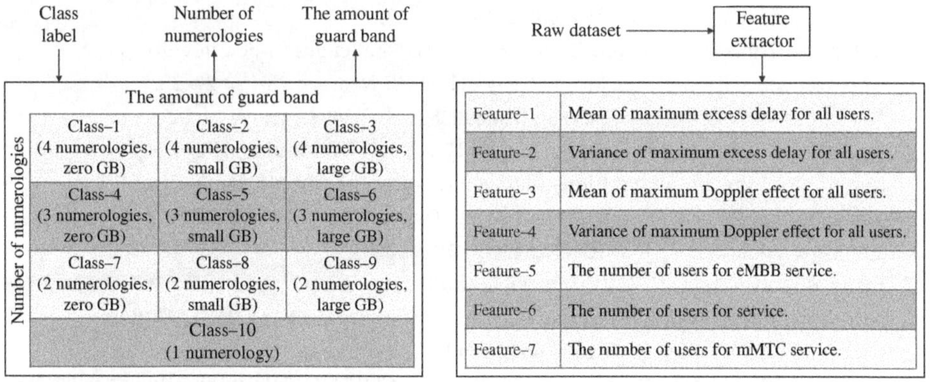

Figure 7.6 An example feature set and class labels for ML-based joint optimization method. Source: Reprinted and adapted with permission from Yazar and Arslan [25]. Copyright 2019, IEEE.

7.4 Conclusion

Until 5G systems, there was not much flexibility on the lattice, pulse shape (filter structure), and frame. Therefore, waveform components were almost fixed and scheduling algorithms were mainly related with the resource allocation between users. The relationship between MAC layer and waveform was limited. However, 5G systems use flexible lattice and pulse shape for OFDM waveform. Hence, frame structures are more flexible than the previous generations. This flexibility increases the importance of scheduling mechanisms at the MAC layer. Waveform parameter assignment is one of the key topics that are related with the scheduling mechanisms.

Standards give freedom to vendors for scheduling related system designs. At that point, analysis of the necessary amount of flexibility and complexity in scheduling mechanisms of 5G is an important, promising research area. Additionally, the role of ML will increase in the next generation cellular networks especially if there are more waveform parameter options. Moreover, the future communications systems can use multi-waveform and multi-numerology structures together in the same frame. There may be different parameter options for different waveforms. Hence, the number of total waveform-related parameter options will be numerous. As can be seen, waveform parameter assignment will be one of the most important topics for scheduling and resource allocation in communications systems.

References

1 Lin, X., Li, J., Baldemair, R. et al. (2018). 5G new radio: unveiling the essentials of the next generation wireless access technology.

2 Ghosh, A., Maeder, A., Baker, M., and Chandramouli, D. (2019). 5G evolution: a view on 5G cellular technology beyond 3GPP release 15. *IEEE Access* 7: 127639–127651. doi: https://doi.org/10.1109/ACCESS.2019.2939938.

3 Yazar, A. and Arslan, H. (2018). A flexibility metric and optimization methods for mixed numerologies in 5G and beyond. *IEEE Access* 6: 3755–3764. doi: https://doi.org/10.1109/ACCESS.2018.2795752.

4 Imran, A., Zoha, A., and Abu-Dayya, A. (2014). Challenges in 5G: how to empower son with big data for enabling 5G. *IEEE Network* 28 (6): 27–33. doi: https://doi.org/10.1109/MNET.2014.6963801.

5 Sahin, A., Guvenc, I., and Arslan, H. (2014). A survey on multicarrier communications: prototype filters, lattice structures, and implementation aspects. *IEEE Communications Surveys Tutorials* 16 (3): 1312–1338. doi: https://doi.org/10.1109/SURV.2013.121213.00263.

6 Yazar, A. and Arslan, H. (2018). Flexible multi-numerology systems for 5G new radio. *River Publishers Journal of Mobile Multimedia* 14 (4): 367–394. doi: https://doi.org/10.13052/jmm1550-4646.1442.

7 Zaidi, A.A., Baldemair, R., Moles-Cases, V. et al. (2018). OFDM numerology design for 5G new radio to support IoT, eMBB, and MBSFN. *IEEE Communications Standards Magazine* 2 (2): 78–83. doi: https://doi.org/10.1109/MCOMSTD.2018.1700021.

8 3GPP (2019). General Aspects for User Equipment (UE) Radio Frequency (RF) for NR. *Technical Report 38.817-01*, ver. 15.5.0.

9 Holma, H., Toskala, A., and Nakamura, T. (2019). *5G Technology: 3GPP New Radio*, 1e. Wiley.

10 Kihero, A.B., Solaija, M.S.J., and Arslan, H. (2019). Inter-numerology interference for beyond 5G. *IEEE Access* 7: 146512–146523. doi: https://doi.org/10.1109/ACCESS.2019.2946084.

11 Zhang, X., Zhang, L., Xiao, P. et al. (2018). Mixed numerologies interference analysis and inter-numerology interference cancellation for windowed OFDM systems. *IEEE Transactions on Vehicular Technology* 67 (8): 7047–7061. doi: https://doi.org/10.1109/TVT.2018.2826047.

12 Bag, T., Garg, S., Shaik, Z., and Mitschele-Thiel, A. (2019). Multi-numerology based resource allocation for reducing average scheduling latencies for 5G NR wireless networks. *2019 European Conference on Networks and Communications (EuCNC)*, June 2019, pp. 597–602. doi: https://doi.org/10.1109/EuCNC.2019.8802009.

13 Chang, B., Zhang, L., Li, L. et al. (2019). Optimizing resource allocation in uRLLC for real-time wireless control systems. *IEEE Transactions on Vehicular Technology* 68 (9): 8916–8927. doi: https://doi.org/10.1109/TVT.2019.2930153.

14 Demir, A.F. and Arslan, H. (2017). The impact of adaptive guards for 5G and beyond. *2017 IEEE 28th Annual International Symposium on Personal, Indoor, and Mobile Radio Communications (PIMRC)*, October 2017, pp. 1–5. doi: https://doi.org/10.1109/PIMRC.2017.8292413.

15 Demmer, D., Gerzaguet, R., Dore, J., and Le Ruyet, D. (2018). Analytical study of 5G NR eMBB co-existence. *2018 25th International Conference on Telecommunications (ICT)*, June 2018, pp. 186–190. doi: https://doi.org/10.1109/ICT.2018.8464938.

16 Gonzalez, A., Kuehlmorgen, S., Festag, A., and Fettweis, G. (2017). Resource allocation for block-based multi-carrier systems considering QoS requirements. *GLOBECOM 2017 - 2017 IEEE Global Communications Conference*, December 2017, pp. 1–7. doi: https://doi.org/10.1109/GLOCOM.2017.8254649.

17 Lagen, S., Bojovic, B., Goyal, S. et al. (2018). Subband configuration optimization for multiplexing of numerologies in 5G TDD new radio. *2018 IEEE 29th Annual International Symposium on Personal, Indoor and Mobile Radio Communications (PIMRC)*, September 2018, pp. 1–7. doi: https://doi.org/10.1109/PIMRC.2018.8580813.

18 Marijanovic, L., Schwarz, S., and Rupp, M. (2018). Optimal numerology in OFDM systems based on imperfect channel knowledge. *2018 IEEE 87th Vehicular Technology Conference (VTC Spring)*, June 2018, pp. 1–5. doi: https://doi.org/10.1109/VTCSpring.2018.8417548.

19 Marijanovic, L., Schwarz, S., and Rupp, M. (2018). Optimal resource allocation with flexible numerology. *2018 IEEE International Conference on Communication Systems (ICCS)*, December 2018, pp. 136–141. doi: https://doi.org/10.1109/ICCS.2018.8689253.

20 Marijanovic, L., Schwarz, S., and Rupp, M. (2019). A novel optimization method for resource allocation based on mixed numerology. *ICC 2019 - 2019 IEEE International Conference on Communications (ICC)*, May 2019, pp. 1–6. doi: https://doi.org/10.1109/ICC.2019.8761921.

21 Marijanovic, L., Schwarz, S., and Rupp, M. (2019). Multi-user resource allocation for low latency communications based on mixed numerology. *2019 IEEE 90th Vehicular Technology Conference (VTC2019-Fall)*, September 2019, pp. 1–7. https://doi.org/10.1109/VTCFall.2019.8891409.

22 Soni, T., Ali, A.R., Ganesan, K., and Schellmann, M. (2018). Adaptive numerology - a solution to address the demanding QoS in 5G-V2X. *2018 IEEE Wireless Communications and Networking Conference (WCNC)*, April 2018, pp. 1–6. doi: https://doi.org/10.1109/WCNC.2018.8377205.

23 Sui, W., Chen, X., Zhang, S. et al. (2019). Energy-efficient resource allocation with flexible frame structure for heterogeneous services. *2019 International Conference on Internet of Things (iThings) and IEEE Green Computing and Communications (GreenCom) and IEEE Cyber, Physical and Social Computing (CPSCom) and IEEE Smart Data (SmartData)*, July 2019, pp. 749–755. doi: https://doi.org/10.1109/iThings/GreenCom/CPSCom/SmartData.2019.00139.

24 Yazar, A. and Arslan, H. (2019). Reliability enhancement in multi-numerology-based 5G new radio using INI-aware scheduling. *EURASIP Journal on Wireless Communications and Networking* 110 (2019): 1–14. doi: https://doi.org/10.1186/s13638-019-1435-z.

25 Yazar, A. and Arslan, H. (2019). Selection of waveform parameters using machine learning for 5G and beyond. *2019 IEEE 30th Annual International Symposium on Personal, Indoor and Mobile Radio Communications (PIMRC)*, September 2019, pp. 1–6. doi: https://doi.org/10.1109/PIMRC.2019.8904153.

26 You, L., Liao, Q., Pappas, N., and Yuan, D. (2018). Resource optimization with flexible numerology and frame structure for heterogeneous services. *IEEE Communications Letters* 22 (12): 2579–2582. doi: https://doi.org/10.1109/LCOMM.2018.2865314.

27 Zhang, J., Xu, X., Zhang, K. et al. (2019). Machine learning based flexible transmission time interval scheduling for eMBB and uRLLC coexistence scenario. *IEEE Access* 7: 65811–65820. doi: https://doi.org/10.1109/ACCESS.2019.2917751.

28 Zaidi, A.A., Baldemair, R., Tullberg, H. et al. (2016). Waveform and numerology to support 5G services and requirements. *IEEE Communications Magazine* 54 (11): 90–98. doi: https://doi.org/10.1109/MCOM.2016.1600336CM.

29 Guan, P., Wu, D., Tian, T. et al. (2017). 5G field trials: OFDM-based waveforms and mixed numerologies. *IEEE Journal on Selected Areas in Communications* 35 (6): 1234–1243. doi: https://doi.org/10.1109/JSAC.2017.2687718.

30 Weitkemper, P., Bazzi, J., Kusume, K. et al. (2016). On regular resource grid for filtered OFDM. *IEEE Communications Letters* 20 (12): 2486–2489. doi: https://doi.org/10.1109/LCOMM.2016.2572183.

31 Ijaz, A., Zhang, L., Grau, M. et al. (2016). Enabling massive IoT in 5G and beyond systems: PHY radio frame design considerations. *IEEE Access* 4: 3322–3339. doi: https://doi.org/10.1109/ACCESS.2016.2584178.

32 Zhang, L., Ijaz, A., Xiao, P. et al. (2017). Subband filtered multi-carrier systems for multi-service wireless communications. *IEEE Transactions on Wireless Communications* 16 (3): 1893–1907. doi: https://doi.org/10.1109/TWC.2017.2656904.

33 Choi, J., Kim, B., Lee, K., and Hong, D. (2019). A transceiver design for spectrum sharing in mixed numerology environments. *IEEE Transactions on Wireless Communications* 18 (5): 2707–2721. doi: https://doi.org/10.1109/TWC.2019.2907239.

34 Chen, H., Hua, J., Li, F. et al. (2019). Interference analysis in the asynchronous F-OFDM systems. *IEEE Transactions on Communications* 67 (5): 3580–3596. doi: https://doi.org/10.1109/TCOMM.2019.2898867.

35 Cheng, X., Zayani, R., Shaiek, H., and Roviras, D. (2019). Inter-numerology interference analysis and cancellation for massive MIMO-OFDM downlink systems. *IEEE Access* 7: 177164–177176. doi: https://doi.org/10.1109/ACCESS.2019.2957194.

36 Cheng, X., Zayani, R., Shaiek, H., and Roviras, D. (2019). Analysis and cancellation of mixed-numerologies interference for massive MIMO-OFDM UL. *IEEE Wireless Communications Letters* 1–1. doi: https://doi.org/10.1109/LWC.2019.2959526.

37 Gokceli, S., Levanen, T., Yli-Kaakinen, J. et al. (2019). PAPR reduction with mixed-numerology OFDM. *IEEE Wireless Communications Letters* 1–1. doi: https://doi.org/10.1109/LWC.2019.2939521.

38 Liu, X., Zhang, X., Zhang, L. et al. (2019). PAPR reduction using iterative clipping/filtering and ADMM approaches for OFDM-based mixed-numerology systems. *IEEE Transactions on Wireless Communications.* 19 (4): 2586–2600. doi: https://doi.org/10.1109/TWC.2020.2966600

39 Liu, X., Zhang, L., Xiong, J. et al. (2019). Peak-to-average power ratio analysis for OFDM-based mixed-numerology transmissions. *IEEE Transactions on Vehicular Technology* 1–1. doi: https://doi.org/10.1109/TVT.2019.2960801.

40 3GPP (2018). NR; Physical Layer; General Description. *Technical Specification 38.201*, ver. 15.0.0.

41 Levanen, T., Tervo, O., Pajukoski, K. et al. (2019). Mobile communications beyond 52.6 GHz: waveforms, numerology, and phase noise challenge, https://arxiv.org/abs/1912.09072.

42 Memisoglu, E., Kihero, A.B., and Arslan, H. (2020). Guard band reduction for 5G and beyond multiple numerologies. *IEEE Communications Letters.* 24 (3): 646–647. doi: https://doi.org/10.1109/LCOMM.2019.2963311.

43 Levanen, T., Pirskanen, J., Pajukoski, K. et al. (2019). Transparent Tx and Rx waveform processing for 5G new radio mobile communications. *IEEE Wireless Communications* 26 (1): 128–136. doi: https://doi.org/10.1109/MWC.2018.1800015.

8

Network Slicing with Spectrum Sharing

Yue Liu, Xu Yang and Laurie Cuthbert

School of Applied Sciences, Macao Polytechnic Institute, Macao SAR, China

8.1 The Need for Spectrum Sharing

Unlike fixed networks where resources can be easily expanded by installing more fibers, wireless networks are limited to a fixed amount of spectrum that is allocated to them. Blocks of spectrum are allocated to types of service by the International Telecommunication Union Radio Communication Sector (ITU-R) and those blocks are allocated, or subdivided and allocated, by national regulators. An example of the allocation (for the UK) is given in Ofcom [1]. Ofcom (Office of Communications) is the UK communications regulator and (under the Wireless Telegraphy Act 2006 and the Communications Act 2003) has the statutory responsibility for allocating spectrum in conformity with the internationally agreed spectrum allocations of the International Telecommunication Union (ITU).

With the increasing use of mobile devices there is greater demand on radio resources, and spectrum sharing is seen as an effective way to utilize the limited frequency resources available. This is different from the traditional allocation of frequencies where each operator can occupy exclusively the allocated block of frequency bands. This sole ownership guarantees simple intra-operator spectrum management and coordination, but the lack of flexibility causes a waste of spectrum. This is illustrated in Figure 8.1. In this scenario there are two network operators (NO_A and NO_B) that are allocated a certain amount of spectrum: in Figure 8.1a both network operators have sufficient resources to serve all their users, but in (b) NO_B has used all its allocated resources and so cannot add any new users, while NO_A still has unused spectrum. Spectrum sharing provides a possible solution to efficiently utilize the whole spectrum. Blocks of spectrum are no longer exclusive to one operator only but are made available to all operators as illustrated in Figure 8.1c where it can be seen, in this simple example, that there would only be unserved users if the total demand exceeds the total allocation.

This example implicitly assumes that there is a central allocation and transmission mechanism to avoid co-channel interference and the easiest way to achieve that is to have one radio access network (RAN) that all operators share. However, the situation and choices

Radio Access Network Slicing and Virtualization for 5G Vertical Industries, First Edition.
Edited by Lei Zhang, Arman Farhang, Gang Feng, and Oluwakayode Onireti.

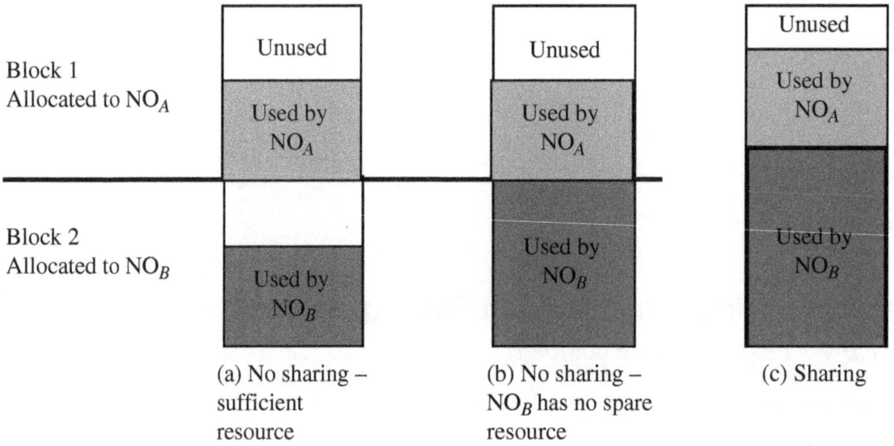

Figure 8.1 Advantages of spectrum sharing.

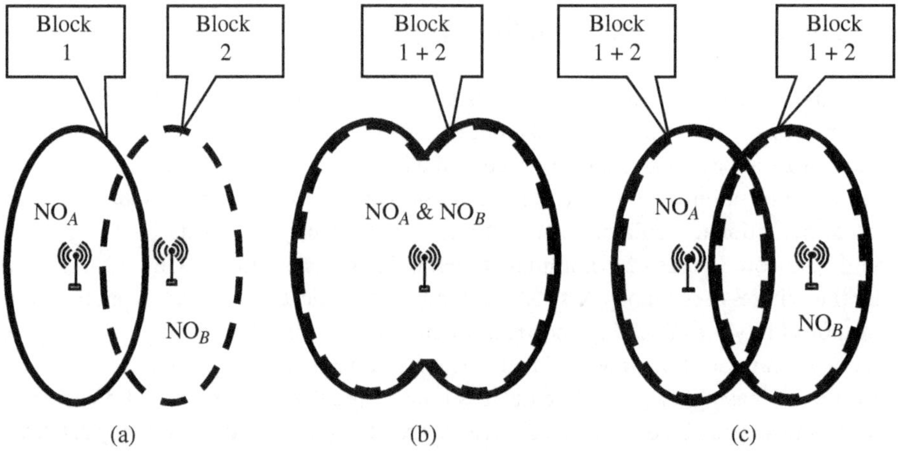

Figure 8.2 Different approaches to sharing. (a) No sharing. (b) Common pool of spectrum and transmission. (c) Common pool of spectrum but separate transmission.

are more complicated than that, as shown simply in Figure 8.2. Figure 8.2a shows the non-shared approach and (b) and (c) show two different approaches to sharing a common pool of spectrum: (b) has one pool of spectrum and both operators share the same transmission equipment so the same spectrum will only be used by a single transmission whereas in (c) the same pool of spectrum is made available to the base stations of both operators but the transmission equipment is not shared so spectrum could be used by several transmissions. Logically these are the same, but a more complex control mechanism is required for (c) as the co-channel interference map is effectively that of the two operators overlaid.

It is clear from Figure 8.2c that operators being allocated the same block of spectrum in the same areas will unavoidably suffer co-channel interference without a high level of cooperation between them – the problem about such a high level of cooperation is that it is likely to lead to a high signaling load. However, there is a trend for infrastructure sharing and this cooperation and deeper integration will have a lower signaling cost, thus making spectrum sharing more practical. For example, the China Tower Company took over the tower assets of the three major network operators in China to reduce redundant construction and the towers will be shared by the three operators [2]. Sharing is now being encouraged by trade bodies and regulators, for example [3, 4].

8.2 Historical Approaches to Spectrum Sharing

Once spectrum has been allocated there may be a need to share that in different ways. For instance, the well-known cellular concept [5, 6] shares spectrum geographically. However, this just allows interference between cells to be mitigated, all the users belonging to the same operator and receiving the same type of service.

A more flexible approach was identified in Mitola and Maguire [7]: cognitive radio (CR). The concept of CR as it has developed is that it allows secondary users (SUs) to sense the "spectrum holes" in the time, space, and frequency domains in an autonomous manner and then make rational use of them in an opportunistic way on the premise of causing limited and tolerable interference to primary users (PUs). The spectrum hole is within the band of frequencies assigned to PUs, but at a particular time and specific geographic location those frequencies are not being utilized by a PU [8]. The concept of CR is attractive because it avoids wasting spectrum that is allocated but unused, at any particular time.

The detection of the holes requires spectrum sensing and although a lot of research has been done on CR, providing rapid and reliable sensing results in all circumstances is difficult and sensing is one of the key factors hindering the development of CR. The authors in [9] noted that problems needing to be addressed include low SNR sensing, the hidden node problem, a quality of service (QoS) guarantee, passive device detection, and challenges in wideband spectrum sensing. This concept of allowing SUs to transmit in the holes is termed *overlay spectrum sharing* [9] (or opportunistic spectrum access [OSA] or dynamic spectrum access [DSA]) and is illustrated in Figure 8.3. The overlay spectrum sharing gives PUs the highest priority on occupying the spectrum and SUs are only allowed to use unused spectrum where there is an absence of PUs so SUs must employ spectrum sensing and keep monitoring as the occupied spectrum must be released by the SU if a PU starts using it. In the IEEE Standard 802.22 this is provided by either a geo-location database or by spectrum sensing [10].

However, as also shown in Figure 8.3, the situation is much more complex when more than one cell is considered because an SU can easily generate inter-cell interference (ICI) to a PU in an adjacent cell, especially when the cell radius is small. Although a lot of work on CR assumes a single cell model (e.g. [11, 12]) the demand by users for more and more bandwidth means that a multi-cell structure is better for providing the required capacity in a denser population area.

Underlay spectrum sharing applies spectrum spreading to the SU's signal in order to spread the transmit power over an ultrawide spectrum so that the transmit power is very low and the interference from the SUs to the PUs does not exceed a certain threshold [13]. This means that the SUs do not have to find specific holes.

Another approach that leads onto spectrum sharing was proposed in Liu et al. [14] that managed spectrum resources to always satisfy PUs, without SUs having to pay attention to the PUs; this was done at the expense of sacrificing the transparency of SUs to the PUs. Such

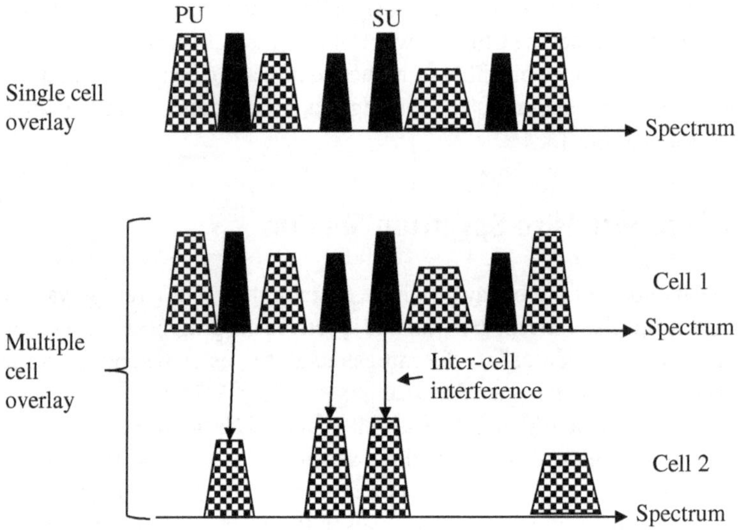

Figure 8.3 Overlay spectrum sharing.

a network requires no sensing hardware as resource allocation is managed across all users and the division between PUs and SUs can be based on the service provider to which they belong, or to different priorities between users for the same service provider. Effectively, the spectrum is shared between two types of users, in this case with absolute priority being given to one type (the PUs). Such sharing improves the efficiency of the whole network.

8.2.1 Classifications of Spectrum Sharing

8.2.1.1 Orthogonality

The different types of user (PU and SU) from CR can be generalized into different network operators and Anchora et al. [15] classified spectrum sharing into *orthogonal spectrum sharing* and *non-orthogonal spectrum sharing*. If the frequency band is allocated to more than one operator simultaneously, it is classified as non-orthogonal sharing and will introduce additional interference into the system that must be mitigated by suitable control. Orthogonal spectrum sharing ensures that, in any given area, no operator can be allocated a frequency that is already in use by another.

In a multi-cell environment, orthogonality, neither the no sharing case (Figure 8.4a) nor orthogonality (Figure 8.4b) will cause interference, but non-orthogonality allows a more dynamic allocation that is different between cells (so better for an uneven load distribution) but could cause interference between network operators (Figure 8.4c).

It is worth noting that in any multi-cell configuration, whether it is orthogonal or non-orthogonal, there exists the possibility of ICI and it is a function of the radio resource management to actually allocate resource blocks (RBs) in such a way as to mitigate the effects of this interference.

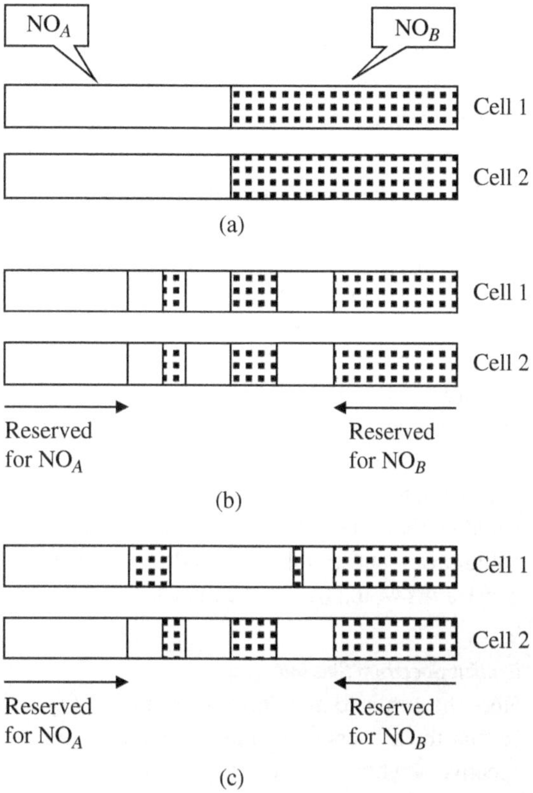

Figure 8.4 Orthogonal and non-orthogonal sharing. (a) No spectrum sharing. (b) Inter-cell orthogonal spectrum sharing. (c) Inter-cell non-orthogonal spectrum sharing.

8.2.1.2 Sharing Rights

In the literature, the type of spectrum sharing between two operators is generally "equal-rights sharing," although CR is a good example where the users of the spectrum do not have equal rights.

Equal rights sharing is where each operator has the same right to frequencies on the shared band. In Jorswieck et al. [16] the full cooperation case is considered with the aim being to maximize the sum of the throughput of two operators. In Singh et al. [17] coordination protocols are proposed to allow negotiations between operators. Equal-rights sharing has been the center of consideration in the literature as this is the simple case and it avoids any priority mechanism between operators. However, such a constraint restricts generality and it is necessary to consider the case where operators could have access to different proportions of the shared spectrum: equal-rights sharing is just a special case of that.

As an example of how this may be applied consider two operators – the method is applicable to any number of operators but using only two (i) is common in the literature and (ii) illustrates the main features.

A simple way of sharing is to pre-allocate *spectrum* (but not specific RBs) between operators in such a way that there is no overlap. A *spectrum sharing ratio (SR)* is then defined as the ratio between the amount of spectrum that each operator can access, However, a better solution is to allow operators to select RBs from anywhere in the whole spectrum and use the SR as a limit on the amount each may receive; this is considered in the following section.

It is important to note in this discussion that spectrum sharing emphasizes the need to allocate sufficient RBs (i.e. sufficient spectrum) to each user so that their connection has a large enough capacity (in the Shannon sense) for them to meet their required QoS: users achieving that QoS are termed *qualified users* (QUs).

8.2.1.3 Allocation of Resources

RBs can be allocated to users belonging to each operator in accordance with these predefined rules (termed *explicit spectrum sharing*) or in such a way that approximates to certain SR: *implicit spectrum sharing*.

The problem with explicit sharing is that it takes no account of the channel conditions and an uneven distribution of users within a cell may mean that those belonging to one network operator may require spectrum to achieve their required QoS than those belonging to the other. Implicit sharing [18] does not allocate RBs according to some SR but chooses how to allocate RBs to users in such a way that it maximizes the number of qualified users, but approximates to a required sharing ratio of the *users* between the operators. An example could be that, in every cell, the users are served in order of their channel conditions. This could mean, in extremis, that all the capacity would be allocated to one operator so suitable constraints would need to be imposed.

Explicit Spectrum Sharing

Here the aim is to maximize the number of qualified users; the RBs are allocated to users so that those users requiring the smallest number of RBs receive the allocation first, irrespective of which operator they belong to.

Users are ranked in ascending order of distance from the BS, distance being a first approximation to the number of RBs required. As each user is considered, the RBs are ordered in terms of the channel conditions, with those providing the best capacity (taking into account the co-channel interference from neighboring cells) being ranked first. Going down the ordered list of users, enough RBs are allocated from the top of the list of RBs to provide the necessary capacity.

However, allocating RBs in particular sectors will affect the co-channel interference, which is why the allocation has to be implemented iteratively taking into account the interference map generated by the allocation.

A running total of the numbers of RBs allocated to users of each operator is required so that once the total that an operator may use (from the sharing ratio) is reached no further users from that operator are served.

Implicit Sharing Mechanism

The overall serving order is adjusted to take into account the share of the spectrum to be taken by each operator. A simple illustration of this principle is given in Figure 8.5 where NO_A is allocated twice the spectrum of NO_B, so two NO_A users are served for every user from NO_B. Within the list for each operator, the serving order is still in ascending order of distance from the BS as illustrated in Figure 8.5.

It is important to note that, with the example of Figure 8.5, this reordering means that the channel conditions may be worse in some of the NO_A users that are served before NO_B

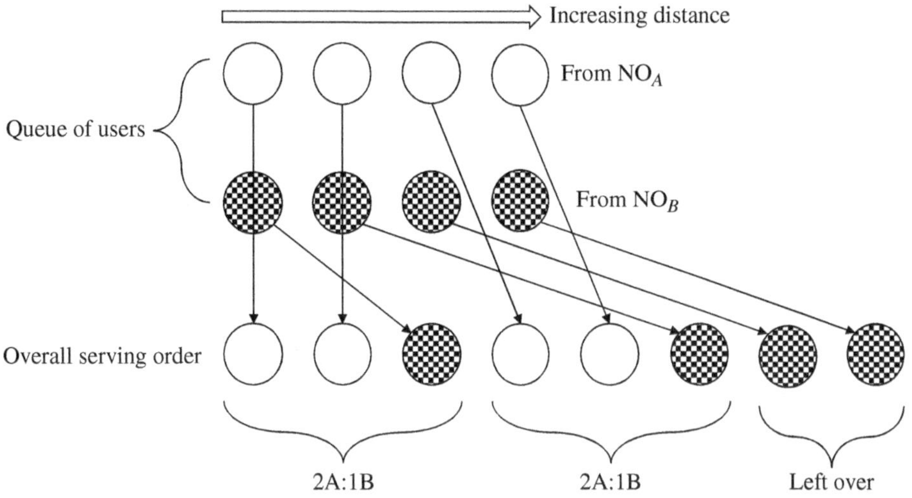

Figure 8.5 Creating non-equal allocation order. Source: Reprinted from Liu et al. 2016 [18] under the terms of the ACM author copyright licence.

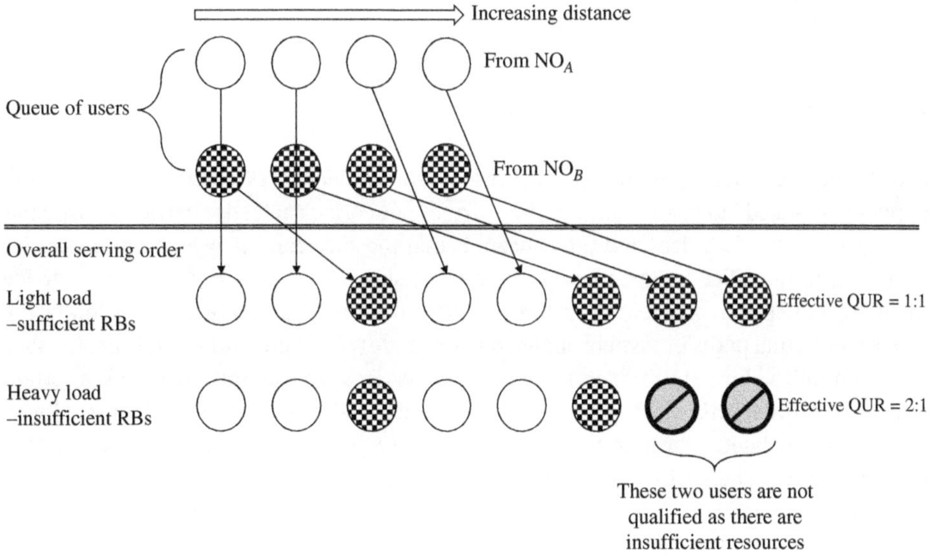

Figure 8.6 Illustration of variability in implicit sharing. Source: Reprinted from Liu et al. 2016 [18] under the terms of the ACM author copyright licence.

users, leading to the case where a user of NO_A cannot be served, but a following user of NO_B has better channel conditions and can be served.

However, sharing spectrum in this way is not a precise share as the ratio of the numbers of users served between the operators will vary depending on the user distribution and channel conditions, as illustrated in the very simple example of Figure 8.6.

In this example, with such a small number of users in the diagram, the variability is exaggerated, but it does illustrate the reasons behind that variability. In the example there is an equal number of users from both operators, but the intention is to allocate twice as much spectrum to NO_A as NO_B, so the serving order is $user_A1$, $user_A2$, $user_B1$, $user_A3$, $user_A4$, $user_B2$, ... The effective qualified user ratio (QUR) in this case gives the ratio of users achieving the required QoS between the two operators.

More Complex Forms of Sharing

While explicit spectrum sharing allocates spectrum between operators, it takes no account of the impact of that sharing on the number of qualified users, and with uneven distributions this may lead to very unequal QURs. On the other hand, implicit spectrum sharing, while concentrating on user priorities, does not consider the actual *spectrum* sharing at all. To balance fairness between spectrum allocation and users served, more complex forms of sharing are required. An example of this was given in Liu et al. [18] where an explicit sharing ratio was used, but the allocation order was adjusted to give users with worse channel conditions a higher priority. This did achieve a better balance, but at the expense of overall throughput because users requiring more RBs were being served before others requiring less. While giving priority to one particular operator would involve determining whether its users are predominantly edge or not, a simple serving order that just alternated between operators gave almost as much mitigation and did not need the user distribution to be determined.

8.3 Network Slicing in the RAN

A technology of interest for future networks is *Network Slicing* [19]. With this concept the network is "sliced" logically into multiple virtual networks; each slice carries a particular service or group of services and the concept is that the slice can manage the resources for users within that slice independently of the other slices. The slices can be used to divide the network between different operators or different services and in many ways follow the concept of virtual paths in asynchronous transfer mode (ATM) networks [20, 21]. Each slice can be optimized for a specific service or group of services, for example, to provide a certain minimum bitrate. This means that each slice has a certain set of attributes (for example, end to end low latency or high bandwidth) that can make it more appropriate for certain type of services. The slices can exist end to end in the network or locally between points within the network.

By splitting the allocated spectrum into these slices, the network elements and functions can be easily configured and reused in each network slice to meet a specific requirement. The implementation of network slicing is conceived to be an end-to-end feature that includes the core network and the RAN [22]. The logical separation provided by network slicing allows networks to be logically separated, with each slice providing customized connectivity, so providing high degree of flexibility in supporting different services and user requirements.

In [23] the authors explain that all slices can run on the same shared infrastructure, or on separate infrastructures as the operator requires, and this is a much more flexible solution than a single physical network.

Points to note about network slicing:

- A network slice provides a set of network capabilities and performance levels that may be suitable for a set of service types.
- A service must be mapped to at least one slice, but it may also be mapped to more than one slice.
- A network slice may be used by only one particular service type, or it may handle several types.
- The service type may represent more than just the normal concept of a service: for example, a mobile virtual network operator (MVNO) may be restricted to one or a limited set of slices by the physical network operator.

In fiber networks the problem of having independent "logical pipes" with no interference between them is fairly easy to solve, but slicing in the RAN [24–27] is much more challenging as (i) that usually provides the constraint on resource availability and (ii) it is more difficult to prevent interference. The RAN can be logically divided to provide flexibility for different services and user requirements, potentially leading to significant cost savings [28, 29].

Figure 8.7 shows the concept of network slicing. Here, for simplicity, the spectrum allocated to an operator is shown as a contiguous block that is divided into slices with no gaps, but realistically a slice could comprise spectrum with gaps and split across different frequency bands.

Notice that the network slicing concept of Figure 8.7 is very similar to spectrum sharing (e.g. Figure 8.4), but the entities sharing are not necessarily different operators: they could be users with different QoS requirements.

The simplest form of slicing is to have separate allocations of spectrum to each slice; if this allocation is fixed, this guarantees resource allocation to the slices but does not allow for dynamic changing demand or for unequal load distributions. Moreover, with multiple cells there will be intra-slice interference if the same spectrum is allocated to the slice in each cell.

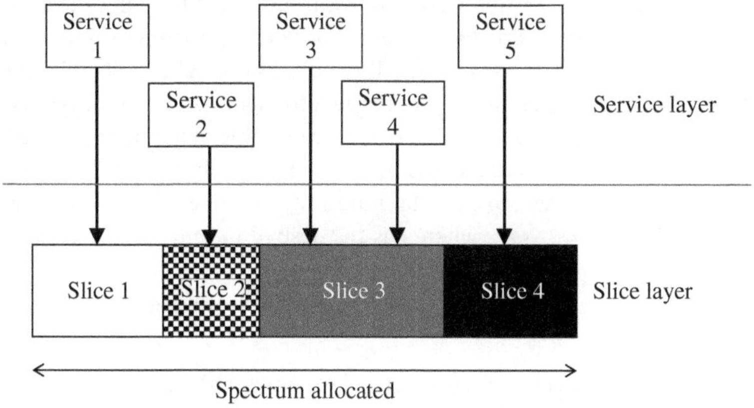

Figure 8.7 Concept of network slicing.

In [25] the authors state that RAN sharing will become more common as studies have shown [30] that resources belonging to an operator are often underutilized. They define a slice in Kokku et al. [25] as "*a group of user flows that belongs to a single entity … and requires a chosen fraction of spectrum resources to be allocated for satisfying the* service level agreements (SLAs)." They do not specify that it is a contiguous block of spectrum in each cell.

In [26], the authors note that it is possible to dynamically allocate the RBs across mutually interfering cells in a multi-cell arrangement, and thus to dynamically assign RBs to slices in a coordinated manner. This is reinforced by the authors in [31] who agree that "*a better method is to allow all operators access to the whole bandwidth but apply a mechanism to manage the fraction of resources each operator consumes on average.*" Allowing such fully dynamic access to all of the spectrum was shown in Yang et al. [32] to not only give higher efficiency but also better fairness between slices.

While this fully dynamic sharing seems at first sight to violate the principles of slicing, logically it provides exactly the same independence for each slice: instead of the slice having a fixed resource allocation, the slice specifies the requirements under an SLA and the RAN is managed to deliver those SLAs. Moreover, fixed slicing is not really sharing of spectrum; it is just allocating different blocks of spectrum to different slices. This arrangement does not allow for changing demand, and ICI means that there can be intra-slice interference.

However, another aspect that needs considering with dynamic RAN slicing is the need to provide *isolation* between slices [23, 33]: what happens in one slice should not affect users in other slices. In the multi-cell RAN scenario, interference reduces isolation because a new user in an adjacent cell may generate interference that will reduce the signal-to-noise-plus-interference ratios (SINR) of an existing user so much that it no longer meets its QoS requirements.

8.4 Radio Resource Allocation that Considers Spectrum Sharing

The definition of QoS can be quite complex considering factors such as bandwidth, throughput, latency, and loss. Cisco has published [34] a detailed assessment of QoS requirements for different types of service. However, this section is about *radio resource* allocation and *spectrum sharing*; such resource allocation can either be blocks of spectrum or time slots. Here, the same approach as found in much of the literature [35] is taken, whereby services are differentiated by the required bitrate and the requirements on different types of QoS requirement must be mapped into allocation of spectrum. This mapping is generally a well-known problem. For example, a simple loss-sensitive sliding-window ARQ protocol needs to have extra bandwidth allocated to allow for retransmission of errored packets; the allocated bitrate is given by $r_{\text{alloc}} = \frac{r}{1-Wp_{we}}$ where r is the desired bit rate, W the window size, and p_{we} the probability of a packet being in error.

Figure 8.8 summarizes the different methods of sharing the spectrum between slices for three slices.

- *Static resource assignment*: A slice operates with a fixed resource. This provides a guaranteed resource allocation to the slices but does not allow for changing demand. ICI means that there can be intra-slice interference.

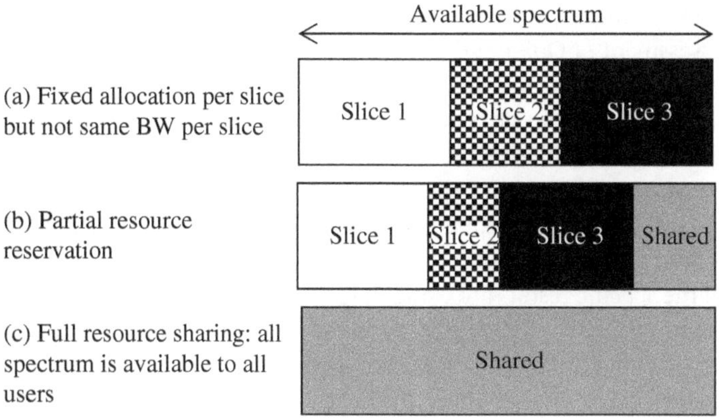

Figure 8.8 Types of network slicing.

- *Partial resource reservation*: This is introduced in Guo and Arnott [31] and combines a fixed allocation with a sharing region (Figure 8.8b). In [36], the authors propose allocating a certain proportion of RBs but in both these papers only a single cell is considered. Zhao et al. [37] discusses a multi-cell problem but models a uniform arrangement so that all cells have the same sharing arrangement. The partial resource reservation is an "in-between solution" and does require complex management. However, it can also be seen as a "mixed solution" where part of the available spectrum is guaranteed to each slice with the remainder being shared.
- *Full resource sharing*: In this model [31], all of the spectrum is available to users from any slice (Figure 8.8c), but management of the sharing will allow users "belonging" to any slice to get the QoS they require. Previous work on spectrum sharing (for instance, [18]) can be extended to network slicing, not only considering a multi-cell environment but also removing any restriction that slicing in each cell would be the same. Limits can be placed on the total amount of bandwidth available to each slice if needed.

Having determined a structure for slices that may, or may not, include sharing, it is necessary to provision resources on a per-slice basis [33].

8.4.1 Example Radio Resource Allocation for Sharing Through Network Slicing

There is a very large amount of literature on spectrum sharing, much of the earlier material considering CR networks but latterly including network slicing in the RAN. A good survey paper is [38] with Fig. 3 in that paper illustrating a framework for spectrum sharing. The authors note that most of the spectrum sharing mechanisms are at layer 2, although a few are at layer 1. Constraints are identified as (i) the requirements of applications (e.g. the type of QoS needed), (ii) business models and cooperation agreements, and (iii) the regulatory framework. Although that paper does not consider network slicing specifically, a lot of the material is relevant.

In Voicu et al. [38] it is also noted that the spectrum sharing in layer 2 can be done in (i) time, (ii) frequency, (iii) code, or (iv) space. From a conceptual point of view, allocating bandwidth in terms of the number of RBs fits the sharing model well. In Szigeti et al. [34] there is a detailed assessment of QoS requirements for different types of service and these can be mapped onto *capacity* (in the Shannon sense) as explained at the start of Section 8.4. Differentiating services by bitrate is quite common (e.g. [35]) and hence a common approach is sharing through considering the frequency domain.

This section is concerned with illustrating the effects that would be expected with dynamic network slicing, optimizing the frequency allocation to maximize the number of users meeting their QoS requirements. Any form of optimization that takes into account the ICI could be used, but the specific results are using the game-theoretic approach in [39] and extended in [18] to apply to network slicing. A different optimization, but still using game theory, is found in [40]. Other authors use different methods, such as auctions [41].

Each run of the game takes place until equilibrium is reached, as explained later, and while this does not guarantee that a global optimization has been reached, the approach stands up well when compared with using a Genetic Algorithm for the optimization [42].

In Liu et al. [39], OFDMA was used with each RB having a bandwidth B_0 depending on the number of subcarriers per RB and bandwidth of each subcarrier. From the well-known Shannon theorem, the capacity C achieved by a user is $B_0 \log_2(1 + {}^{S}/_{N})$ where S is the signal power and N the total noise power including interference. For SINR of more than a few dB this can be approximated by the well-known $C = B_0 \times \text{SINR}/3.01$.

Hence, the total capacity of a user with L RBs is $C = \frac{B_0}{3.01} \sum_1^L \text{SINR}_l$ and $\sum_1^L \text{SINR}_l$ can be denoted as the "sum of signal-to-noise-plus-interference ratios" (*sSINR*).

This approximation to Shannon's theorem is reasonably valid for an SINR of >5 dB (Figure 8.9) but it is a *pessimistic* approximation as it gives a lower capacity than the actual value; *allocating* resource based on the approximation (easier and faster to implement) will give more resources than the minimum required, but *the achieved capacity* is always taken from the full equation.

The problem then is to determine how best to allocate the minimum number of RBs to each user so they achieve the necessary sSINR to meet their QoS requirements, subject to

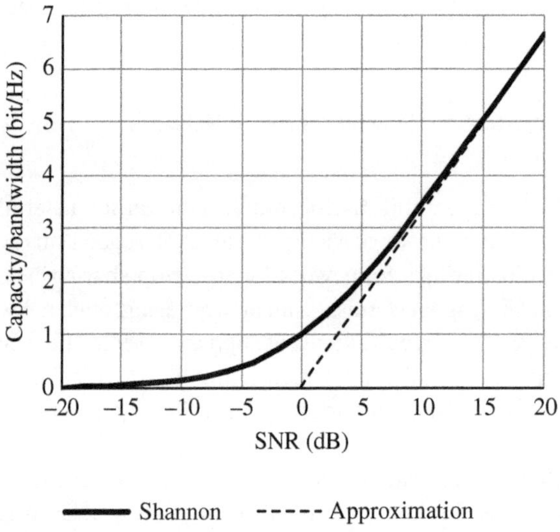

Figure 8.9 Approximation to Shannon.

(i) the interference seen by that user, (ii) the channel conditions to that user, and (iii) users in other cells being allocated resources at the same time.

In Liu et al. [39] the entity allocating the resource was the cell but here the situation is more complex, depending on which model of sharing is used: it may be the slice rather than the cell as explained later. The term *resource allocation entity* (RAE) is used as a more general description that may refer to the cell or the slice.

The game allows each RAE to maximize its number of qualified users by choosing the best allocation of RBs, while competing against the allocation schemes of other RAE. The maximum is reached when all RAEs maintain their current positions. Non-cooperative game theory is suitable for modeling this type of multi-person problem characterized by strategic interdependency [43] although the solution may be suboptimal [44]. This allows an RAE to be fully in charge of resource allocation independently of other RAEs (although dependent on the interference generated by the other RAEs) and so avoids the overhead of message exchange between RAEs. Compared with other optimization techniques, it can achieve much faster convergence and can be easily applied in practice although it might cause suboptimal solutions. The *Nash Equilibrium* [44] is regarded as the solution to a non-cooperative game; it consists of every player's *best response* against all others' strategies. In other words, it is a steady-state point where none of the players has an incentive to change their strategy since none of them can unilaterally increase their utility function given that the other players stick to their current strategies. In the example here, the game has converged to reach the equilibrium.

A static scenario is considered: multiple users request bandwidth resource from RAEs in a multi-cell OFDMA network with wrap-around [45]. The *players* are the N RAEs and the *action space* is the combination of all possible resource allocation schemes of all players.

To implement the game for the situation being considered, the utility function needs to be mapped to that situation. In general, the utility function is as follows:

Maximize the number of users who achieve a capacity $C = B_0 \times \text{sSINR}/3.01 \geq C_{\text{min}}$ where C_{min} is the minimum capacity needed for that user to meet its QoS requirements.

Subject to (i) the total transmission power of a single RB not exceeding P_{max}, (ii) in any cell a particular RB can only be allocated to one user, and (iii) the maximum number of RBs allocated to a particular slice may be limited by a factor such as a predefined sharing ratio.

This is the general approach, but the application to the static and fully dynamic network slicing types is different:

Fixed slices: A portion of RBs is reserved for each slice according to some predetermined ratio. Each slice is only allowed to use its reserved RBs. In this case, it is straightforward to apply the game theory as each slice is an RAE and independent games can be implemented, one for each slice. The players are the instances in each cell of a particular slice, so in the multi-cell environment the players compete to maximize the number of qualified users.

Fully dynamic slicing: In this case the RAE is a cell. All the RBs are available to all the users and the allocation is done in such a way that the overall sharing between slices is as specified in the overall design. Sharing is not orthogonal since each slice may have users in different parts of the spectrum in neighboring cells and this leads to inter-slice interference, unless other actions are taken.

A consideration is how to specify some limits to the sharing between slices in this case for the admission control to use. The following approaches are feasible ([18] and Section 1.2.3):

- *A defined spectrum sharing ratio (SR)*: this is more usefully specified as a minimum share for each slice.
- *Number of qualified users*: sharing *spectrum* does not guarantee that any required number of qualified users is achieved as that depends on channel conditions and user distribution. An alternative is to consider the number of qualified users per slice, or at least the ratio of qualified users between slices. To do this, the *implicit spectrum sharing* concept of Liu et al. [18] can organize the allocation to users to achieve the required ratio between qualified users in the slices.

The steps in the algorithm are as follows:

(i) *Initial prioritized RB allocation*: In every cell, the sectors allocate RBs to their users in the appropriate serving order [14]. As no ICI can be obtained beforehand, the number of RBs required by each user is estimated by taking into account noise, channel conditions, and fading, but assuming there is no ICI. The lack of consideration of interference means that many users become unqualified after the first round, but this can be mitigated by adding extra RBs as a "guard band."

(ii) *RB release*: Releases the RBs of the users who do not reach C_{min} (these are called unqualified users) and puts those released RBs back into the unallocated RB pool.

(iii) *RB reallocation*: Reallocates the unallocated RB pool to unqualified users and keeps a record of the number of which RBs are used.
As RBs are allocated to unqualified users, the interference map will change so that some previously qualified users may become unqualified. In this case, extra RBs must be allocated to those users, or the actual RBs may be changed to those suffering a lower interference.

(iv) *RB release*: As in step (ii)

(v) *Go back to RB reallocation* and continue the process until the RB allocation result converges to the Nash Equilibrium, that is, when the BSs do not change their current allocation strategies.

To illustrate the effectiveness of using a more intelligent form of slicing than simply assigning fixed amounts of bandwidth, the game theory model was implemented for three slices with static reservation in Yang et al. [32], and also for the full resource sharing approach. In both cases a seven-cell cluster with wrap-around was implemented for an OFDMA downlink using a suburban channel model. Three types of service (*S*) were included and following the approach of Jiang et al. [35] these were specified as requiring different bandwidths – here $S1 = 1$ Mbit/s, $S2 = 2$ Mbit/s, and $S3 = 3$ Mbit/s. Each service was allocated to a separate slice, the fixed slices being allocated 30 RBs each (90 RBs overall per cell).

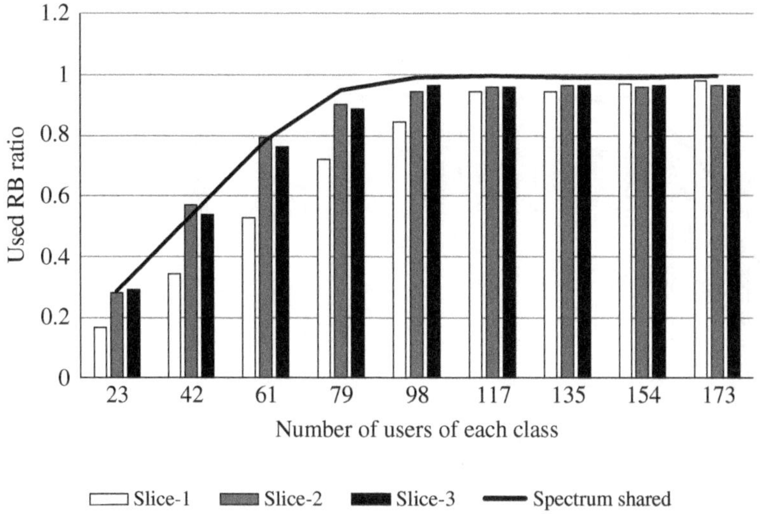

Figure 8.10 Comparison between fixed slices and overall sharing (even load) ©IEEE 2017.

As a benchmark, the users are distributed uniformly and one third of the users are of each service class.[1]

In Figure 8.10 [32] the usage of RBs is compared between the fixed slice model and the spectrum sharing approach as the offered load on the system increases. This is expressed as the ratio of used RBs to total RBs for that RAE – i.e. for the slice in the fixed allocation and for the cell in the fully dynamic case.

It is clear from these results that, as the load on the system starts to become heavy, the fixed model does not use all available resources, especially in Slice 1 where the bandwidth demands are lower. As the load increases, the whole system becomes saturated and there are almost no free RBs.

The same tests, but with different distributions of service class, are shown in Figure 8.11: in case (a) 55%, 30%, 15% for *S*1, *S*2, *S*3 respectively and for case (b) 15%, 30%, 55%.

With fixed slices, Figure 8.11 shows that there are RBs unused even as the network saturates, thus wasting resources. This is most marked in case (b) for the fixed slices where there is a smaller number of lower bitrate users requiring service so that Slice 1 is underused.

The effect of this waste is shown in Figure 8.12 by considering the QUR: there is a significant benefit to the total number of users served by allocating resources from the whole block of spectrum rather than as fixed slices.

Fairness between service classes is shown in Figure 8.13 using the well-known Raj Jain index [46] for the overall QUR: the fairness obtained with sharing the whole spectrum is essentially 1 (i.e. maximum) for both cases, but for the fixed slice approach the fairness becomes quite low.

Clearly these results show the benefit of sharing the spectrum using dynamic network slicing, particularly under uneven load; what the results do not show is whether isolation

1 The results are from [32] and reprinted under the author copyright agreement of the IEEE.

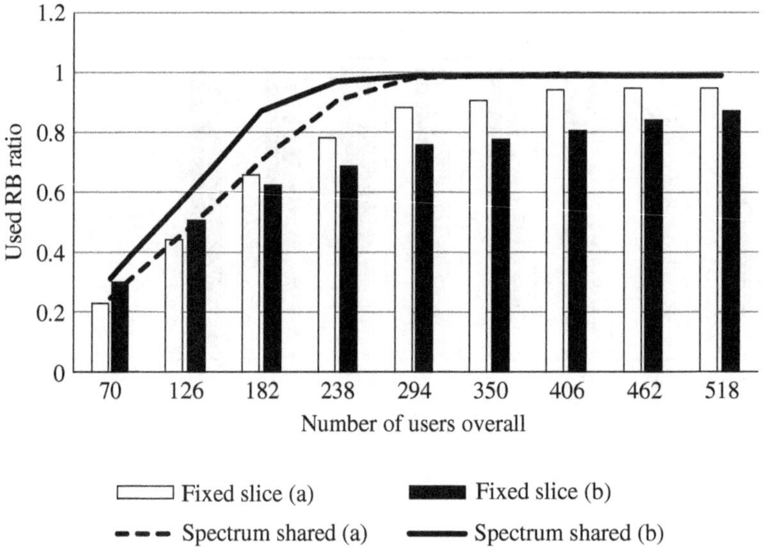

Figure 8.11 Comparison between fixed slices and overall sharing (uneven load) ©IEEE 2017.

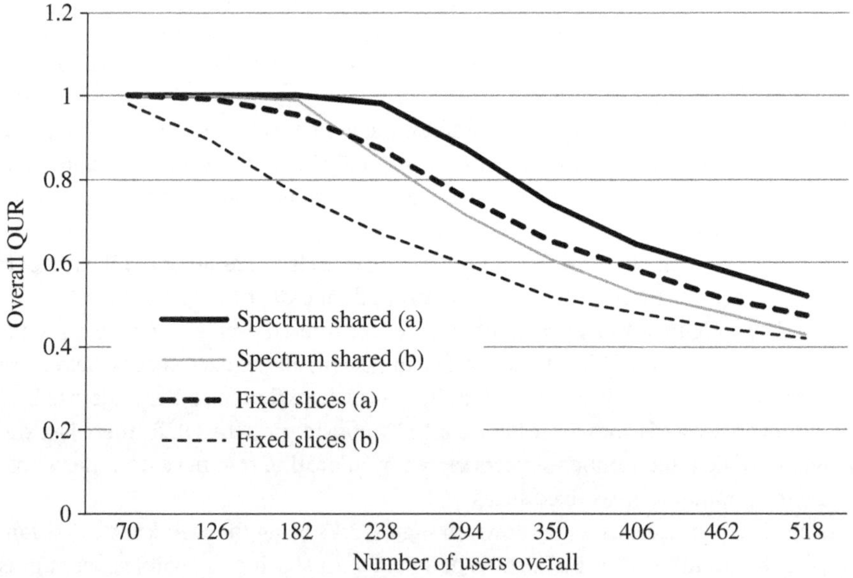

Figure 8.12 Overall results for qualified user ratio ©IEEE 2017.

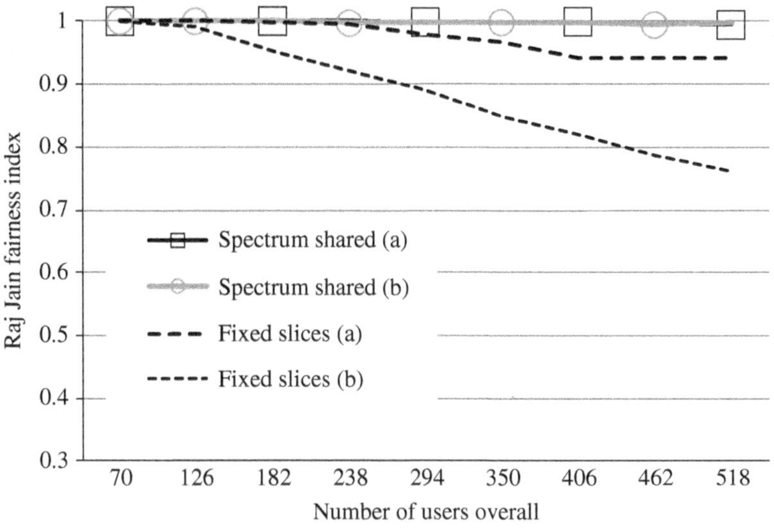

Figure 8.13 Raj Jain fairness index ©IEEE 2017.

is maintained between slices. Since that is a very important aspect of network slicing, it is considered in detail in Section 8.6.

8.4.2 Other Considerations

Since the use of game theory in the example is to implement an optimization process for resource allocation, it is possible to add different constraints to that optimization. As an example, consider adding in the need to keep transmitted power to a minimum, compared with the previous example where each RB had the same transmission power.

Since the power transmitted to user $c_{n,\,i}$ on RB m is $p_{n,\,i}(m)$ the total power is

$$p_{\text{tot}} = \sum_{n \in N} \sum_{i \in I_n} p_{n,i}(m).$$

This total can be minimized by minimizing the power to each qualified user *subject to the constraint that each QU remains qualified* – i.e. it still has sufficient sSINR to meet the QoS objective.

However, there is an additional constraint in that the received power (at the mobile terminal) $p^R_{n,i}$ on RB m for user $c_{n,\,i}$ must be above a certain value $p^R_{\min}$ for satisfactory detection and the relationship between received power and transmitted power is given as

$$p^R_{n,i}(m) = h^n_{n,i}(m) \times p_{n,i}(m).$$

From this it is clear that the optimization needs to maximize the number of users that are qualified, yet at the same time minimize the power as expressed above; this is a multifunction optimization problem with the variables being the RB allocation and the power.

One general approach for solving multi-objective optimization problems is to combine the individual objective functions into a single composite function such as the weighted sum method [47]. This is simple yet probably the most widely used classical approach but does suffer from the need to find suitable weights and can be difficult to specify when one objective is to be maximized and another minimized. The authors in [48] carried out a very comprehensive survey of multi-objective optimization methods and came up with a helpful concluding remark: "*In general, multi-objective optimization requires more computational effort than single-objective optimization. Unless preferences are irrelevant or completely understood, solution of several single objective problems may be necessary to obtain an acceptable final solution.*"

This separation into separate phases was the approach adopted in Yang et al. [49]: that work did not include sharing between slices as the interest was only in the overall power. However, a scenario where users "belong" to different slices could be considered as the slice owning a user has no effect on the power allocation.

Phase 1 is the allocation of RBs: power allocation is set at the maximum power for each RB, denoted as p_{max}, and the allocation of RBs was carried out as described in Section 8.5.1 to maximize the QUR.

Phase 2 is the power reduction. This can be structured as a game but by applying the constraints that (i) RB allocation is unchanged and (ii) no user that is already qualified is allowed to become unqualified, it is possible to simplify the process because none of the users is actually competing against any other. This is because reducing the power to a user reduces the co-channel interference and hence improves the SINR for users allocated that RB in adjacent cells – no user is adversely affected. However, it should be noted that although there is only one variable for each RB, the transmit power $p_{n,i}(m)$, there are two constraints to be met:

- The received power: $p^R_{n,i} \geq p^R_{min}$
- The sSINR: $y_{n,i} \geq y^{min}_{n,i}$

The candidates for power reduction are those users that meet the sSINR requirement and have more received power than the minimum for detection. The implementation is as follows:

Step 1: Initialize by calculating the interference map and the sSINR ($y_{n,i}$) for each RB allocated to each qualified user using the RBs already allocated and set the power on each RB $p_{n,i}$ to be the maximum transmit power for an RB: p_{max}. Note that (i) unqualified users have power set to 0 and no RBs allocated and (ii) that it is assumed here that for any given user the power on each RB is the same and the consequences of that assumption are considered later.

Step 2: For each qualified user:

(i) Calculate the transmit power required to reduce the sSINR to $y^{min}_{n,i}$ and the equivalent receive power $p^{R(\text{sSINR target})}_{n,i}$.

(ii) Check whether the received power $p^{R(\text{sSINR target})}_{n,i}$ satisfies the receive power constraint p^R_{min}.

(iii) If $p^{R(\text{sSINR target})}_{n,i} \geq p^R_{min}$ then the receive power is sufficient so set the value of $p_{n,i}$ for the next loop to be $p^{R(\text{sSINR target})}_{n,i}/h^n_{n,i}$ using the worst value of $h^n_{n,i}(m)$ for the RBs allocated to that user. Go back to Step 2(i) for the next user.

(iv) If $p_{n,i}^{R(\text{sSINR target})} < p_{\min}^{R}$ then the receive power is insufficient so set the transmission power as $p_{\min}^{R}/h_{n,i}^{n}$ (again using the worst value of $h_{n,i}^{n}(m)$) and compare with $p_{\max}$.

(v) If it is $\leq p_{\max}$ then use this value as $p_{n,i}$ for the next loop. Go back to Step 2(i) for the next user.

(vi) If it is $>p_{\max}$ then use $p_{\max}$ as $p_{n,i}$ for the next loop. Go back to Step 2(i) for the next user.

Step 3: When all users have been considered in Step 2, calculate the new interference map using the new transmit powers for the next loop. Loop back to Step 2 until convergence is reached.

Because a user may have been allocated several RBs and the channel conditions for each will be different there is another inherent optimization problem on how to reduce the power to each RB separately but here that is simplified by assuming (i) that the transmit power for each RB used by a particular user is the same and (ii) that the worst channel condition is used for all RBs. This will not lead to the lowest possible power but is conservative in that the sSINR will actually be a bit higher than the minimum and the last condition in Step 2(iv) prevents the worst channel condition leading to transmit power values higher than $p_{\max}$.

The example results in Figure 8.14 show the effectiveness in terms of (a) reducing the overall power through (b) removing excess sSINR from the system: there is no point in allocating excess to any user since the requirement is to allocate just sufficient (plus a safety margin) so that each user is qualified.

Other objective functions could, of course, be added and that demonstrates the flexibility of spectrum sharing: any form of suitable optimization (not just game theory) can be used to tailor the allocation of spectrum to what is required.

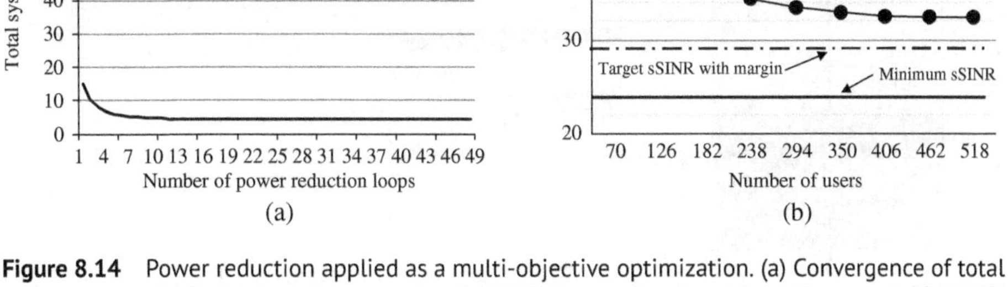

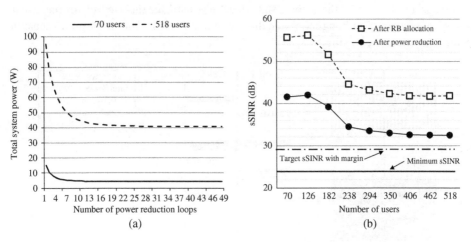

Figure 8.14 Power reduction applied as a multi-objective optimization. (a) Convergence of total system power. (b) Reduction of excess overall sSINR. Source: Reprinted from Yang et al. 2017 [49] under the author copyright agreement of the IEEE.

8.5 Isolation

Isolation is seen as the key in providing effective slicing and is defined as "The operator shall be able to operate different network slices in parallel with isolation that, for example, prevents one slice's data communication to negatively impact services in other slices" [23].

In the RAN, isolation is affected by interference. Unless a frequency resource factor ≥ 1 is used, there will be co-channel ICI and, depending on the slicing structure, this may generate co-channel interference between slices. If the ICI reduces the SINR of an existing user to such an extent that can no longer meet its QoS requirements, then there is no isolation. Hence, it is necessary to consider the effect of ICI on the different resource allocation approaches. This can be exacerbated by movement of users and changing channel conditions leading to changes in ICI.

With fixed allocation there is no impact between slices but there will be ICI that leads to intra-slice interference (Figure 8.15a). However, with fully dynamic allocation inter-slice interference may occur as well (Figure 8.15b) as the co-channel interference may be between users allocated to different slices. Although this is not a problem in lightly loaded networks, as more users are admitted it becomes difficult to allocate spectrum so that there is no ICI, and the purpose of this section is to show how the effects of the ICI can be mitigated in order to achieve isolation.

The literature generally considers *inter-slice isolation*: the isolation of a slice from the effects of other slices in another cell, but there is also *implicit isolation* that is applicable to a sharing environment. This means that instead of a physical separation [23], suitable radio resource management can be used to ensure that any change in one slice should not lead to deterioration in the QoS of users in that slice or in other slices in neighboring cells. It is noted in Foukas et al. [50] (in their Table 1) that "*shared resources*" lead to "*more efficient use of radio resources*" but require "*more sophisticated techniques to ensure isolation of radio resources.*" In Li et al. [51] the authors state that "*In general, the slice control function can be shared among slices,*" although they add the caveat that for some communications separate control functions are required.

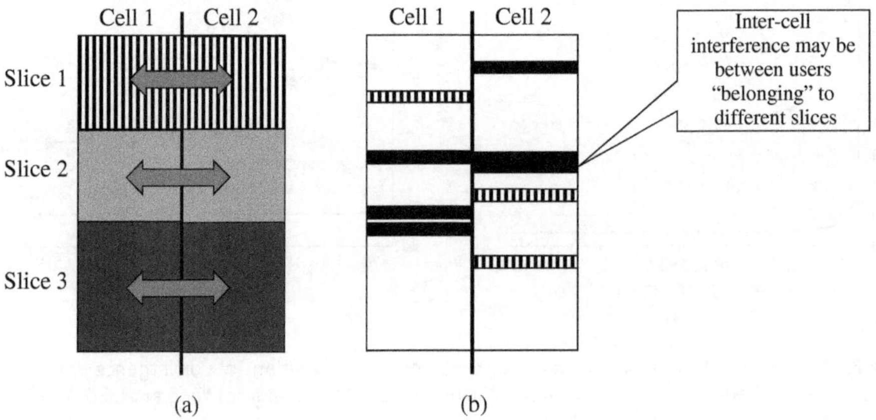

Figure 8.15 Reduction in isolation between slices through inter-cell interference (a) Fixed sharing. (b) Dynamic sharing.

In fact, the concept of implicit isolation has been around for a very long time in the form of connection admission control (CAC), which is indeed a control-plane function. For example, quoting from Cuthbert and Sapanel [20] (for an ATM network), "*A connection request is accepted only when sufficient resources are available to establish the call through the whole network at its required quality of service and to maintain the agreed quality of service of existing calls.*" It is important to note that it is the "*maintain the agreed quality of service of existing calls*" that is the essential factor to give implicit isolation. The problem becomes one of implementing a suitable CAC mechanism as a shared control-plane mechanism rather than finding a form of separation at the radio layer. Of course, the CAC mechanism has to be integrated with the resource allocation to ensure that is as efficient as possible. CAC has also been used in other areas of mobile networks, for instance in Canales et al. [52], Cho et al. [53], and Gelenbe et al. [54].

8.5.1 Example Isolation Results Using CAC

This work is based on that in Yang et al. [55]: this uses the game theory allocation approach described earlier but considers how to use CAC to force isolation.

CAC is applied whenever a new user request is received to determine whether sufficient resources can be made available to make the new user qualified without causing any of the existing users to become unqualified as a result of the extra interference. The scenario considered is the traditional "blocked calls cleared": those users that are not qualified have their RBs released and the connections marked as rejected and not considered further. (It would also be possible to set the scenario as "blocked calls queued" but for simplicity that is not considered here.)

The basic steps of CAC are as follows, but the details of each step vary with the implementation chosen, as explained later:

Whenever a new user comes,

1. store the current RB allocation and co-channel interference map;
2. mark all existing QUs as PrUs (protected users);
3. use the game-theoretic approach to allocate resources to **all the unqualified users** (RB reallocation);
4. if the existing (protected) users have become unqualified, provide extra resources to those users to meet their desired SINRs (PrU compensation);
5. if one or more PrUs is still unqualified, loop back to step 3 until the result converges or the maximum number of iterations is reached, or all the users become qualified;
6. In the end,
   - if one or more PrUs still cannot meet the desired SINR, reject the new user and roll back to position at step 1 and end;
   - if all PrUs are qualified, the new RB allocation will be recorded and the new user is accepted.

The concept of PrUs is derived from the PUs in CR networks [56] but it should be noted that **all existing users** are PrUs at the start of each optimization round so the scenario generally contains many PrUs, but only one new user (equivalent to the SU in CR). As the emphasis of this example is on demonstrating the effectiveness of the CAC in providing isolation, for simplicity only one type of user is considered. However, mixing users that have different characteristics (i.e. belonging to different slices) is straightforward as explained in Yang et al. [32].

This example again uses a seven-cell cluster with wrap-around for an OFDMA uplink using a suburban channel model, but the one type of service included had a target bitrate of 500 kbit/s.

Different approaches to CAC were implemented for further discussion and comparison.

8.5.1.1 Type A: Baseline – CAC Without Network Isolation and Without Protection for Existing Users

This model is used to show the number of users that would be qualified if there were no CAC constraints on existing users. This means that, overall, those users with the best channel conditions were allocated resources, irrespective of whether they were existing or new users. It would be expected that this approach would lead to the highest number of qualified users but there would be no enforced isolation.

To implement the model, the steps used are shown below; changes to the basic steps are shown in bold.

Whenever a new user comes,

1. store the current RB allocation and interference map;
2. **omitted** (Marking all existing QUs as PrUs);
3. use the game-theoretic approach to allocate resources to **all users**;
4. **omitted** (PrU compensation);
5. if one or more **users** are still unqualified, loop back to step 3 until the result converges or the maximum number of iterations is reached, or all the users become qualified;
6. reject any unqualified user, including unqualified PrUs.

8.5.1.2 Type B: Optimum Types – B1 and B2

This approach includes compensation for the PrUs (after the extra interference from the new user has occurred) by allocating extra RBs to any unqualified PrU. Here, two versions are shown (B1 and B2). B1 also includes a roll back to the previous state if 100% protection is not there for the PrUs, but this is omitted in B2 to show the penalty (in terms of number of qualified users).

The steps are as follows:

1. Store the current RB allocation and interference map.
2. Mark all existing QUs as PrUs.
3. Use the game-theoretic approach to allocate resources to **all the unqualified users** (RB reallocation).
4. If the existing (protected) users have become unqualified, provide extra resource to those users (PrU compensation).
5. If one or more PrUs is still unqualified, loop back to step 3 until the result converges or the maximum number of iterations is reached, or all the users become qualified.
6. If one or more PrUs is unqualified, reject the new user and roll back to position at step 1 and end (**B1 only**).

8.5.1.3 Type C: Without Compensation – C1 and C2

The inclusion of compensation in Type B adds complexity to the process since extra resources must be allocated and this increases the implementation time. However, the simple step of using roll back guarantees the operation of the CAC so Type C omits compensation: version C1 implements roll back and, for comparison, version C2 does not roll back and rejects all unqualified users.

1. Store the current RB allocation and interference map.
2. Mark all existing QUs as PrUs.
3. Use the game-theoretic approach to allocate resources to **all the unqualified users** (RB reallocation).
4. **Omitted** (PrU compensation).
5. If one or more PrUs is still unqualified, loop back to step 3 until the result converges or the maximum number of iterations is reached, or all the users become qualified.
6. If one or more PrUs is still unqualified, reject new user and roll back to position at step 1 and end (**C1 only**); reject any unqualified user, including unqualified PrUs **(C2).**

The comparison was carried out with users being distributed with a random uniform distribution across all seven cells, with no constraint on maintaining the same number of users in each cell.

Each set of results[2] shown in Figures 8.16–8.18 below is the average of multiple runs of the simulation. At each measurement point a new user request occurs (chosen randomly to be in any one of the seven cells) and the algorithms run: if the new user cannot be added then it is rejected and "cleared" so that it is not considered again. The results show the performance as more users are added until the system becomes saturated.

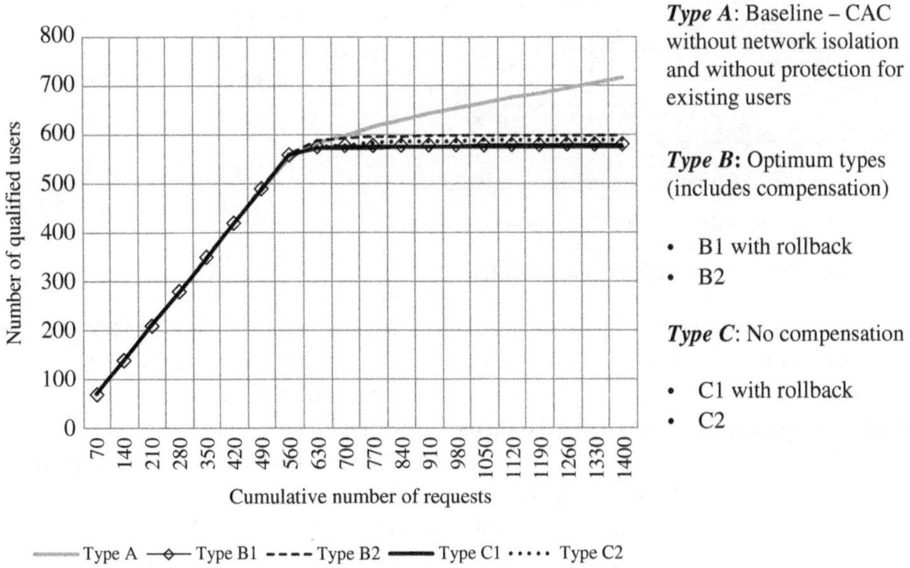

Figure 8.16 Number of qualified users as requests increase ©IEEE 2019.

2 The results are from [55] and reprinted under the author copyright agreement of the IEEE.

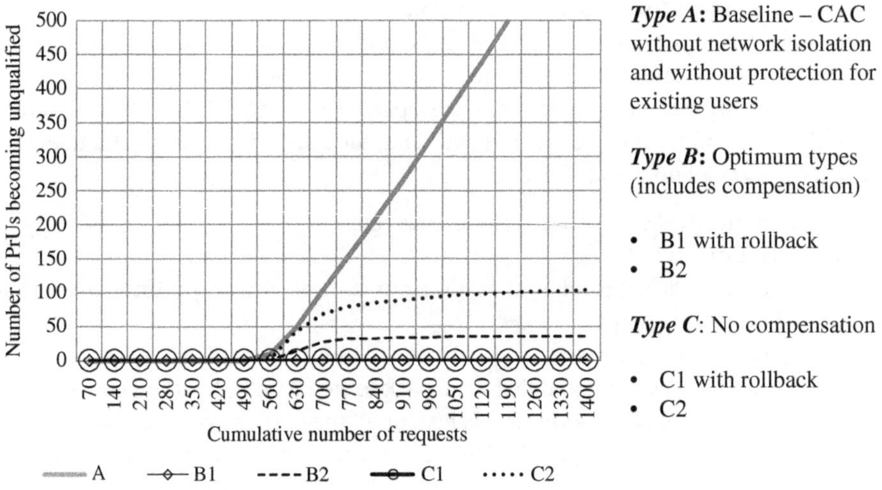

Figure 8.17 Cumulative number of protected users dropped ©IEEE 2019.

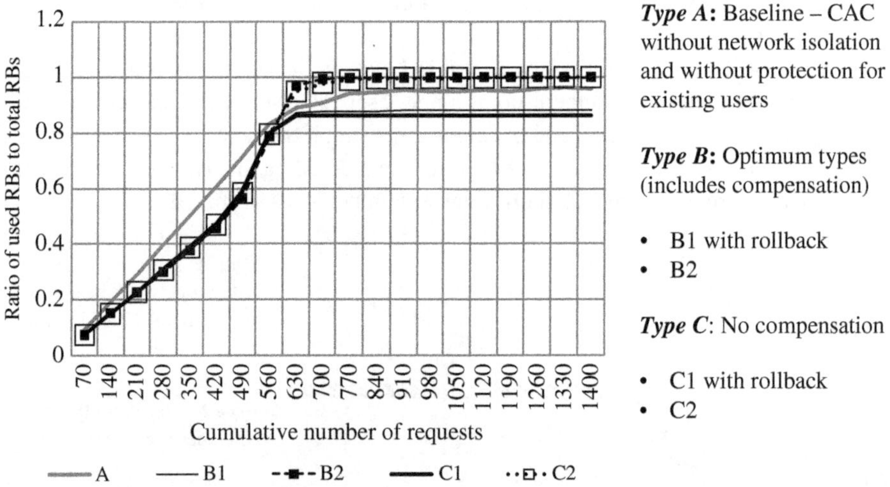

Figure 8.18 Ratio of used RBs to total RBs ©IEEE 2019.

Figure 8.16 shows the results (in terms of qualified user number, QU) of applying the algorithms to the uniform distribution. It is clear that the QU number for the case without any constraints on CAC is the highest because new users with better channel conditions can get resources at the expense of existing users that are being rejected.

There is a definite saturation shown with types B and C because the CAC is trying to ensure that PrUs are not affected by the addition of new users. The saturated region also indicates that there is a distinct, albeit small, difference with B2 and C2 allowing a greater total number of QUs. However, this needs to be considered along with Figure 8.17, which shows the number of PrUs that become unqualified as new users are added.

Type A, as expected, has no protection for the PrUs; types B1 and C1 provide full protection for PrUs with none becoming unqualified, i.e. they provide complete *implicit isolation*.

This is achieved through the roll-back mechanism that strictly enforces protection in that in the event of a PrU becoming unqualified, the new user is rejected and the system rolled back to the conditions before the new request was received.

Type B2 uses the PrU compensation approach of Liu et al. [56], but it can be seen that this protection is incomplete; although the loss of PrUs is small it is still finite so isolation is not achieved completely. The reason this performs much worse, in that sense, than the PrU compensation mechanism in Liu et al. [56] is that all the users are priority users, apart from the single new user being added, so there are insufficient free RBs to perform the compensation.

Type C2 is provided purely as a comparison with C1 by omitting rollback. As with B, not including rollback means that the implicit isolation is not 100%, although there is a greater number of QUs overall since some of the new users will have better channel conditions than the PrUs.

Another effect of the protection is shown in Figure 8.18 where the need to ensure isolation shows there is a waste of RBs – the B1/C1 types with rollback prioritize isolation at the expense of allocating all the RBs. However, the waste is relatively small, of the order of 12%, but it is a price that must be paid for the isolation.

These results are for cases where, once a user is allocated enough RBs to achieve its desired SINR, it is never removed from the network. In a realistic case, users only have a finite connection time, so at some point they will terminate their connection. This is demonstrated in Figure 8.19 for a single example run using Type C1 where after 700 requests from 2% of the PrUs are randomly chosen to terminate their connection so that the RBs released can then be used for new users as shown.

The conclusion of this example is that CAC is a very effective means of maintaining isolation between slices, the rollback mechanism (i.e. not allowing any changes) achieving

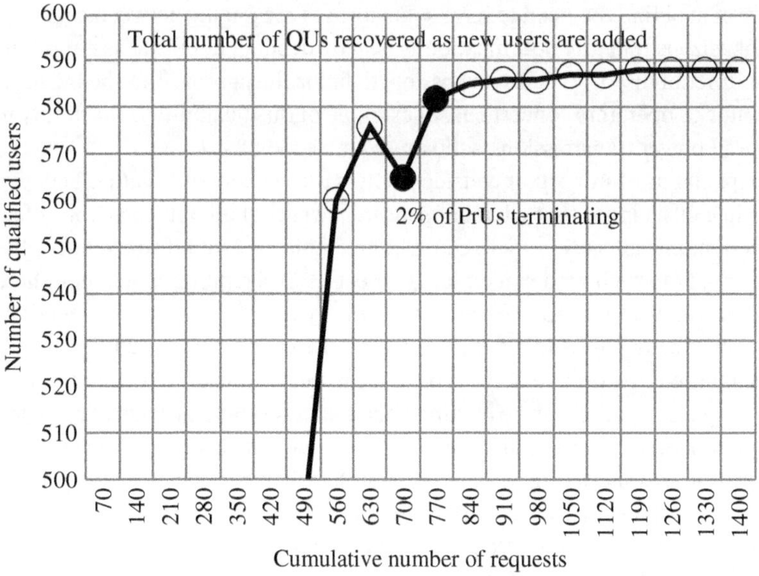

Figure 8.19 Example of dynamic change to number of PrUs ©IEEE.

100% isolation between slices despite any inter-slice or ICI. Method C1 provides the best overall implementation for maintaining complete isolation without adding the complexity of compensation and the rollback can be applied independently of the resource allocation, so it is not specific to the game-theory approach. However, this strict approach means that fewer users achieve the required QoS: in B2 for example, there is no rollback and so there is a high QUR overall, but a small number of existing PrUs are dropped. It should also be noted that the compensation in B2 is included as part of the game, so something equivalent would need to be implemented in any other optimization technique.

8.6 Conclusions

This chapter has sought to highlight the advantages and the issues surrounding network slicing when it is implemented in a dynamic spectrum-sharing manner. These have been considered in a generic way, but the merits of the dynamic approach have been demonstrated using results from research simulations. Of course, the detail of the results would have changed with different implementations, but the principles remain the same.

Clearly, the use of dynamic sharing of a whole block of spectrum leads to improvements in the number of users meeting their QoS requirements and allows flexible changes to the load between operators/slices. However, allowing all users access to all the spectrum does raise issues about ICI being not only between cells but also between different slices whereas fixed allocation limits interference to being within a slice. This means that an interference mitigation strategy needs to be implemented between cells so that the SINR seen by users is sufficient to allow them to meet their QoS requirements. The approach used in the example results shown is to use game theory to perform the necessary multi-cell optimization of resources; while other optimization techniques could be used, non-cooperative game theory has the advantage of avoiding the need to have some form of messaging between cells.

A point to emphasize is that any optimization is heavily dependent on the assumptions and the objective function used, so that it is possible to tailor the approach to the business needs of the operator or operators concerned. An example of this flexibility is the ability to minimize the overall power transmission as illustrated in Section 8.4.2.

However, perhaps the most important consideration is the concept of isolation between slices as the principle of slicing is that what happens in one slice shall not negatively affect other slices. With dynamic allocation there could be inter-slice ICI and if this reduces the SINR of an existing user to such an it can no longer meet its QoS requirements, then there is no isolation. However, using an appropriate CAC strategy can prevent new connections requested in one cell affecting users in others, such an approach having the additional benefit of being implemented irrespective of the type of resource optimization technique. This guarantees that complete isolation can be obtained but does have the drawback of reducing slightly the overall number of QUs. It is also possible to include controls within the resource allocation optimization and these might, as in the example shown, allow a slightly higher QUR but at the expense of having less than 100% isolation.

Overall, it is clear that dynamic resource allocation has many advantages over fixed allocation, as noted in the literature, and that the major potential issues with dynamic allocation can be mitigated by using suitable resource allocation and isolation strategies.

Acknowledgments

Part of the work leading to the results described here was supported by the Science and Technology Development Fund (Fundo para o Desenvolvimento das Ciências e da Tecnologia) of the Macao SAR under grant numbers 074/2015/A and 272/2017/A.

The authors would also like to express their appreciation for the implementation work done by Ka Seng Chou and Ieok Cheng Wong.

References

1 Ofcom (2017). The United Kingdom frequency allocation table [WWW document]. https://www.ofcom.org.uk/spectrum/information/uk-fat (accessed 14 October 2019).

2 Waring, J. (2015). China's operators sell tower assets to China Tower [WWW document]. http://www.mobileworldlive.com/asia/asia-news/chinas-operators-sell-tower-assets-to-china-tower (accessed 20 September 2010).

3 GSMA (2018). Spectrum sharing [WWW document]. Public Policy Position. https://www.gsma.com/spectrum/wp-content/uploads/2019/09/Spectrum-Sharing-PPP.pdf (accessed 20 June 2020).

4 Ofcom (2016). A framework for spectrum sharing [WWW document]. https://www.ofcom.org.uk/__data/assets/pdf_file/0032/79385/spectrum-sharing-framework.pdf (accessed 14 October 2019).

5 Donald, V.H.M. (1979). Advanced mobile phone service: the cellular concept. *Bell System Technical Journal* 58: 15–41.

6 Ring, D.H. (1947). Mobile telephony-wide area coverage. Bell Teleph. Lab. Tech. Rep. 20564.

7 Mitola, J. and Maguire, G. (2001). *Software Radio Technologies: Selected Readings*, 413–418. Wiley-IEEE Press.

8 Haykin, S. (2005). Cognitive radio: brain-empowered wireless communications. *IEEE Journal on Selected Areas in Communications* 23: 201–220.

9 Venkataraman, H., Purohit, A., Pareek, R., and Muntean, G.-M. (2012). Radio resource allocation for cognitive radio based ad hoc wireless networks. *Lecture Notes in Electrical Engineering* 116 LNEE: 287–305.

10 Stevenson, C.R., Chouinard, G., Lei, Z. et al. (2009). IEEE 802.22: the first cognitive radio wireless regional area network standard. *IEEE Communications Magazine* 47: 130–138.

11 Bhattacharjee, S., Konar, A., and Nagar, A.K. (2011). Channel allocation for a single cell cognitive radio network using genetic algorithm. In: *Proceedings – 2011 5th International Conference on Innovative Mobile and Internet Services in Ubiquitous Computing, IMIS 2011*, 258–264. IEEE.

12 Gharavol, E.A., Liang, Y.-C., and Mouthaan, K. (2010). Robust downlink beamforming in multiuser MISO cognitive radio networks with imperfect channel-state information. *IEEE Transactions on Vehicular Technology* 59: 2852–2860.

13 Kanth, V.U., Chandra, K.R., and Kumar, R.R. (2013). Spectrum sharing in cognitive radio networks. *International Journal of Engineering Trends and Technology (IJETT)* 1: 1172–1175.

14 Liu, Y., Cuthbert, L., Yang, X., and Wang, Y. (2012). QoS-aware resource allocation for multimedia users in a multi-cell spectrum sharing radio network. In: *Proceedings of the 7th ACM Workshop on Performance Monitoring and Measurement of Heterogeneous Wireless and Wired Networks – PM2HW2N '12, PM2HW2N '12*, 45. New York: ACM.

15 Anchora, L., Badia, L., Karipidis, E., and Zorzi, M. (2012). Capacity gains due to orthogonal spectrum sharing in multi-operator LTE cellular networks. In: *Proceedings of the International Symposium on Wireless Communication Systems (ISWCS)*, 286–290. IEEE.

16 Jorswieck, E.A., Badia, L., Fahldieck, T. et al. (2014). Spectrum sharing improves the network efficiency for cellular operators. *IEEE Communications Magazine* 52: 129–136.

17 Singh, B., Hailu, S., Koufos, K. et al. (2015). Coordination protocol for inter-operator spectrum sharing in co-primary 5G small cell networks. *IEEE Communications Magazine* 53: 34–40.

18 Liu, Y., Yang, X., and Chou, K.S. (2016). Flexible spectrum sharing in OFDMA cellular networks. In: *Proceedings of the 14th ACM International Symposium on Mobility Management and Wireless Access – MobiWac '16, MobiWac '16*, 67–74. New York: ACM.

19 Rost, P., Mannweiler, C., Michalopoulos, D.S. et al. (2017). Network slicing to enable scalability and flexibility in 5G mobile networks. *IEEE Communications Magazine* 55: 72–79.

20 Cuthbert, L.G. and Sapanel, J.-C. (1993). *ATM: The Broadband Telecommunications Solution*. IET.

21 ITU-T (1997). *Vocabulary of Terms for Broadband Aspects of ISDN*, Series I: Integrated Services Digital. International Telecommunication Union.

22 Huawei Technologies (2018). 5G network architecture: a high-level perspective, Huawei Whitepaper. https://www.huawei.com/uk/industry-insights/outlook/mobile-broadband/insights-reports/5g-network-architecture (accessed 14 October 2019).

23 5G Americas (2016). *Network Slicing for 5G Networks and Services*. 5G Americas.

24 Chang, C.Y. and Nikaein, N. (2018). RAN runtime slicing system for flexible and dynamic service execution environment. *IEEE Access* 6: 34018–34042.

25 Kokku, R., Mahindra, R., Zhang, H., and Rangarajan, S. (2013). CellSlice: cellular wireless resource slicing for active RAN sharing. 2013 5th International Conference on Communication Systems and Networks, COMSNETS, Bangalore, India (7–10 January 2013). IEEE.

26 Sallent, O., Perez-Romero, J., Ferrus, R., and Agusti, R. (2017). On radio access network slicing from a radio resource management perspective. *IEEE Wireless Communications* 24: 166–174.

27 Vo, P.L., Nguyen, M.N.H., Le, T.A., and Tran, N.H. (2018). Slicing the edge: resource allocation for RAN network slicing. *IEEE Wireless Communications Letters* 7: 970–973.

28 Costa-Perez, X., Swetina, J., Guo, T. et al. (2013). Radio access network virtualization for future mobile carrier networks. *IEEE Communications Magazine* 51: 27–35.

29 Meddour, D.-E., Rasheed, T., and Gourhant, Y. (2011). On the role of infrastructure sharing for mobile network operators in emerging markets. *Computer Networks* 55: 1576–1591.

30 Paul, U., Subramanian, A.P., Buddhikot, M.M., and Das, S.R. (2011). Understanding traffic dynamics in cellular data networks. 2011 Proceedings IEEE INFOCOM, Shanghai, China (10–15 April 2011). IEEE.

31 Guo, T. and Arnott, R. (2013). Active LTE RAN sharing with partial resource reservation. In: *2013 IEEE 78th Vehicular Technology Conference (VTC Fall)*, 1–5. Las Vegas, NV: IEEE.

32 Yang, X., Liu, Y., Chou, K.S., and Cuthbert, L. (2017). A game-theoretic approach to network slicing. In: *2017 27th International Telecommunication Networks and Applications Conference (ITNAC)*, 1–4. Melbourne, Australia: IEEE.

33 Kokku, R., Mahindra, R., Zhang, H., and Rangarajan, S. (2012). NVS: a substrate for virtualizing wireless resources in cellular networks. *IEEE/ACM Transactions on Networking* 20: 1333–1346.

34 Szigeti, T., Hattingh, C., Barton, R., and Briley, K. Jr., (2013). *End-to-End QoS Network Design: Quality of Service for Rich-Media & Cloud Networks.* Cisco Press.

35 Jiang, M., Condoluci, M., and Mahmoodi, T. (2016). Network slicing management & prioritization in 5G mobile systems. In: *Proceedings: European Wireless 2016; 22th European Wireless Conference*, 1–6. VDE.

36 Kamel, M.I., Le, L.B., and Girard, A. (2014). LTE wireless network virtualization: dynamic slicing via flexible scheduling. In: *2014 IEEE 80th Vehicular Technology Conference (VTC2014-Fall)*, 1–5. Vancouver, BC: IEEE.

37 Zhao, L., Li, M., Zaki, Y. et al. (2011). LTE virtualization: from theoretical gain to practical solution. In: *Proceedings of the 23rd International Teletraffic Congress, ITC '11*, 71–78. International Teletraffic Congress.

38 Voicu, A.M., Simić, L., and Petrova, M. (2019). Survey of spectrum sharing for inter-technology coexistence. *IEEE Communication Surveys and Tutorials* 21: 1112–1144.

39 Liu, Y., Cuthbert, L., Yang, X., and Wang, Y. (2012). QoS-aware radio resource allocation for multi-cell OFDMA network. In: *2012 IEEE International Conference on Communication Systems, ICCS 2012*, 408–412. IEEE.

40 Halabian, H. (2019). Distributed resource allocation optimization in 5G virtualized networks. *IEEE Journal on Selected Areas in Communications* 37: 627–642.

41 Jiang, M., Condoluci, M., and Mahmoodi, T. (2017). Network slicing in 5G: an auction-based model. IEEE International Conference on Communications, Paris, France (21–25 May 2017). IEEE.

42 Liu, Y., Yang, X., Wong, I.C., et al. (2019). Evaluation of game theory for centralized resource allocation in multi-cell network slicing. 2019 IEEE 30th Annual International Symposium on Personal, Indoor and Mobile Radio Communications (PIMRC) Workshops – W5: Workshop on Radio Access Network Slicing for 5G and Beyond (IEEE PIMRC'19 WS on RAN Slicing). Istanbul, Turkey (8–11 September 2019). IEEE.

43 Aguirre, I. (2009). *Notes on Non-cooperative Game Theory – Microeconomic Theory IV*. Universidad del País Vasco.

44 Liu, K.J.R. and Wang, B. (2010). *Cognitive Radio Networking and Security: A Game-Theoretic View*. Cambridge University Press.

45 Chu, X., López-Pérez, D., Yang, Y., and Gunnarsson, F. (2011). *Heterogeneous Cellular Networks: Theory, Simulation and Deployment*. Cambridge University Press.

46 Jain, R. (1991). *The Art of Computer Systems Performance Analysis: Techniques for Experimental Design, Measurement, Simulation, and Modeling*. Wiley.

47 Deb, K. (2001). *Multi-objective Optimization Using Evolutionary Algorithms*. Wiley.

48 Marler, R.T. and Arora, J.S. (2004). Survey of multi-objective optimization methods for engineering. *Structural and Multidisciplinary Optimization* 26: 369–395.

49 Yang, X., Liu, Y., Chou, K.S., and Cuthbert, L. (2017). QoS-aware power allocation for spectrum sharing. In: *2017 International Conference on Selected Topics in Mobile and Wireless Networking, MoWNeT 2017*, 119–124. Avignon, France: IEEE.

50 Foukas, X., Patounas, G., Elmokashfi, A., and Marina, M.K. (2017). Network slicing in 5G: survey and challenges. *IEEE Communications Magazine* 55: 94–100.

51 Li, X., Samaka, M., Chan, H.A. et al. (2017). Network slicing for 5G: challenges and opportunities. *IEEE Internet Computing* 21: 20–27.

52 Canales, M., Gállego, J.R., Hernández-Solana, Á., and Valdovinos, A. (2009). QoS provision in mobile ad hoc networks with an adaptive cross-layer architecture. *Wireless Networks* 15: 1165–1187.

53 Cho, S., Akyildiz, I.F., Bender, M.D., and Uzunalioğlu, H. (2002). A new connection admission control for spotbeam handover in LEO satellite networks. *Wireless Networks* 8: 403–415.

54 Gelenbe, E., Sakellari, G., and D'Arienzo, M. (2008). Admission of QoS aware users in a smart network. *ACM Transactions on Autonomous and Adaptive Systems* 3 https://doi.org/10.1145/1342171.1342175.

55 Yang, X., Liu, Y., Wong, I.C., et al. (2019). Effective isolation in dynamic network slicing. IEEE Wireless Communications and Networking Conference (WCNC), Marrakesh, Morocco (15–18 April 2019). IEEE.

56 Liu, Y., Yang, X.X., Chou, K.S.K.S., and Cuthbert, L. (2017). Cognitive radio using spectrum-sharing and power minimisation. In: *18th IEEE International Symposium on a World of Wireless, Mobile and Multimedia Networks, WoWMoM 2017 – Conference*, 1–6. IEEE.

9

Access Control and Handoff Policy Design for RAN Slicing

Yao Sun[1], *Lei Zhang*[1], *Gang Feng*[2] *and Muhammad Ali Imran*[1]

[1] *University of Glasgow, James Watt School of Engineering, UK*
[2] *University of Electronic Science and Technology of China, National Key Laboratory on Communications, China*

In radio access networks (RANs) slicing, resource management (including access control and handoff policy design) should be exploited to improve network efficiency and flexibility while meeting quality of service (QoS) requirements. However, the hierarchical network architecture, diversified use cases, as well as various QoS requirements make resource management a challenging issue. In this chapter, we propose to solve two problems: user access control and mobile user handoff policy design. Let us start with the discussions on user access control.

9.1 A Framework of User Access Control for RAN Slicing

To take the benefits of network slicing technology, access control is of paramount importance to be investigated in network slice (NS)-based networks. An appropriate access control scheme can improve both network and users performance by flexibly allocating inter-/intra NS resources. Moreover, driven by the explosive growth of wireless data traffic, improving resource utilization is crucial for 5G-and-beyond communication systems [1–4]. Therefore, from both network performance and resource utilization perspectives, it is imperative to develop efficient access control schemes for radio access network (RAN) slicing.

In RAN slicing, user equipment (UE) access control and resource allocation are fundamentally different from that in conventional mobile networks due to the introduction of NS [5, 6]. First, from the network architecture aspect, slices are logically virtualized and isolated over shared physical networks. Hence, both physical and virtual resource constraints need to be considered to form a function chain for a specific service. Second, from the user association aspect, UE should be associated with an NS via a specific physical access point (AP), such as base station (BS). Not all APs are able to provide the specific slice/service due to either a missing functionality or limited resources. Hence, joint optimization of NS and BS selection should be addressed. Third, from the service aspect, NS-based networks provide guaranteed QoS for all serving UEs [7] instead of the traditional *best effort* model [8]. Owing to the aforementioned differences, applying traditional resource management

Radio Access Network Slicing and Virtualization for 5G Vertical Industries, First Edition.
Edited by Lei Zhang, Arman Farhang, Gang Feng, and Oluwakayode Onireti.

mechanisms to RAN slicing may lead to low resource utilization, poor QoS provisioning, and frequent NS reconfigurations. Therefore, designing new resource management mechanisms dedicated for RAN slicing to optimize network performance and resource utilization becomes an essential yet challenging issue.

In this chapter, we aim for a unified framework for access control in RAN slicing. Owing to resource limitation, not all users' QoS requirements can be satisfied simultaneously. Hence, we design two user access control policies to select admissible users for optimizing the QoS and the number of admissible users respectively. Numerical results show that in typical scenarios, our proposed access control policies can significantly outperform traditional policies in terms of the number of admissible users.

9.1.1 System Model for RAN Slicing

We consider an NS-based mobile network shown in Figure 9.1, which consists of core and radio access networks. Some network function (NF) modules and access function (AF) modules are deployed to form an end-to-end network slice with specific service provisioning. NFs and AFs are related to some specific logic functions, such as connection management, mobility management, security, etc., in core network and RAN respectively. Detailed descriptions of the network architecture can be found in [7]. Here we focus on the resource management aspect of the system.

From Figure 9.1, we see that slices can share resource in both core and radio access networks. Specifically, in the core network, the slices using the same link should share link bandwidth resource, computing resource, and NFs. In RAN, the slices deployed in the same AP will share wireless resource as well as AFs. The slice information is broadcast by the APs, and not all NSs will be accessible via every AP. In the example of Figure 9.1, AP 3 broadcasts the information of NS 2 and NS 4, and thus UEs can access NS 2 or NS 4 via AP 3.

Focusing on RAN slicing, we consider a multi-slice and multi-AP model shown in Figure 9.2, where the BSs used by multiple slices are deployed in the area (we use BS to

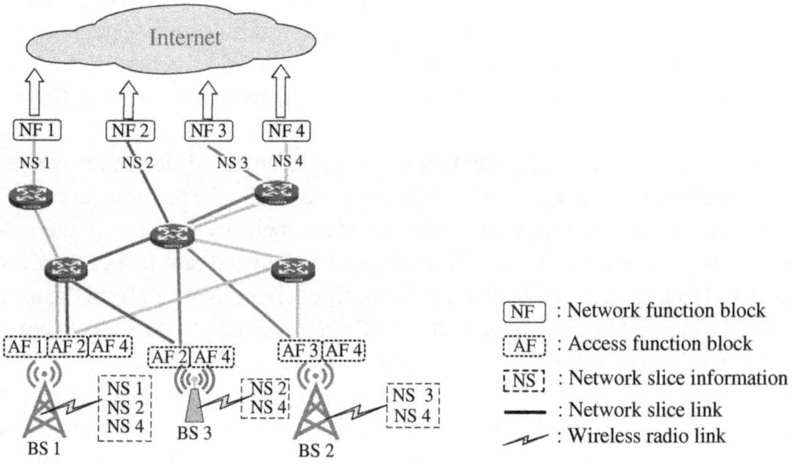

Figure 9.1 Network slice-based network architecture.

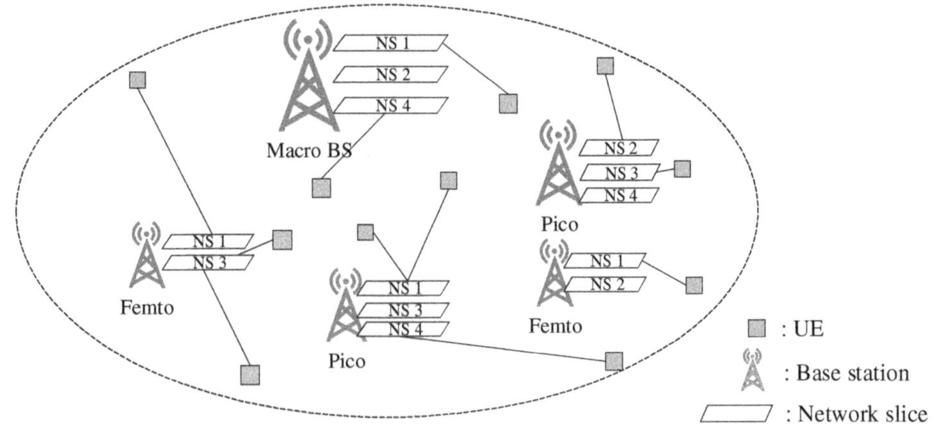

Figure 9.2 Multi-slice and multi-AP RAN slicing.

denote all kinds of APs throughout this chapter). Each BS can support multiple NSs with different provisioned QoS, and each NS may also be covered by multiple BSs (i.e. each slice information is broadcast by several BSs). Multiple UEs are randomly distributed in this area with different QoS requirements. They can access a specific NS via a BS in the coverage of the NS, thus forming a three layer association relationship. Let $\mathcal{B}$, $\mathcal{S}$, and $\mathcal{U}$ denote the set of BSs, NSs, and UEs, respectively. For a specific BS, say BS k, we use $\mathcal{S}_k$ to denote the set of NSs supported by BS k.

We identify a specific NS, say NS j, by transmission rate, delay, and the resource allocation in core and access networks. Specifically, besides the slice ID, four elements $(R_j, D_j, \Lambda_j, \vec{B}_j)$ are used to identify the j-th slice, where R_j and D_j respectively denote the minimum transmission rate and the maximum delay that NS j can provide to its serving UEs, Λ_j denotes the bandwidth allocated to NS j in the core network, and $\vec{B}_j$ is a vector denoting the wireless bandwidth allocation of NS j from all BSs. Let $b_j^{(k)}$ be the k-th element of vector $\vec{B}_j$ denoting the bandwidth of NS j allocated by BS k. $b_j^{(k)} = 0$ when BS k is not in the coverage of NS j.

For a specific UE, say UE n with q_n volume data to transmit, the QoS can be described by two metrics: transmission rate $\bar{r}_n$ and delay $\bar{d}_n$ [9]. Thus, NS j is admissible for UE n only if $R_j \geq \bar{r}_n$ and $D_j \leq \bar{d}_n$. We now study the two QoS metrics respectively. Let $r_n^{j,k}$ be the transmission rate of UE n served by NS j via BS k. For simplicity, we use Shannon theory to define the transmission rate, i.e. $r_n^{j,k} = w_n^{j,k}\log_2(1 + SINR_n^k)$, where $w_n^{j,k}$ is the wireless bandwidth that BS k allocates to the UE n served by NS j, and SINR_n^k is the signal-to-interference-plus-noise-ratio (SINR) between UE n and BS k. We use $d_n^{j,k} = q_n/r_n^{j,k}$ to denote the delay in RAN of UE n served by NS j via BS k. Thus, combined with the delay in core network, the end-to-end delay can be approximately calculated as $d_n^{j,k} + D_j$. Note that more sophisticated and accurate transmission rate and delay models can be used here. However, these do not affect the following derivations. This is because the models in this work may affect the absolute value of bandwidth consumption, but do not invalidate the relative performance enhancement of our proposed policies.

Traditional networks usually provide the *best effort* service model to users [8]. In this model, the network allocates resources among all of the active UEs and attempts to serve all of them without making any explicit commitment on rate or any other service quality [8], while NS-based networks need to guarantee the different and heterogeneous QoS of UEs without interference among one another [10]. Hence, the best effort service model cannot be applicable directly to RAN slicing since it cannot guarantee the individual user's QoS. In the following, we will formulate the joint access control and bandwidth allocation problem in RAN slicing to meet individual UE requirements.

9.1.2 UE Association Problem Description

The optimization problem in this work can be stated as: minimizing the wireless bandwidth consumption subject to QoS requirements, NS and BS resource constraints through joint access control and bandwidth allocation for UEs. To formulate this problem, we define a binary variable $x_n^{j,k} \in \{0,1\}, \forall (n,j,k) \in \mathcal{U} \times \mathcal{S} \times \mathcal{B}$, where $x_n^{j,k} = 1$ indicates that UE n is served by NS j via BS k. We can now formulate the access control and bandwidth allocation problem as **P1**:

$$\mathbf{P1}: \min \sum_{n\in\mathcal{U}} \sum_{j\in\mathcal{S}} \sum_{k\in\mathcal{B}} x_n^{j,k} w_n^{j,k}, \tag{9.0}$$

$$\text{s.t.} \sum_{n\in\mathcal{U}} \sum_{k\in\mathcal{B}} x_n^{j,k} r_n^{j,k} \le \Lambda_j, \quad \forall j \in \mathcal{S} \tag{9.1-1}$$

$$\sum_{n\in\mathcal{U}} x_n^{j,k} w_n^{j,k} \le b_j^{(k)}, \quad \forall j \in \mathcal{S}, \quad \forall k \in \mathcal{B} \tag{9.1-2}$$

$$\sum_{j\in\mathcal{S}} \sum_{k\in\mathcal{B}} x_n^{j,k} r_n^{j,k} \ge \bar{r}_n, \quad \forall n \in \mathcal{U} \tag{9.1-3}$$

$$\sum_{j\in\mathcal{S}} \sum_{k\in\mathcal{B}} x_n^{j,k} R_j \ge \bar{r}_n, \quad \forall n \in \mathcal{U} \tag{9.1-4}$$

$$\sum_{j\in\mathcal{S}} \sum_{k\in\mathcal{B}} x_n^{j,k} (d_n^{j,k} + D_j) \le \bar{d}_n, \quad \forall n \in \mathcal{U} \tag{9.1-5}$$

$$\sum_{j\in\mathcal{S}} \sum_{k\in\mathcal{B}} x_n^{j,k} = 1, \quad \forall n \in \mathcal{U} \tag{9.1-6}$$

$$x_n^{j,k} \in \{0,1\}, \quad \forall (n,j,k) \in \mathcal{U} \times \mathcal{S} \times \mathcal{B} \tag{9.1-7}$$

where $x_n^{j,k}$ and $w_n^{j,k}$ are the optimization variables. Constraint (9.1-1) refers to the wired link resource constraint to guarantee that the total transmission rate offered by an NS does not exceed the link budget. Constraint (9.1-2) states the wireless bandwidth constraint, ensuring that the total bandwidth allocated to UEs by NS j via BS k does not exceed the wireless bandwidth budget of NS j allocated from BS k. Note that constraint (9.1-2) also ensures that UEs cannot access an NS via the BSs that cannot provide such a service. Constraints (9.1-3)–(9.1-5) guarantee that the QoS (rate and delay) of UEs can be satisfied by its serving BS and NS. Constraints (9.1-6) and (9.1-7) ensure that each UE can only access one NS via one BS at a time.

9.1.3 Admission Control Mechanisms Design for RAN Slicing

We first design AC schemes to find a set of admissible UEs whose QoS can be satisfied simultaneously. The main idea is that if we cannot meet the QoS of all UEs, we will reject some of the UEs to give more resources to others, thus avoiding significant overall network performance degradation. Before studying this problem, we first give the following assumption and definition.

Assumption 9.1 *Network slice allocates the minimal required wireless bandwidth to UEs to satisfy the UEs' QoS requirements.* □

According to Assumption 9.1, the original constraints (9.1-3)–(9.1-5) about the QoS of UEs become equality constraints, and the admissible UEs obtained under Assumption 9.1 are still admissible for the original problem. Moreover, as we use this assumption, the allocated bandwidth $w_n^{j,k}$ is not the optimal variable for AC design. We will optimize $w_n^{j,k}$ in Section 9.1.3.2 by solving **P1** for those admissible UEs.

Definition 9.1 *Subset $\mathcal{A}$ is an admissible UE set (AUS) if problem* ***P1*** *is feasible when $\mathcal{U}$ is replaced by $\mathcal{A}$.* □

Definition 9.1 describes the feasibility of a UE subset. Hence, for a specific AUS, the RAN slicing can simultaneously guarantee the QoS of all the UEs in this AUS. However, it is not a sufficient condition to achieve good overall network performance. For example, it is meaningless to choose an AUS that contains only one UE. Therefore, we need to design an approach to find a UE subset that is not only feasible for problem **P1** but also achieves good overall network performance. We will thus develop two AC schemes under Assumption 9.1 to optimize the QoS and the number of admissible UEs respectively.

9.1.3.1 Optimal QoS AC Mechanism

We first consider the AC scheme from the UE QoS perspective. As not all UEs could be admissible, there is a gap between the achieved QoS and the required QoS of UEs (we use QoS degradation to represent this gap). Our main idea is to serve the UEs whose QoS can be satisfied simultaneously while minimizing the overall QoS degradation. To this end, we formulate a new optimization problem based on **P1**, and then design an AC policy based on the solution to the new problem. Let us illustrate the process in detail in the following.

First, we introduce two elastic variables $\check{r}_n$ and $\check{d}_n$ for UE n to describe the rate and delay degradation respectively. We restrict that $0 \leq \check{r}_n \leq \bar{r}_n$ and $0 \leq \check{d}_n \leq \bar{D}$, where $\bar{D}$ is a very large parameter. Therefore, the rate and delay requirement of UE n can be referred to as $\bar{r}_n - \check{r}_n$ and $\bar{d}_n + \check{d}_n$ respectively. UE n is admissible when $\check{r}_n = 0$ and $\check{d}_n = 0$. Then we give the following definition to describe a subset of UEs whose QoS can be simultaneously satisfied while the QoS degradation of others is minimum.

Definition 9.2 *Subset $\mathcal{A}$ is a QoS-admissible UE set (QoS-AUS) if $\mathcal{A}$ is an AUS with the minimum achievable value of $\sum_{n \in \mathcal{U} \setminus \mathcal{A}} \left(\frac{\check{r}_n}{\bar{r}_n} + \frac{\check{d}_n}{\bar{d}_n} \right)$.* □

Here we use the normalized degradation of transmission rate (i.e., $\check{r}_n / \bar{r}_n$) and delay ($\check{d}_n / \bar{d}_n$). This definition describes both the feasibility and the QoS performance of a UE subset. In the following, we design an AC scheme, called QoS-AC, to find such a QoS-AUS in Definition 9.2. By introducing elastic variables $\check{r}_n$ and $\check{d}_n$, we formulate problem **P2** as follows.

$$\textbf{P2}: \min \sum_{n\in\mathcal{U}} \left(\frac{\check{r}_n}{\bar{r}_n} + \frac{\check{d}_n}{\bar{d}_n} \right), \tag{9.1}$$

$$\text{s.t.} \sum_{n\in\mathcal{U}} \sum_{k\in\mathcal{B}} x_n^{j,k} r_n^{j,k} \le \Lambda_j, \quad \forall j \in \mathcal{S} \tag{9.2-1}$$

$$\sum_{n\in\mathcal{U}} x_n^{j,k} w_n^{j,k} \le b_j^{(k)}, \quad \forall j \in \mathcal{S}, \forall k \in \mathcal{B} \tag{9.2-2}$$

$$\sum_{j\in\mathcal{S}} \sum_{k\in\mathcal{B}} x_n^{j,k} r_n^{j,k} = \bar{r}_n - \check{r}_n, \quad \forall n \in \mathcal{U} \tag{9.2-3}$$

$$\sum_{j\in\mathcal{S}} \sum_{k\in\mathcal{B}} x_n^{j,k} R_j = \bar{r}_n - \check{r}_n, \quad \forall n \in \mathcal{U} \tag{9.2-4}$$

$$\sum_{j\in\mathcal{S}} \sum_{k\in\mathcal{B}} x_n^{j,k} (d_n^{j,k} + D_j) = \bar{d}_n + \check{d}_n, \quad \forall n \in \mathcal{U} \tag{9.2-5}$$

$$\sum_{j\in\mathcal{S}} \sum_{k\in\mathcal{B}} x_n^{j,k} = 1, \quad \forall n \in \mathcal{U} \tag{9.2-6}$$

$$x_n^{j,k} \in \{0,1\}, \quad \forall (n,j,k) \in \mathcal{U} \times \mathcal{S} \times \mathcal{B} \tag{9.2-7}$$

In **P2**, the objective is to minimize the normalized QoS degradation of all UEs. Compared with the constraints in **P1**, the only difference is using equalities in constraints (9.2-3)-–(9.2-5) to replace the inequalities in (9.1-3)–(9.1-5) by introducing elastic variables. In **P2**, the optimization variables are binary indicators $x_n^{j,k}$ as well as the continuous elastic variables $\check{r}_n$ and $\check{d}_n$. Hence, **P2** is a mixed integer linear programming (MILP). As we introduce the elastic variables into the MILP, the QoS of UEs can vary with the elastic variables, and thus problem **P2** is always feasible. In the following, we solve **P2**.

Using Lagrange decomposition theory [11], we first introduce constraints (9.2-3)–(9.2-5) into the optimization objective by associating Lagrange multipliers λ_n, υ_n, and μ_n. Let $\boldsymbol{\lambda}, \boldsymbol{\upsilon}, \boldsymbol{\mu}, \check{\boldsymbol{r}}, \check{\boldsymbol{d}}$, and $\boldsymbol{x}$ be the corresponding vectors of $\lambda_n, \upsilon_n, \mu_n, \check{r}_n, \check{d}_n$, and $x_n^{j,k}$, respectively. For **P2**, we give Lagrange dual problem **P3** with respect to constraints (9.2-3)–(9.2-5):

$$\begin{aligned} \textbf{P3}: g(\boldsymbol{\lambda}, \boldsymbol{\upsilon}, \boldsymbol{\mu}) &\triangleq \inf_{\check{\boldsymbol{r}}, \check{\boldsymbol{d}}, \boldsymbol{x}} L(\boldsymbol{\lambda}, \boldsymbol{\upsilon}, \boldsymbol{\mu}, \check{\boldsymbol{r}}, \check{\boldsymbol{d}}, \boldsymbol{x}) \\ &\text{s.t. Constraints (9.2-1), (9.2-2), (9.2-6) and (9.2-7)} \end{aligned} \tag{9.3}$$

where Lagrangian $L(\boldsymbol{\lambda}, \boldsymbol{\upsilon}, \boldsymbol{\mu}, \check{\boldsymbol{r}}, \check{\boldsymbol{d}}, \boldsymbol{x})$ is expressed as

$$\begin{aligned} &L(\boldsymbol{\lambda}, \boldsymbol{\upsilon}, \boldsymbol{\mu}, \check{\boldsymbol{r}}, \check{\boldsymbol{d}}, \boldsymbol{x}) \\ &= \sum_{n\in\mathcal{U}} \left(\frac{\check{r}_n}{\bar{r}_n} + \frac{\check{d}_n}{\bar{d}_n} \right) + \sum_{n\in\mathcal{U}} \lambda_n \left(\sum_{j\in\mathcal{S}} \sum_{k\in\mathcal{B}} x_n^{j,k} r_n^{j,k} - (\bar{r}_n - \check{r}_n) \right) \\ &\quad + \sum_{n\in\mathcal{U}} \upsilon_n \left(\sum_{j\in\mathcal{S}} \sum_{k\in\mathcal{B}} x_n^{j,k} R_j - (\bar{r}_n - \check{r}_n) \right) \\ &\quad + \sum_{n\in\mathcal{U}} \mu_n \left(\sum_{j\in\mathcal{S}} \sum_{k\in\mathcal{B}} x_n^{j,k} (d_n^{j,k} + D_j) - (\bar{d}_n + \check{d}_n) \right) \end{aligned}$$

$$= \sum_{n\in\mathcal{U}} \left(\frac{\check{r}_n}{\bar{r}_n} + \frac{\check{d}_n}{\bar{d}_n} \right) + \sum_{n\in\mathcal{U}} (\lambda_n \check{r}_n + \upsilon_n \check{r}_n - \mu_n \check{d}_n) + \sum_{n\in\mathcal{U}} \lambda_n \left(\sum_{j\in\mathcal{S}} \sum_{k\in\mathcal{B}} x_n^{j,k} r_n^{j,k} - \check{r}_n \right)$$
$$+ \sum_{n\in\mathcal{U}} \upsilon_n \left(\sum_{j\in\mathcal{S}} \sum_{k\in\mathcal{B}} x_n^{j,k} R_j - \check{r}_n \right) + \sum_{n\in\mathcal{U}} \mu_n \left(\sum_{j\in\mathcal{S}} \sum_{k\in\mathcal{B}} x_n^{j,k} (d_n^{j,k} + D_j) - \check{d}_n \right). \quad (9.4)$$

Here, strong duality holds. Therefore, we first solve **P3** with the fixed Lagrange multipliers $\boldsymbol{\lambda}$, $\boldsymbol{\nu}$, and $\boldsymbol{\mu}$, and then maximize $g(\boldsymbol{\lambda}, \boldsymbol{\nu}, \boldsymbol{\mu})$ to find the optimal solution to **P2**.

Owing to the independence of $\check{\boldsymbol{r}}, \check{\boldsymbol{d}}$, and $\boldsymbol{x}$, $L(\boldsymbol{\lambda}, \boldsymbol{\nu}, \boldsymbol{\mu}, \check{\boldsymbol{r}}, \check{\boldsymbol{d}}, \boldsymbol{x})$ can be decoupled into two sub-functions $L^1(\boldsymbol{\lambda}, \boldsymbol{\nu}, \boldsymbol{\mu}, \check{\boldsymbol{r}}, \check{\boldsymbol{d}})$ and $L^2(\boldsymbol{\lambda}, \boldsymbol{\nu}, \boldsymbol{\mu}, \boldsymbol{x})$, i.e. $L(\boldsymbol{\lambda}, \boldsymbol{\nu}, \boldsymbol{\mu}, \check{\boldsymbol{r}}, \check{\boldsymbol{d}}, \boldsymbol{x}) = L^1(\boldsymbol{\lambda}, \boldsymbol{\nu}, \boldsymbol{\mu}, \check{\boldsymbol{r}}, \check{\boldsymbol{d}}) + L^2(\boldsymbol{\lambda}, \boldsymbol{\nu}, \boldsymbol{\mu}, \boldsymbol{x})$, where

$$\begin{aligned} L^1(\boldsymbol{\lambda}, \boldsymbol{\nu}, \boldsymbol{\mu}, \check{\boldsymbol{r}}, \check{\boldsymbol{d}}) &\triangleq \sum_{n\in\mathcal{U}} \left(\frac{\check{r}_n}{\bar{r}_n} + \frac{\check{d}_n}{\bar{d}_n} + \lambda_n \check{r}_n + \upsilon_n \check{r}_n - \mu_n \check{d}_n \right) \\ &= \sum_{n\in\mathcal{U}} \left[\left(\frac{1}{\bar{r}_n} + \lambda_n + \upsilon_n \right) \check{r}_n + \left(\frac{1}{\bar{d}_n} - \mu_n \right) \check{d}_n \right], \end{aligned} \quad (9.5)$$

and

$$\begin{aligned} L^2(\boldsymbol{\lambda}, \boldsymbol{\nu}, \boldsymbol{\mu}, \boldsymbol{x}) \triangleq& \sum_{n\in\mathcal{U}} \lambda_n \left(\sum_{j\in\mathcal{S}} \sum_{k\in\mathcal{B}} x_n^{j,k} r_n^{j,k} - \bar{r}_n \right) + \sum_{n\in\mathcal{U}} \upsilon_n \left(\sum_{j\in\mathcal{S}} \sum_{k\in\mathcal{B}} x_n^{j,k} R_j - \bar{r}_n \right) \\ &+ \sum_{n\in\mathcal{U}} \mu_n \left(\sum_{j\in\mathcal{S}} \sum_{k\in\mathcal{B}} x_n^{j,k} (d_n^{j,k} + D_j) - \bar{d}_n \right). \end{aligned} \quad (9.6)$$

Hence, **P3** is decomposed into two subproblems:

$$\begin{aligned} &\textbf{P3(1)}: g^1(\boldsymbol{\lambda}, \boldsymbol{\nu}, \boldsymbol{\mu}) \triangleq \inf_{\check{\boldsymbol{r}}, \check{\boldsymbol{d}}} L^1(\boldsymbol{\lambda}, \boldsymbol{\nu}, \boldsymbol{\mu}, \check{\boldsymbol{r}}, \check{\boldsymbol{d}}) \\ &\text{s.t. Constraints (9.2-1), (9.2-2), (9.2-6) and (9.2-7)} \end{aligned} \quad (9.7)$$

and

$$\begin{aligned} &\textbf{P3(2)}: g^2(\boldsymbol{\lambda}, \boldsymbol{\nu}, \boldsymbol{\mu}) \triangleq \inf_{\boldsymbol{x}} L^2(\boldsymbol{\lambda}, \boldsymbol{\nu}, \boldsymbol{\mu}, \boldsymbol{x}) \\ &\text{s.t. Constraints (9.2-1), (9.2-2), (9.2-6) and (9.2-7)} \end{aligned} \quad (9.8)$$

When $\boldsymbol{\lambda}, \boldsymbol{\nu}, \boldsymbol{\mu}$ are fixed, we solve **P3(1)** and **P3(2)** respectively.

Focusing on **P3(1)**, we find that the constraints (9.2-1), (9.2-2), (9.2-6), and (9.2-7) are unrelated to the optimization variables $\check{\boldsymbol{r}}, \check{\boldsymbol{d}}$. Hence, **P3(1)** is actually an unconditional optimization problem that can be easily solved. Given the fixed Lagrange multipliers $\boldsymbol{\lambda}$, $\boldsymbol{\nu}$, and $\boldsymbol{\mu}$, the partial derivative of $\check{r}_n$ and $\check{d}_n$ are constant, which means that the objective function $L^1(\boldsymbol{\lambda}, \boldsymbol{\nu}, \boldsymbol{\mu}, \check{\boldsymbol{r}}, \check{\boldsymbol{d}})$ is monotone increasing or decreasing with respect to $\check{r}_n$ and $\check{d}_n$. Therefore, the optimal solution is

$$\check{r}_n^* = \begin{cases} 0, & \frac{1}{\bar{r}_n} + \lambda_n + \upsilon_n \le 0 \\ \bar{r}_n, & \frac{1}{\bar{r}_n} + \lambda_n + \upsilon_n > 0 \end{cases}, \quad (9.9)$$

and

$$\check{d}_n^* = \begin{cases} 0, & \frac{1}{\bar{d}_n} - \mu_n \leq 0 \\ \bar{D}, & \frac{1}{\bar{d}_n} - \mu_n > 0 \end{cases}. \tag{9.10}$$

We now study the solution to **P3(2)**. We give the Lagrange dual problem of **P3(2)** with respect to constraints (9.2-6) and (9.2-7) as follows:

$$\begin{aligned} &\mathbf{P4:} \; h(\lambda, \boldsymbol{v}, \boldsymbol{\mu}, \boldsymbol{\eta}, \boldsymbol{\theta}) \triangleq \inf_{\boldsymbol{x}} L^3(\lambda, \boldsymbol{v}, \boldsymbol{\mu}, \boldsymbol{\eta}, \boldsymbol{\theta}, \boldsymbol{x}) \\ &\text{s.t. Constraints (9.2-1), (9.2-6) and (9.2-7)} \end{aligned} \tag{9.11}$$

where

$$\begin{aligned} L^3(\lambda, \boldsymbol{v}, \boldsymbol{\mu}, \boldsymbol{\eta}, \boldsymbol{\theta}, \boldsymbol{x}) = {} & L^2(\lambda, \boldsymbol{v}, \boldsymbol{\mu}, \boldsymbol{x}) - \sum_{j \in \mathcal{S}} \eta_j \left(\sum_{n \in \mathcal{U}} \sum_{k \in \mathcal{B}} x_n^{j,k} r_n^{j,k} - \Lambda_j \right) \\ & - \sum_{j \in \mathcal{S}} \sum_{k \in \mathcal{B}} \theta_{j,k} \left(\sum_{n \in \mathcal{U}} x_n^{j,k} w_n^{j,k} - b_j^{(k)} \right), \end{aligned} \tag{9.12}$$

and vectors $\boldsymbol{\eta}$ and $\boldsymbol{\theta}$ are Lagrange multipliers containing the elements $\eta_j \geq 0$ and $\theta_{j,k} \geq 0$ respectively. Here, **P3(2)** is a linear programming, and thus strong duality holds again, i.e. $\inf_{\boldsymbol{x}} L^2(\lambda, \boldsymbol{v}, \boldsymbol{\mu}, \boldsymbol{x}) = \sup_{\eta,\theta} h(\lambda, \boldsymbol{v}, \boldsymbol{\mu}, \boldsymbol{\eta}, \boldsymbol{\theta})$. Based on (9.12), we rewrite $L^3(\lambda, \boldsymbol{v}, \boldsymbol{\mu}, \boldsymbol{\eta}, \boldsymbol{\theta}, \boldsymbol{x})$ as

$$L^3(\lambda, \boldsymbol{v}, \boldsymbol{\mu}, \boldsymbol{\eta}, \boldsymbol{\theta}, \boldsymbol{x}) = \sum_{n \in \mathcal{U}} \Phi_n(\lambda, \boldsymbol{v}, \boldsymbol{\mu}, \boldsymbol{\eta}, \boldsymbol{\theta}, \boldsymbol{x}) + \sum_{j \in \mathcal{S}} \eta_j \Lambda_j + \sum_{j \in \mathcal{S}} \sum_{k \in \mathcal{B}} \theta_{j,k} b_j^{(k)}, \tag{9.13}$$

where

$$\begin{aligned} \Phi_n(\lambda, \boldsymbol{v}, \boldsymbol{\mu}, \boldsymbol{\eta}, \boldsymbol{\theta}, \boldsymbol{x}) = {} & \lambda_n \left(\sum_{j \in \mathcal{S}} \sum_{k \in \mathcal{B}} x_n^{j,k} r_n^{j,k} - \bar{r}_n \right) + v_n \left(\sum_{j \in \mathcal{S}} \sum_{k \in \mathcal{B}} x_n^{j,k} R_j - \bar{r}_n \right) \\ & - \sum_{j \in \mathcal{S}} \sum_{k \in \mathcal{B}} \theta j, k x_n^{j,k} w_n^{j,k} + \mu_n \left(\sum_{j \in \mathcal{S}} \sum_{k \in \mathcal{B}} x_n^{j,k} (d_n^{j,k} + D_j) - \bar{d}_n \right) - \sum_{j \in \mathcal{S}} \eta_j \sum_{k \in \mathcal{B}} x_n^{j,k} r_n^{j,k}. \end{aligned} \tag{9.14}$$

As there is no cross-term of $\boldsymbol{x}$ in $L^3(\lambda, \boldsymbol{v}, \boldsymbol{\mu}, \boldsymbol{\eta}, \boldsymbol{\theta}, \boldsymbol{x})$, we can change the computation order as

$$h(\lambda, \boldsymbol{v}, \boldsymbol{\mu}, \boldsymbol{\eta}, \boldsymbol{\theta}) = \sum_{n \in \mathcal{U}} \inf_{\boldsymbol{x}} \Phi_n(\lambda, \boldsymbol{v}, \boldsymbol{\mu}, \boldsymbol{\eta}, \boldsymbol{\theta}, \boldsymbol{x}) + \sum_{j \in \mathcal{S}} \eta_j \Lambda_j + \sum_{j \in \mathcal{S}} \sum_{k \in \mathcal{B}} \theta_{j,k} b_j^{(k)}. \tag{9.15}$$

Hence, solving **P4** is equivalent to solving the following subproblem **P4(n)** for each UE n separately.

$$\begin{aligned} &\mathbf{P4(n):} \; h_n(\lambda, \boldsymbol{v}, \boldsymbol{\mu}, \boldsymbol{\eta}, \boldsymbol{\theta}) \triangleq \inf_{\boldsymbol{x}} \Phi_n(\lambda, \boldsymbol{v}, \boldsymbol{\mu}, \boldsymbol{\eta}, \boldsymbol{\theta}, \boldsymbol{x}) \\ &\text{s.t. Constraints (9.2-6) and (9.2-7)} \end{aligned} \tag{9.16}$$

Rewriting function $\Phi_n(\lambda, \boldsymbol{v}, \boldsymbol{\mu}, \boldsymbol{\eta}, \boldsymbol{\theta}, \boldsymbol{x})$, we have

$$\begin{aligned} \Phi_n(\lambda, \boldsymbol{v}, \boldsymbol{\mu}, \boldsymbol{\eta}, \boldsymbol{\theta}, \boldsymbol{x}) = {} & \sum_{j \in \mathcal{S}} \sum_{k \in \mathcal{B}} x_n^{j,k} [\lambda_n r_n^{j,k} + v_n R_j + \mu_n (d_n^{j,k} + D_j) - \eta_j r_n^{j,k} - \theta_{j,k} w_n^{j,k}] \\ & - (\lambda_n + v_n) \bar{r}_n - \mu_n \bar{d}_n. \end{aligned} \tag{9.17}$$

Since we want to find a binary solution of $x_n^{j,k}$ for fixed Lagrange multipliers $\boldsymbol{\lambda}, \boldsymbol{v}, \boldsymbol{\mu}, \boldsymbol{\eta}$, $\boldsymbol{\theta}$, solving **P4(n)** can be described as: for UE n, we choose an NS j^* and BS k^* from the admissible set of NSs and BSs respectively to maximize the value of $(\lambda_n - \eta_{j^*})r_n^{j^*,k^*} + v_n R_{j^*} + \mu_n(d_n^{j^*,k^*} + D_{j^*}) - \theta_{j^*,k^*} w_n^{j^*,k^*}$.

Then we design a QoS-AC policy based on the solution to **P2** to find the QoS-AUS in Definition 9.2. Let $\check{r}_n^*$ and $\check{d}_n^*$ be the optimal solution of UE n. If $\check{r}_n^* = 0$ and $\check{d}_n^* = 0$, it means that there is no QoS degradation of UE n. In other words, this UE can be admitted for the network. Hence, based on this observation, we design the QoS-AC policy, where the UEs with $\check{r}_n^* = 0$ and $\check{d}_n^* = 0$ can be accepted by the network, and others are rejected due to limited resources. Hence, the admissible set of UEs can be expressed by $\mathcal{A}_{Q-A} = \{n : \check{r}_n^* = 0, \check{d}_n^* = 0, n \in \mathcal{U}\}$.

Theorem 9.1 Subset $\mathcal{A}_{Q-A}$ is a QoS-AUS.

Proof: To prove Theorem 9.1, we should prove the feasibility and QoS performance of $\mathcal{A}_{Q-A}$ respectively.

(1) *feasibility*: Let $x_n^{j^*k^*}$ be the optimal solution $\boldsymbol{x}^*$ of UE n to **P2**. We denote $\boldsymbol{y}$ as an $|\mathcal{A}_{Q-A}|$-dimensional vector with elements y_n, where $y_n = x_n^{j^*,k^*}$ for $n \in \mathcal{A}_{Q-A}$. As $\boldsymbol{x}^*$ is the optimal solution to **P2**, it is also a feasible solution. Moreover, $\forall n \in \mathcal{A}_{Q-A}$, we have $\check{r}_n^* = 0, \check{d}_n^* = 0$. Hence, it is easy to verify that $\boldsymbol{y}$ can satisfy all constraints in **P1** when we use $\mathcal{A}_{Q-A}$ to replace $\mathcal{U}$. In other words, $\mathcal{A}_{Q-A}$ is an AUS.
(2) *QoS performance*: According to Definition 9.2, we need to prove that the achievable value of $\sum_{n\in\mathcal{U}\setminus\mathcal{A}_{Q-A}} \left(\frac{\check{r}_n}{\bar{r}_n} + \frac{\check{d}_n}{\bar{d}_n}\right)$ is no greater than that of any other AUS $\mathcal{H}$. Let us start from set $\mathcal{A}_{Q-A}$. According to the definition of $\boldsymbol{y}$, we have $\check{r}_n^* = 0, \check{d}_n^* = 0$ for all $n \in \mathcal{A}_{Q-A}$, and thus $\sum_{n\in\mathcal{U}\setminus\mathcal{A}_{Q-A}} \left(\frac{\check{r}_n^*}{\bar{r}_n} + \frac{\check{d}_n^*}{\bar{d}_n}\right) = \sum_{n\in\mathcal{U}} \left(\frac{\check{r}_n^*}{\bar{r}_n} + \frac{\check{d}_n^*}{\bar{d}_n}\right)$. For any other AUS $\mathcal{H}$, let $\check{\boldsymbol{r}}^{\mathcal{H}}$, $\check{\boldsymbol{d}}^{\mathcal{H}}$, and $\boldsymbol{x}^{\mathcal{H}}$ be the optimal solution of $\check{\boldsymbol{r}}$, $\check{\boldsymbol{d}}$, and $\boldsymbol{x}$, respectively, and thus the minimum achievable value is $\sum_{n\in\mathcal{U}} \left(\frac{\check{r}_n^{\mathcal{H}}}{\bar{r}_n} + \frac{\check{d}_n^{\mathcal{H}}}{\bar{d}_n}\right)$. As $\mathcal{H}$ is an AUS, we have $\check{r}_n^{\mathcal{H}} = 0, \check{d}_n^{\mathcal{H}} = 0$ for all $n \in \mathcal{H}$, and thus $\sum_{n\in\mathcal{U}\setminus\mathcal{H}} \left(\frac{\check{r}_n^{\mathcal{H}}}{\bar{r}_n} + \frac{\check{d}_n^{\mathcal{H}}}{\bar{d}_n}\right) = \sum_{n\in\mathcal{U}} \left(\frac{\check{r}_n^{\mathcal{H}}}{\bar{r}_n} + \frac{\check{d}_n^{\mathcal{H}}}{\bar{d}_n}\right)$. As $\{\check{\boldsymbol{r}}^*, \check{\boldsymbol{d}}^*, \boldsymbol{x}^*\}$ is the optimal solution to **P2**, $\sum_{n\in\mathcal{U}} \left(\frac{\check{r}_n^*}{\bar{r}_n} + \frac{\check{d}_n^*}{\bar{d}_n}\right) \le \sum_{n\in\mathcal{U}} \left(\frac{\check{r}_n^{\mathcal{H}}}{\bar{r}_n} + \frac{\check{d}_n^{\mathcal{H}}}{\bar{d}_n}\right)$, and thus $\sum_{n\in\mathcal{U}\setminus\mathcal{A}_{Q-A}} \left(\frac{\check{r}_n^*}{\bar{r}_n} + \frac{\check{d}_n^*}{\bar{d}_n}\right) \le \sum_{n\in\mathcal{U}\setminus\mathcal{H}} \left(\frac{\check{r}_n^{\mathcal{H}}}{\bar{r}_n} + \frac{\check{d}_n^{\mathcal{H}}}{\bar{d}_n}\right)$.

Therefore, according to the above proof of feasibility and QoS performance, we can conclude that subset $\mathcal{A}_{Q-A}$ is a QoS-AUS. □

According to Theorem 9.1, QoS-AC policy guarantees both QoS performance and feasibility. Moreover, this AC policy also guides network operators to re-allocate bandwidth in the NS reconfiguration phase thus to satisfy QoS of all the UEs in set $\mathcal{U}$ with the minimum bandwidth consumption. However, network slice reconfiguration is beyond the scope of this work. QoS-AC policy is focused on network QoS while the performance in terms of the number of admissible UEs cannot be guaranteed. In Section 9.1.3.2, we will design another AC scheme Num-AC to maximize the number of admissible UEs.

9.1.3.2 Num-AC Mechanism

In the proposed QoS-AC, we find that some UEs with only unsatisfied rate or delay requirement (i.e. $\check{r}_n^* > 0, \check{d}_n^* = 0$ or $\check{d}_n^* > 0, \check{r}_n^* = 0$) should be rejected. This means that some unviolated constraints are deleted in **P1**, implying that the network may have some spare resources to accept more UEs. Hence, from the number of admissible UEs viewpoint, the performance of QoS-AC policy may not be good. Moreover, the number of admissible UEs is also one of the key performance measures for UE admission control policy. Therefore, we propose Num-AC policy based on the solution to **P2** to further optimize the number of admissible UEs.

By analyzing the optimal solution to **P2**, we find that the smaller $\check{r}_n^*$ or $\check{d}_n^*$ is, the more likely the rate or delay of UE n can be satisfied. Based on this observation, we develop Num-AC policy to find the admissible UE subset, which is denoted by $\mathcal{A}_{N-A}$. The basic idea of this policy is trying to add the UEs with small value of $\check{r}_n^*$ and $\check{d}_n^*$ into set $\mathcal{A}_{N-A}$.

First of all, the UEs with $\check{r}_n^* = 0$ and $\check{d}_n^* = 0$ are definitely admissible to the network. Hence, $\mathcal{A}_{Q-A} \subseteq \mathcal{A}_{N-A}$. We then try to find more admissible UEs from set $\mathcal{U} \backslash \mathcal{A}_{Q-A}$, and add these UEs into $\mathcal{A}_{N-A}$. The details of Num-AC policy are summarized as Algorithm 9.1.

Algorithm 9.1: Algorithm of Num-AC policy.

Require: problem **P2** formulated in (2).
Ensure: set of admissible UEs $\mathcal{A}_{N-A}$.

Initialization Stage:

1: $\mathcal{A}_{N-A} = \emptyset, \mathcal{A}_{temp} = \emptyset$
2: obtain the optimal solution $\check{\boldsymbol{r}}^*$, $\check{\boldsymbol{d}}^*$ and $\boldsymbol{x}^*$ by solving **P2**
3: add all UEs with $\check{r}_n^* = 0$ and $\check{d}_n^* = 0$ into $\mathcal{A}_{N-A}$

Search Stage:

4: find UE i: $\min_{i \in \mathcal{U} \backslash \mathcal{A}_{N-A}} \left(\frac{\check{r}_i^*}{\bar{r}_i} + \frac{\check{d}_i^*}{\bar{d}_i} \right)$
5: $\mathcal{A}_{temp} = \{\mathcal{A}_{N-A} \cup \text{UE } i\}$
6: obtain the optimal solution $\check{\boldsymbol{r}}^*$, $\check{\boldsymbol{d}}^*$ and $\boldsymbol{x}^*$ by solving **P2** with respect to $\mathcal{A}_{temp}$
7: **if** $\sum_{n \in \mathcal{A}_{temp}} \left(\frac{\check{r}_n^*}{\bar{r}_n} + \frac{\check{d}_n^*}{\bar{d}_n} \right) = 0$ **then**
8: $\quad \mathcal{A}_{N-A} = \mathcal{A}_{temp}$
9: $\quad$ Go back to line 4
10: **else**
11: $\quad$ break
12: **end if**
13: **output** $\mathcal{A}_{N-A}$

In the initialization stage, we add the definitely admissible UEs (i.e. the UEs with $\check{r}_n^* = 0$ and $\check{d}_n^* = 0$). Then in the search stage, we check the feasibility of other UEs one by one. The smaller the value of $\left(\frac{\check{r}_n^*}{\bar{r}_n} + \frac{\check{d}_n^*}{\bar{d}_n} \right)$ is, the more likely the UE is admissible. Hence, we check the UEs with the smallest value of $\left(\frac{\check{r}_n^*}{\bar{r}_n} + \frac{\check{d}_n^*}{\bar{d}_n} \right)$ first. To reduce the computational complexity, once a UE is unfeasible for the network, we terminate the check, and then obtain the set $\mathcal{A}_{N-A}$. Therefore, this policy needs to solve **P2** at most $|\mathcal{U}|$ times in the worst case.

9.1.4 Experiments, Results, and Discussions

In this section, we evaluate the performance of the proposed resource management framework. We compare with two access control schemes: (i) NS prior association (NSA) and (ii) BS prior association (BSA) based on traditional maximum SINR scheme [12]. Owing to the introduction of NS, some necessary modifications should be done in NSA and BSA. In detail, for a specific UE, the NSA scheme first finds the NS that satisfies the QoS requirement of the UE, and then finds the BS covered by this NS with sufficient bandwidth. BSA scheme first finds the BS with the maximum SINR for the UE, and then finds the NS deployed in this BS with the QoS guarantee satisfied. In both NSA and BSA schemes, if such a pair of NS and BS is found, the UE is admissible and associated with the NS and BS.

We consider a network that consists of a macro BS (MBS) located at the center of a circular area with a radius of 500 m and varying number of pico BSs (PBSs), femto BSs (FBSs), NSs, and UEs. The transmit power of MBS (PBS) and FBS is set to 46, 30, and 20 dBm, respectively. We use $L(d) = 34 + 40\log(d)$ and $L(d) = 37 + 30\log(d)$ to model the pass loss for the MBS, PBSs, and FBSs, respectively [13]. All the BSs share 20 MHz bandwidth. Each NS randomly covers 4 BSs, and provides different rate and delay performance. UEs are randomly distributed in this area with different rate and delay requirements.

In the first experiment, we compare the number of admissible UEs of the four AC policies QoS-AC, Num-AC, NSA and BSA. In this experiment, we fix the number of NSs and BSs as 20 and 21 (including one MBS) respectively. Figure 9.3 shows the number of admissible UEs for the four AC policies with different UE densities. From Figure 9.3, we can see that the number of admissible UEs of QoS-AC and Num-AC is always higher than that of the other two traditional policies that do not consider the characteristics of NS. Specifically, when the number of UEs is 200, the admissible number of UEs for Num-AC, QoS-AC, BSA, and NSA is 173, 142, 118, and 92, respectively. These results show that the proposed

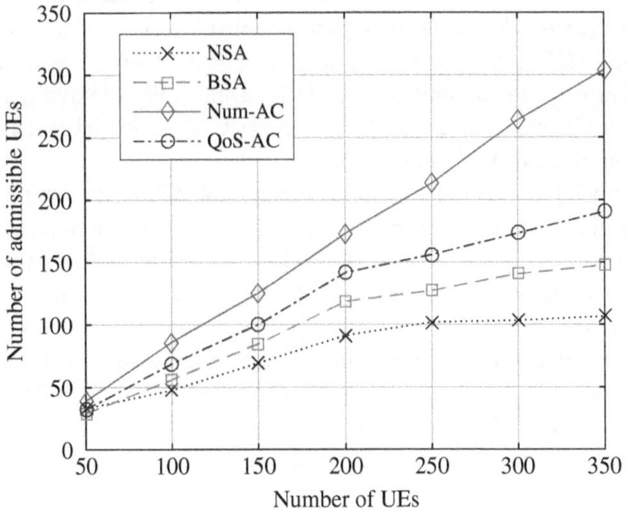

Figure 9.3 Comparisons of the number of admissible UEs with different UE density. (The number of NSs is 20, and the number of BSs is 21.)

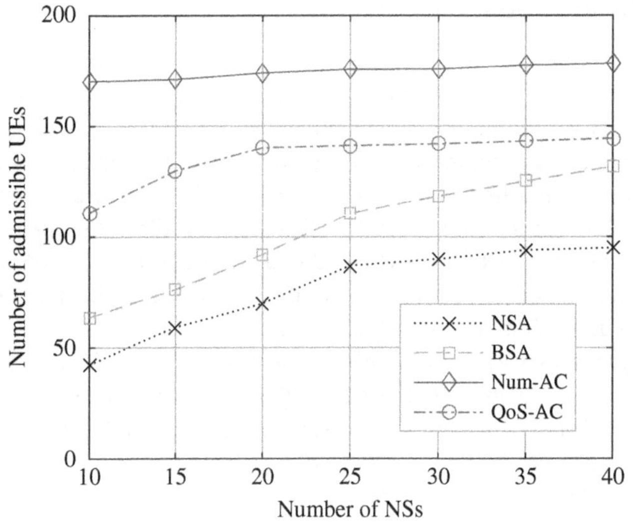

Figure 9.4 Comparisons of the number of admissible UEs vs. number of NSs. (The number of UEs 200, and the number of BSs 21.)

Num-AC policy can serve 47% and 88% more UEs when compared with NSA and BSA respectively.

In the second experiment, we evaluate the number of admissible UEs of the four AC policies for varying number of NSs while using fixed number of UEs 200. Figure 9.4 shows the number of admissible UEs for the four policies as a function of the number of NSs. From Figure 9.4, we can see that the number of admissible UEs of all four policies increases with the number of NSs. This is because the more the number of NSs deployed the more the association choices for UEs, and thus the more UEs can be admitted. Specifically, the number of admissible UEs of QoS-AC, NSA, and BSA increases rapidly under the low NS density, and when the number of NSs is larger than 25, the number of admissible UEs of all the four policies increases slowly due to the limited bandwidth resource. We also find that the number of admissible UEs of Num-AC and QoS-AC is always significantly higher than that of NSA and BSA under all NS density circumstances.

We next examine the relationship between the number of admissible UEs and the number of BSs with the same parameters as those in the first experiment, and fix the number of UEs to 200. Figure 9.5 shows the number of admissible UEs for the four policies as a function of the number of BSs. From this figure, we can see that the number of admissible UEs of Num-AC, QoS-AC, and NSA monotonically increases with the number of BSs while that of BSA decreases, which is due to the decreasing number of NSs deployed in each BS. Moreover, the number of admissible UEs of Num-AC and QoS-AC is always much higher than that of the other two traditional AC policies. These results clearly demonstrate the performance gain of the proposed Num-AC and QoS-AC policies in terms of the number of admissible UEs.

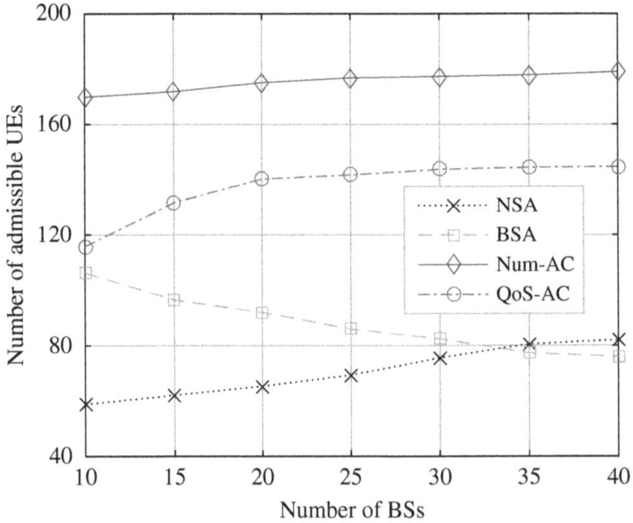

Figure 9.5 Comparisons of the number of admissible UEs vs. number of BSs. (The number of UEs is 200, and the number of NSs 20.)

9.2 Smart Handoff Policy Design for RAN Slicing

Owing to the introduction of network slicing into radio access networks forming a UE-BS-NS three-layer association, handoff becomes very complicated and cannot be resolved by conventional policies. In this section, we propose a multi-agent reinforcement LEarning based Smart handoff policy, named LESS, to reduce handoff cost while maintaining user QoS requirements in RAN slicing. LESS is based on a modified distributed Q-learning algorithm with small action space to make handoff decisions. Numerical results show that in typical scenarios, LESS can significantly reduce the handoff cost when compared with traditional handoff policies without learning.

9.2.1 RAN Slice Based Mobile Network Model

We consider an NS-based mobile network architecture shown in Figure 9.6, which consists of multiple end-to-end NSs, BSs, as well as UEs. These NSs share the physical resources in both core networks and RAN. Each NS has different NF modules, such as connection management, mobility management, security, etc., to provide a specific service for UEs. The detailed descriptions of the network architecture can be found in [7]. Here we focus on RAN slicing from the mobility management perspective.

We consider a multi-BS and multi-NS RAN model shown in Figure 9.6. Let $\mathcal{B}$, $\mathcal{N}$, and $\mathcal{U}$ be the set of BSs, NSs, and UEs, respectively. We assume that UEs in the system move at a random speed and in a random direction. Similar to that in [14], we use two parameters to describe QoS requirements: minimum threshold of transmission rate $\gamma_i^{\min}$ and endurable

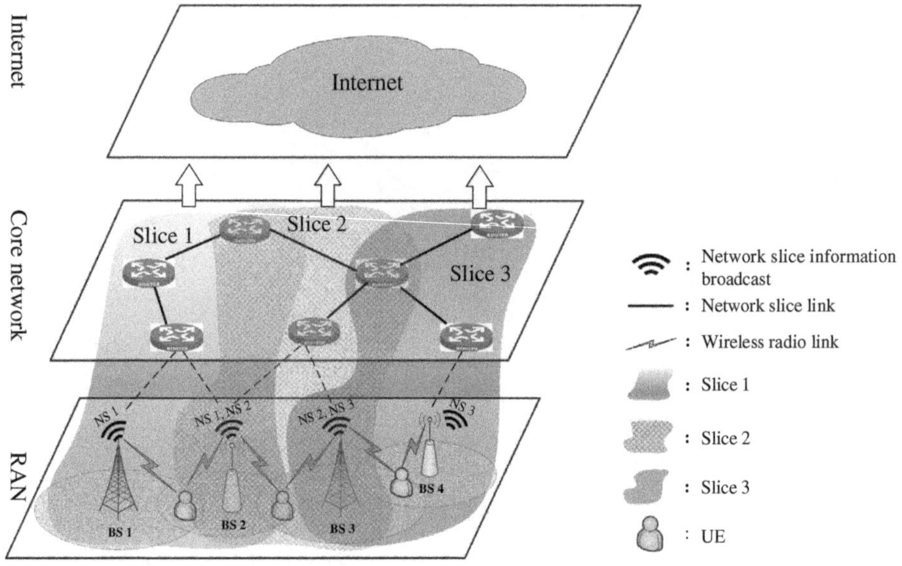

Figure 9.6 NS-based mobile network architecture.

time τ_i, which is the maximum time a UE is allowed to have a transmission rate lower than the minimum threshold. Let $\mathcal{T} = \{T_1, T_2, \dots, T_L\}$ be the set of all service types, and $\psi_i \in \mathcal{T}$ be the service type of UE i. We say $\psi_i = T_n$ when both $\gamma_i^{\min}$ and τ_i can fulfill the requirement of the service type T_n.

We identify a specific NS, say NS j, by the two elements $(\mathcal{T}_j, \mathbf{B}_j)$, where $\mathcal{T}_j$ is the set of service types that NS j can provide, and $\mathbf{B}_j$ is a vector denoting the bandwidth allocation of NS j from all BSs. Let $\bar{b}_j^{(k)}$ be the k-th element of vector $\mathbf{B}_j$ denoting the bandwidth of NS j allocated by BS k. $\bar{b}_j^{(k)} = 0$ when BS k is not in the coverage of NS j. UEs can access the NS via only the covered BSs. In the example of Figure 9.6, UEs can access NS 1 via only BSs 1 and 2.

We describe the handoff model from two aspects: handoff trigger condition and handoff cost. Handoff should occur once the QoS of the UE cannot be satisfied [14, 15]. Based on the definition of QoS, the handoff trigger condition for UE n can be written as

$$\forall t_0 \in [t - \tau_n, t], r_n(t_0) < \gamma_n^{\min}, \tag{9.18}$$

where $r_n(t_0)$ is the achievable transmission rate of UE n at time t_0. This condition states that UE n cannot achieve the minimum rate requirement $\gamma_n^{\min}$ in the last τ_n time.

Once the handoff trigger condition is met, UEs need to select suitable target BSs and NSs. As aforementioned, each type of handoff corresponds to a specific handoff cost. Here we define four handoff costs generated by the four handoff types: (i) C_{NS}, switch NS only; (ii) C_{BS}, switch BS only; (iii) $C_{\text{NS-BS}}$, switch both NS and BS; (iv) C_{New}, deploy a new NS (this type can be seen as a special handoff in RAN slicing); with the relationship $C_{\text{NS}} < C_{\text{BS}} < C_{\text{NS-BS}} < C_{\text{New}}$. Based on this, we design a handoff mechanism to minimize the overall handoff cost through target BS and NS selections while guaranteeing the QoS of UEs.

9.2.2 Multi-Agent Reinforcement Learning Based Handoff Framework

In this section, we first formulate the handoff decision problem as a multi-agent reinforcement learning (RL) model, and then propose an intelligent handoff mechanism based on the learning model, LESS.

Once the handoff trigger condition for a UE is met, it should choose an appropriate serving NS and BS in order to maintain the desired QoS. We model this target BS and NS selection problem as a multi-agent RL consisting of four main elements: agents, states, actions, and reward. In detail, each UE is an agent to make handoff decisions. The states are defined as the available bandwidth levels of NSs. Let $s_j^k(t)$ denote the available bandwidth level of NS j via BS k at time t after discretization. The environment state can be written as $S(t) = (s_j^k(t))_{(|\mathcal{B}||\mathcal{N}|)\times 1}$ at time t.

An action means selecting both target BS and NS when a handoff occurs. In detail, we denote by $\boldsymbol{a}_i(t) = (x_i(t), y_i(t))$ the action taken by UE i at time t, where $x_i(t)$ and $y_i(t)$ is the target BS and NS respectively. If $y_i(t) \notin \mathcal{N}$, the action denotes deploying a new NS. Let $\mathcal{A}$ be the action space for a UE, and thus the system action space for all UEs is $\mathcal{A}^{|\mathcal{U}|}$. The reward denoted by $r_i(S(t), \boldsymbol{a}_i(t))$ is the handoff cost for UE i in state $S(t) \in \mathcal{S}$ with action $\boldsymbol{a}_i(t) \in \mathcal{A}$ at time t, which can be expressed as

$$r_i(S(t), \boldsymbol{a}_i(t)) = \begin{cases} C_{\text{NS}}, & \text{if } x_i(t) = x_i(t-1),\ y_i(t) \neq y_i(t-1), \\ C_{\text{BS}}, & \text{if } x_i(t) \neq x_i(t-1),\ y_i(t) = y_i(t-1), \\ C_{\text{NS-BS}}, & \text{if } x_i(t) \neq x_i(t-1),\ y_i(t) \neq y_i(t-1), \\ C_{\text{New}}, & \text{if } y_i(t) \notin \mathcal{N}. \end{cases} \tag{9.19}$$

Our objective is to minimize the total long-term handoff cost $\sum_{t=1}^{\infty} \sum_{i=1}^{|\mathcal{U}|} r_i(S(t), \boldsymbol{a}_i(t))$ by designing an intelligent handoff mechanism. Traditional Q-learning algorithm [16] is widely used to get the optimal solution to RL problems. However, considering that the system action space $|\mathcal{A}^{|\mathcal{U}|}|$ is very large, Q-learning algorithm needs a long time to converge. To address this issue, we propose the distributed learning based LESS handoff mechanism in the following.

9.2.3 LESS Algorithm for Target BS and NS Selection

Q-learning is a simple yet effective algorithm for solving RL problems, and it can be briefly described as follows. Denote by vector $\mathbf{A} = [\boldsymbol{a}_1(t), \ldots, \boldsymbol{a}_{|\mathcal{U}|}(t)] \in \mathcal{A}^{|\mathcal{U}|}$ the actions for all UEs. $Q_t(S, \mathbf{A})$ and $r(S, \mathbf{A})$ are respectively the Q-value and reward for the state–action pair $(S, \mathbf{A}) \in \mathcal{S} \times \mathcal{A}^{|\mathcal{U}|}$, where $r(S, \mathbf{A}) = \sum_{i=1}^{|\mathcal{U}|} r_i(S(t), \boldsymbol{a}_i(t))$. The update rule of Q-value can be expressed as

$$\begin{aligned} &Q_0(S, \mathbf{A}) = M, \text{for all } \mathbf{A} \in \mathcal{A}^{|\mathcal{U}|} \text{ and } S \in \mathcal{S}, \\ &Q_{t+1}(S, \mathbf{A}) = \\ &\begin{cases} Q_t(S, \mathbf{A}), & \text{if } \mathbf{A}(t) \neq \mathbf{A} \text{ or } S(t) \neq S, \\ r(S, \mathbf{A}) + \beta \min\limits_{\mathbf{A}' \in \mathcal{A}^{|\mathcal{U}|}} Q_t(S(t+1), \mathbf{A}'), & \text{otherwise}, \end{cases} \end{aligned} \tag{9.20}$$

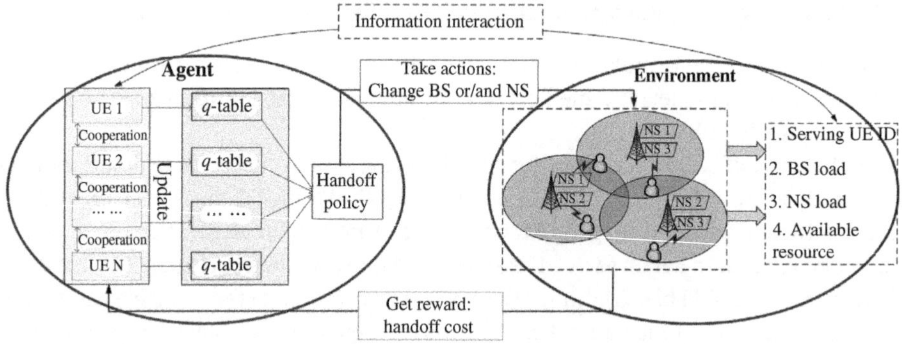

Figure 9.7 The framework of LESS-DL.

where $S(t)$ and $\boldsymbol{A}(t)$ are respectively the state and the action vector at time t, M is a large constant for initialization, and $\beta(0 < \beta < 1)$ is the discount factor. The target BS and NS selection policy is to choose the NS–BS pair with the smallest Q-value with respect to ϵ-greedy policy.

However, applying traditional Q-learning to solving our problem requires a large action space (i.e. $|\mathcal{A}^{|\mathcal{U}|}|$), which takes a long time to converge. Moreover, it requires all UEs to make handoff decisions simultaneously, which is unrealistic. Thus, to overcome these issues, we develop a distributed learning algorithm, LESS-DL, shown in Figure 9.7 to select target BS and NS for each individual UE. The main idea of LESS-DL is that each UE only needs to maintain a reduced Q-table where the action space is composed of its own actions without distinguishing the actions from other UEs.

9.2.3.1 *q*-Value Update Policy

Denote by q^i-table the reduced Q-table maintained by UE i, and $q_t^{(i)}(S, \boldsymbol{a}_i)$ the Q-value of UE i at time t with the state–action pair $(S, \boldsymbol{a}_i)$. For convenience, we use q-value and Q-value to denote the value in the reduced and original Q-table respectively. Using a similar idea of [17], $q_t^{(i)}(S, \boldsymbol{a}_i)$ can be updated as

$$\begin{aligned} & q_0^{(i)}(S, \boldsymbol{a}_i) = M, \text{for all } \boldsymbol{a}_i \in \mathcal{A} \text{and } S \in \mathcal{S}, \\ & q_{t+1}^{(i)}(S, \boldsymbol{a}_i) = \\ & \begin{cases} q_t^{(i)}(S, \boldsymbol{a}_i), \text{ if } \boldsymbol{a}_i(t) \neq \boldsymbol{a}_i \text{ or } S(t) \neq S, \\ \min\{q_t^{(i)}(S, \boldsymbol{a}_i), r_i(S, \boldsymbol{a}_i) + \beta \min\limits_{\boldsymbol{a}' \in \mathcal{A}} q_t(\boldsymbol{a}', \delta(S, \mathbf{A}))\}, \\ \text{otherwise.} \end{cases} \end{aligned} \tag{9.21}$$

By using this update method, we can get the reduced q-tables for all UEs. Although the reduced q-table of all UEs cannot construct the original big Q-table, the following proposition gives a very good property of the reduced q-table, which makes it possible to take actions in a distributed manner.

Proposition 9.1 *The value of $q_t^{(i)}(S, \boldsymbol{a}_i)$ in the reduced q^i-table is the minimum value in the original Q-table defined in (9.20) when the action of UE i is $\boldsymbol{a}_i$, i.e.*

$$q_t^{(i)}(S, \boldsymbol{a}_i) = \min_{\mathbf{A} \in \mathcal{A}^{|\mathcal{U}|} \boldsymbol{a}^{(i)} = \boldsymbol{a}_i} Q_t(S, \mathbf{A}) \tag{9.22}$$

where $\boldsymbol{a}^{(i)}$ denotes the i-th element in action vector $\mathbf{A}$.

Proof: Similar to that in [17], we can easily obtain our Proposition 9.1 by replacing all the notation "max" to "min." □

Proposition 9.1 states that by using the q-value update method in (9.21), we can obtain the minimum value $Q_t(S, \mathbf{A})$, which makes it possible to design an optimal NS-BS selection policy for UEs in a distributed manner. In the following, we illustrate how to use the q-value to obtain the optimal target BS-NS selection policy.

9.2.3.2 Optimal Action Policy

For traditional Q-learning, we know that once we get a proper converged Q-table, the policy that we choose the action with the smallest Q-value can guarantee the optimality [16]. However, for the reduced q^i-table, if we choose the action with the smallest value $q_t^{(i)}(S, \boldsymbol{a}_i)$ for each UE, it cannot guarantee the optimal policy. In other words, choosing action $\boldsymbol{a}_i{}^*$ with the smallest value $q_t^{(i)}(\boldsymbol{a}_i{}^*, S)$ for each UE may not provide the optimal action-vector $\mathbf{A}^*$ of all UEs with the smallest Q-value $Q_t(\mathbf{A}^*, S)$ [17], i.e. we cannot guarantee that

$$[\boldsymbol{a}_1{}^*, \boldsymbol{a}_2{}^*, \dots, \boldsymbol{a}_{|\mathcal{U}|}{}^*] = \mathbf{A}^*. \tag{9.23}$$

To overcome this issue, we use the following policy to choose actions. The main idea is to update and store an action policy in parallel with $q_t^{(i)}(S, \boldsymbol{a}_i)$ update. Once the value of $\min q_t^{(i)}(S, \boldsymbol{a}_i)$ decreases, we update our action policy as we can find a better action, and then we store the better one as the current optimal action. When the q-table converges, implying that the value of $\min q_t^{(i)}(S, \boldsymbol{a}_i)$ stays unchanged, our action policy is stable, and the current stored policy is optimal. The update rule of stored action policy $\pi_t^{(i)}(S)$ of UE i is stated as

$$\begin{aligned}
&\pi_0^{(i)}(S) \in \mathcal{A}, \text{ arbitrarily,}\\
&\pi_{t+1}^{(i)}(S) =\\
&\begin{cases}
\pi_t^{(i)}(S), \text{if } S \neq S_t \text{ or } \min\limits_{\boldsymbol{a}_i \in \mathcal{A}} q_t^{(i)}(S, \boldsymbol{a}_i) = \min\limits_{\boldsymbol{a}_i \in \mathcal{A}} q_{t+1}^{(i)}(S, \boldsymbol{a}_i),\\
\boldsymbol{a}_i(t), \text{ otherwise,}
\end{cases}
\end{aligned} \tag{9.24}$$

where $\boldsymbol{a}_i(t)$ is the action of UE i at time t.

From Ref. [17] we can get the corollary that for an arbitrary state S, we have

$$[\pi_t^{(1)}(S), \pi_t^{(1)}(S), \dots, \pi_t^{(|\mathcal{U}|)}(S)] = \arg \min_{\mathbf{A} \in \mathcal{A}^{|\mathcal{U}|}} Q_t(S, \mathbf{A}). \tag{9.25}$$

Thus, when we get the converged q-table, choosing the current stored action $\pi_t^{(i)}(S)$ for each individual UE can guarantee the minimum handoff cost.

In general, based on ϵ-greedy policy LESS-DL can be described as: before q-value is converged, we choose the current stored policy $\pi_t^{(i)}(S)$ as the target NS–BS pair for each UE respectively with probability $p = (1 - \epsilon)$, and choose other pairs randomly with probability $p = \epsilon$. Then, we update the q-values in a distributed manner according to (9.21) by using the obtained handoff cost. Finally, we update the current stored action policy based on (9.24) for the next handoff decision. Once we get the converged q-tables, we always choose the current stored action as the target NS–BS pair for each individual UE.

9.2.4 Experiment, Results, and Discussions

In this section, we compare the performance of LESS with three other handoff mechanisms: Max-SINR, NS-Prior, and LESS-DL. In detail, Max-SINR first selects the BS with the maximum SINR for UEs [12], and then finds the NS deployed in this BS with satisfied QoS provisioning. If such a BS–NS pair is found, select them as the target; otherwise, deploy a new NS that satisfies the UE's QoS. NS-Prior mechanism first selects the NS that satisfies the QoS requirement of UEs, and then finds the BS covered by this NS with sufficient bandwidth. Lastly, by comparing with LESS-DL, we can verify the effectiveness of LESS-DS data sharing policy. The handoff trigger condition is the same for all the four mechanisms in (9.18).

We consider a network that consists of an MBS located at the center of a circular area with a radius of 1000 m and varying number of PBSs, FBSs, and UEs. The number of NSs deployed is 40. Each NS covers 8 BSs randomly, and provides different rate and delay (in term of τ_n in our model) performance. The transmit power of MBS, PBS, and FBS is set to

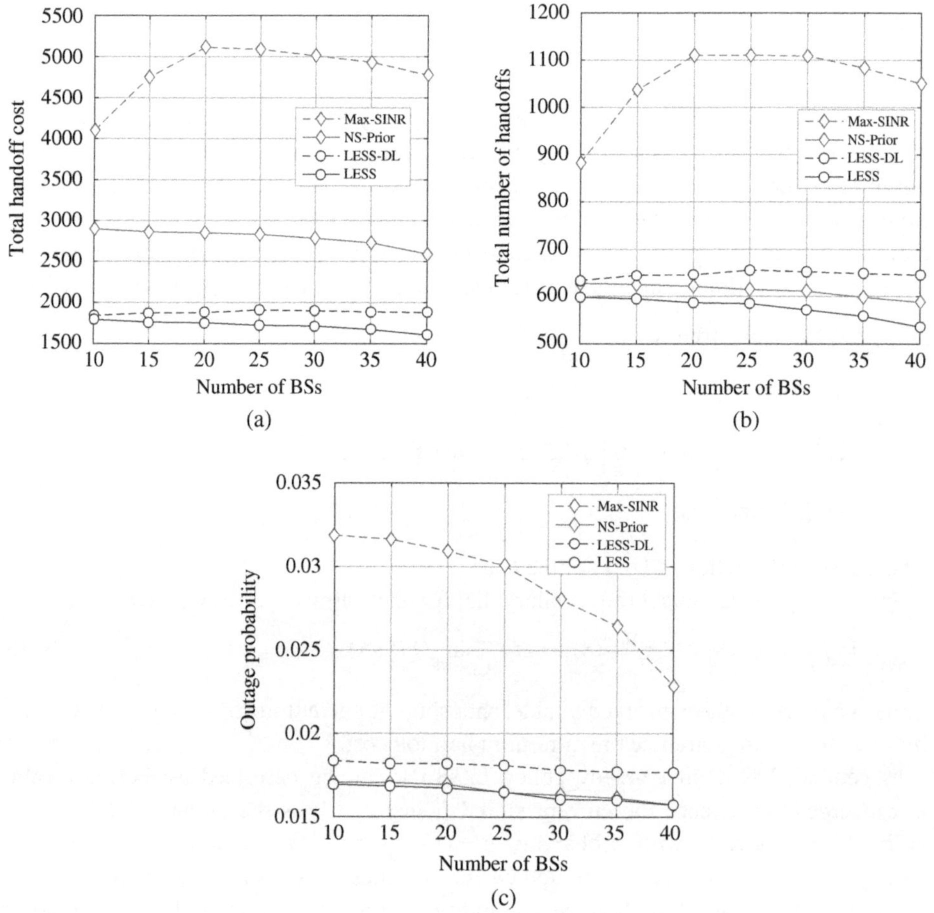

Figure 9.8 Comparisons of handoff performance for the four handoff mechanisms with different number of BSs. (a) Handoff cost, (b) number of handoffs, and (c) outage probability of UEs.

46, 30, and 20 dBm, respectively [13]. All the BSs share a 20 MHz bandwidth, and allocate them to the deployed NSs based on the NS QoS provisioning. UEs are randomly distributed in the area with different rate and delay requirements. In the following, we examine the performance of the proposed LESS handoff mechanism.

In the first experiment, we compare the handoff cost, the number of handoffs, and the UE outage probability for the four handoff mechanisms when the number of BSs varies from 10 to 40 as shown in Figure 9.9. As expected, we find that the handoff cost of the two learning based mechanisms LESS and LESS-DL is much lower than that of the other two traditional mechanisms shown in Figure 9.9a. In particular, when compared with Max-SINR, NS-Prior, and LESS-DL, the handoff cost gain of LESS is about 51%, 40%, and 10%, respectively when the number of BS is 25. From Figure 9.9b, we can see that LESS achieves the smallest number of handoffs, while the number of handoffs in LESS-DL is higher than that in NS-Prior due to the lack of data gathered from the LESS-DS policy. Finally, in terms of UE outage probability, Figure 9.9c shows that NS-Prior achieves a performance as good as LESS due to the prior consideration of NS service provisioning.

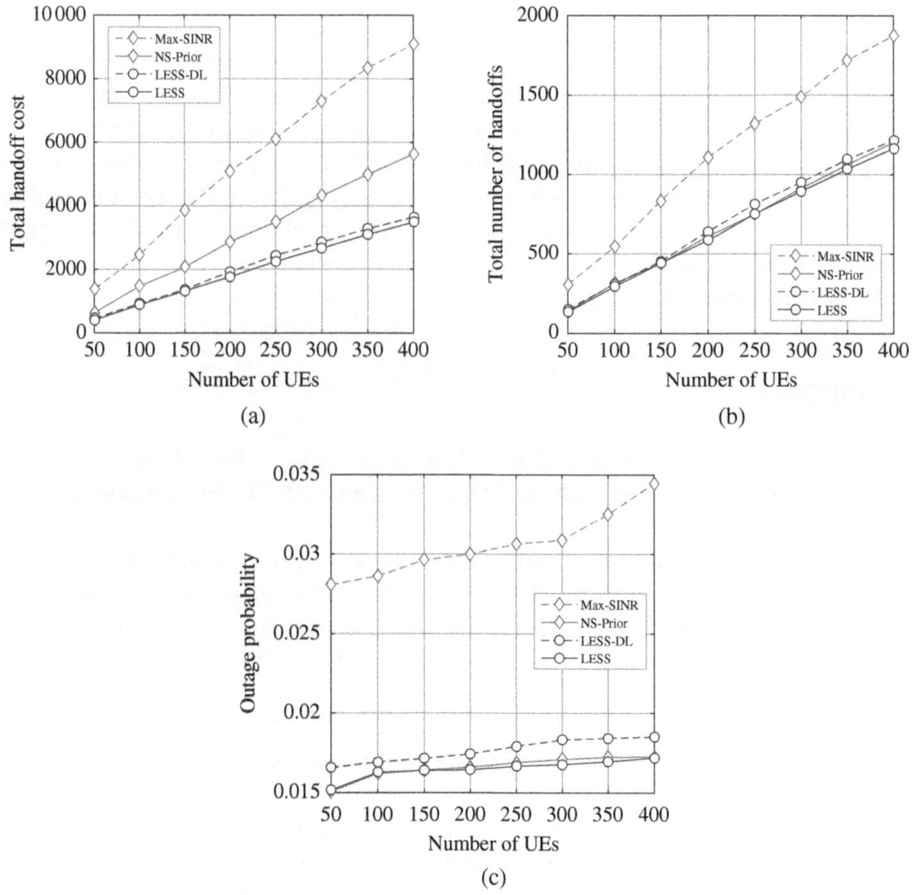

Figure 9.9 Comparisons of handoff performance for the four handoff mechanisms with different number of UEs. (a) Handoff cost, (b) number of handoffs, and (c) outage probability of UEs.

Next, we compare the handoff performance for the four handoff mechanisms with different number of UEs, as shown in Figure 9.8. Figure 9.8a shows that the learning based mechanisms LESS and LESS-DL significantly outperform the other two in terms of handoff cost. As the learning objective is handoff cost, the performance in terms of the number of handoffs and UE outage probability of LESS is very close to that of NS-Prior mechanism, which considers NS service type when making handoff decisions. This can be concluded from Figure 9.8b,c.

9.3 Summary

In this chapter, we discuss the access control and handoff policy design for RAN slicing. First, we propose a unified framework for access control in RAN slicing. Owing to resource limitation, not all users' QoS requirements can be satisfied simultaneously. Hence, we design two user access control policies to select admissible users for optimizing the QoS and the number of admissible users respectively. Numerical results show that in typical scenarios, our proposed access control policies can significantly outperform traditional policies in terms of the number of admissible users. Next, we investigate the handoff issue for mobile users in RAN slicing. We propose a multi-agent reinforcement LEarning based Smart handoff policy, named LESS, to reduce handoff cost while maintaining user QoS requirements in RAN slicing. LESS is based on a modified distributed Q-learning algorithm with small action space to make handoff decisions. Numerical results show that in typical scenarios, LESS can significantly reduce the handoff cost when compared with traditional handoff policies without learning.

Bibliography

1 Zhang, H., Liu, N., Chu, X. et al. (2017). Network slicing based 5G and future mobile networks: mobility, resource management, and challenges. *IEEE Communications Magazine* 55 (8): 138–145.

2 Zhang, L., Ijaz, A., Mao, J. et al. (2017). Multi-service signal multiplexing and isolation for physical-layer network slicing (PNS). *VTC2017-Fall Proceedings, Toronto, Canada.*

3 Zhang, L., Ijaz, A., Xiao, P. et al. (2017). Subband filtered multi-carrier systems for multi-service wireless communications. *IEEE Transactions on Wireless Communications* 16 (3): 1893–1907.

4 Ijaz, A., Zhang, L., Grau, M. et al. (2016). Enabling massive IoT in 5G and beyond systems: PHY radio frame design considerations. *IEEE Access* 4: 3322–3339.

5 Sun, Y., Feng, G., Zhang, L. et al. (2019). Distributed learning based handoff mechanism for radio access network slicing with data sharing. *ICC 2019–2019 IEEE International Conference on Communications (ICC)*,Shanghai, China, IEEE, pp. 1–6.

6 Sun, Y., Feng, G., Zhang, L. et al. (2019). User access control and bandwidth allocation for slice-based 5G-and-beyond radio access networks. *ICC 2019–2019 IEEE International Conference on Communications (ICC)*, IEEE, pp. 1–6.

7 An, X., Zhou, C., Trivisonno, R. et al. (2016). On end to end network slicing for 5G communication systems. *Transactions on Emerging Telecommunications Technologies* 28 (4): e3058.

8 Clark, D.D. and Fang, W. (1998). Explicit allocation of best-effort packet delivery service. *IEEE/ACM Transactions on Networking* 6 (4): 362–373.

9 Liang, C. and Yu, F.R. (2015). Wireless network virtualization: a survey, some research issues and challenges. *IEEE Communications Surveys & Tutorials* 17 (1): 358–380.

10 Nakao, A., Du, P., Yaoshiaki, K. et al. (2017). End-to-end network slicing for 5G mobile networks. *Journal of Information Processing* 25: 153–163.

11 Boyd, S.P. and Vandenberghe, L. (2004). *Convex Optimization.* Cambridge, UK: Cambridge university press.

12 3GPP TS 36.331 (2016). E-UTRA Radio Resource Control (RRC); Protocol specification (Release 9).

13 Ye, Q., Rong, B., Chen, Y. et al. (2013). User association for load balancing in heterogeneous cellular networks. *IEEE Transactions on Wireless Communications* 12 (6): 2706–2716.

14 Sun, Y., Feng, G., Qin, S., and Sun, S. (2018). Cell association with user behavior awareness in heterogeneous cellular networks. *IEEE Transactions on Vehicular Technology* 67 (5): 4589–4601.

15 Sun, Y., Feng, G., Qin, S. et al. (2017). The smart handoff policy for millimeter wave heterogeneous cellular networks. *IEEE Transactions on Mobile Computing* 17 (6): 1456–1468.

16 Watkins, C.J.C.H. and Dayan, P. (1992). Q-learning. *Machine Learning* 8 (3–4): 279–292.

17 Lauer, M. and Riedmiller, M. (2000). An algorithm for distributed reinforcement learning in cooperative multi-agent systems. *Proceedings of the 17th International Conference on Machine Learning, Orlando, FL, USA*, pp. 535–542.

10

Robust RAN Slicing

Ruihan Wen[1,2] and Gang Feng[2]

[1] *Southwest Minzu University, School of Computer Science and Technology, Chengdu, China*
[2] *University of Electronic Science and Technology of China, Chengdu, China*

10.1 Introduction

The next generation mobile network (5G) is envisioned to support a wide range of use cases including extreme mobile broadband (eMBB), massive machine-type communications (mMTC), and ultra-reliable low-latency communications (URLLC), where eMBB provides both extremely high throughputs and low-latency communications, mMTC provides wireless connectivity for dozens of billions of network-enabled devices, and URLLC provides ultra-reliable low-latency and/or resilient communication links for network services with extreme requirements on availability, latency, and reliability [1, 2]. To address diverse application demands, network slicing has been viewed as one of the most promising architectural technologies for the forthcoming 5G era [3, 4]. A network slice is a logical (virtualized) end-to-end network provided to third parties in an on-demand manner to realize a use case or service model. A network slice is configured and lifecycle managed through optimal network functions placement and chaining over the infrastructure network, which encompasses user devices, radio access network (RAN), transport network, and core network (CN) [4].

Software defined networking (SDN) and network function virtualization (NFV) will enable the launch of network slices to meet the requirements of various 5G use cases in a more cost efficient way [5]. There are already prototype implementations of RAN slicing in wireless mobile networks, which enable the placement of virtual CN functions including virtual PDN gateway, serving gateway, mobility management entity [6–8], and RAN network functions including virtual base station (vBS) and base band units (BBUs) [9–11] in the framework of SDN. However, the reliability problems in SDN/NFV supported RAN slices have been poorly addressed [12]. We assume that a slice is constituted by a set of VNFs and virtual links. Leveraging NFV technology, multiple types of VNFs (e.g. vBSs and vSGWs) of a slice are run as instances on servers in data centers or network nodes [13, 14], in which failures may occur occasionally due to servers being powered down for maintenance, software faults (e.g. firmware errors), or misconfiguration of servers [15–17]. Therefore, it is imperative to develop effective mechanisms for link/node monitoring and

Radio Access Network Slicing and Virtualization for 5G Vertical Industries, First Edition.
Edited by Lei Zhang, Arman Farhang, Gang Feng, and Oluwakayode Onireti.

slice recovery to relieve RAN slices from substrate link and node failures, especially for the use cases of ultra-reliable machine-type communications [5].

Besides link and network function failures in SDN/NFV supported RAN slices, which will trigger a RAN slice recovery process, the uncertain traffic demands of users in a RAN slice may also trigger the reconfiguration of a RAN slice, as the link resources allocated to a RAN slice are assumed to be deterministic parameters and shared among the users within the same RAN slice. However, variation in the traffic demands of these users results in uncertain (stochastic) aggregated demands of the link resources of a RAN slice, which may in turn violate the assumption that the link resources are deterministic parameters.

The aforementioned observations inspire us to investigate effective robust RAN slicing mechanisms to address the RAN slice recovery issue against link and network function failures and the reconfiguration issue against the uncertain traffic demands in a unified framework. In this chapter, our contributions are three-fold. We first formulate a recovery problem as a baseline, to solve remapping of a slice with deterministic demands, where remapping is a process of re-selecting VNFs and links to accommodate the failed ones. Based on that, we propose two robust RAN slicing algorithms for slice recovery and reconfiguration under stochastic demands: the first algorithm is an optimal one with slow convergence, which is developed by leveraging the robust optimization theory by transforming the problem into its computational tractable equivalent, which is called robust failure recovery problem (ROBUST) [18]. ROBUST can recover the failed RAN slice and also tackle the uncertain traffic demands by introducing redundancies. However, ROBUST is a mixed integer problem (MIP) that is proved to be $\mathcal{NP}$-hard [19] and convergence is very slow using the existing solver to obtain an optimal solution [20]. Nevertheless, the result of ROBUST can be used as the performance upperbound for evaluating other solutions. To address these problems, we propose a heuristic robust RAN slicing algorithm based on variable neighborhood search (VNS). The remainder of this chapter is organized as follows. Section 10.2 elaborates the terminology of RAN slicing and system model. In Section 10.3, we first formulate and solve the non-robust RAN slicing problem. Then, we propose two robust RAN slicing algorithms: the exact algorithm, i.e. ROBUST, and the heuristic algorithm, i.e. VNS. In Section 10.4, we discuss the simulation results to evaluate the performance of the proposed robust RAN slicing algorithms. We draw a conclusion and point out some future work in Section 10.5.

10.2 Network Model

10.2.1 Slice Failure Detection Process

The main processes during the lifecycle of a RAN slice should include deployment, operation, and termination [21], as shown in Figure 10.1. We depict the process where a RAN slice failure is detected. The controller receives a slice reconfiguration request that contains service specific requirements, e.g. quality-of-service (QoS) (latency and guaranteed bandwidth), mobility, security, reliability, and connectivity requirements. Owing to this, different technologies and primitives are involved in NFV and packet forwarding, and the controller decomposes a slice reconfiguration request into a VNF reconfiguration request and a path reconfiguration request in Step 5a and 5b. By analyzing the VNF requirements, the VNF

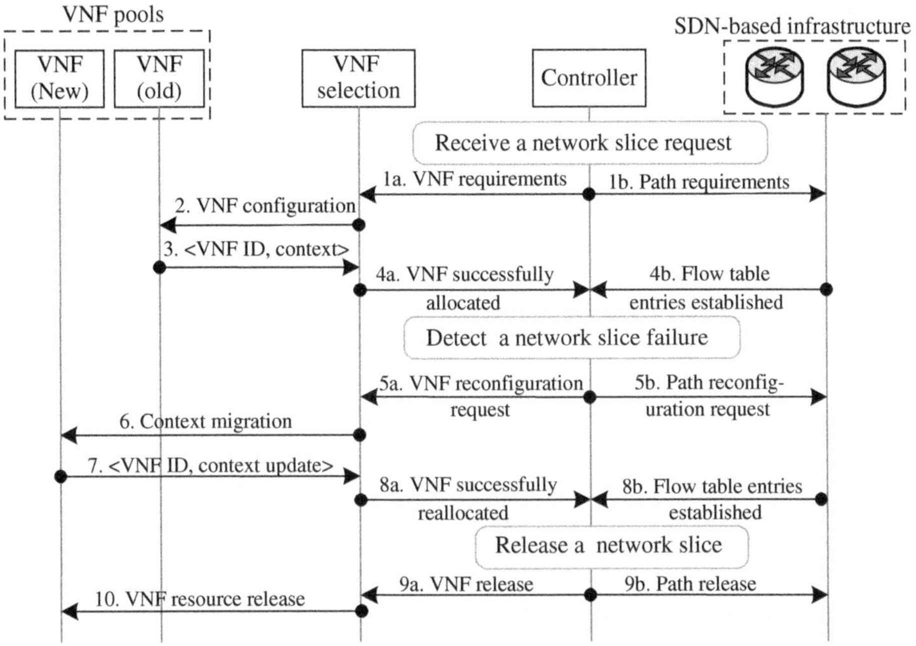

Figure 10.1 The main processes of the lifecycle of a RAN slice.

selection module reconfigures an appropriate set of VNFs in the VNFs pool in Step 6. The VNF selection module migrates the context to the new VNF, where the context includes VNF type, capability, property, and requirement following the TOSCA standard [22]. The VNF ID and context update confirmation massage are sent back to the VNF selection module in Step 7. A RAN slice failure is recovered once the controller receives feedbacks from the VNF selection modules and open flow switches in the SDN-based infrastructure in Step 8a and 8b, respectively. In Section 10.2.2, the system model for robust RAN slicing mechanisms are presented.

10.2.2 System Model

In this section, we present the system model for robust RAN slicing. We model the substrate network (SN) as a weighted directed graph $\mathcal{G}^S = (\mathcal{N}^S, \mathcal{L}^S)$, where $\mathcal{N}^S$ and $\mathcal{L}^S$ denote the set of substrate nodes and links respectively. A substrate node can be a commodity server or a physical network equipment that provides (virtual) network functions. A substrate link can be softwarized into several virtual links using ports virtualization [23]. Let $\mathcal{N}^S_\mathcal{F}$ and $\mathcal{L}^S_\mathcal{F}$ denote the set of failed substrate nodes and links respectively. Here, failed (resp. available) substrate nodes and links are those malfunctioning (resp. properly functioning) commodity severs and physical network interfaces, respectively. Similarly, the set of available substrate nodes and links is denoted by $\mathcal{N}^S_\mathcal{A}$ and $\mathcal{L}^S_\mathcal{A}$ respectively, where $\mathcal{N}^S_\mathcal{A} = \mathcal{N}^S - \mathcal{N}^S_\mathcal{F}$ and $\mathcal{L}^S_\mathcal{A} = \mathcal{L}^S - \mathcal{L}^S_\mathcal{F}$. We use a binary vector $\boldsymbol{\beta}^i$ to indicate whether substrate node i possesses the corresponding network functions, where $\beta^k_i = 1$ if substrate node i possesses network function k, and $\beta^k_i = 0$ otherwise. The dimension of vector $\boldsymbol{\beta}^i$ is denoted by L. The elements of vector $\boldsymbol{\beta}_i$ represent various network functions. For simplicity, we use ij to denote the substrate link between substrate node i and j. The bandwidth of substrate link ij is denoted by b^S_{ij}. We define an indicator $a_{w,ij}$, $w \in \mathcal{N}^S$, $ij \in \mathcal{L}^S$ to reflect adjacency of the directed graph $\mathcal{G}^S = (\mathcal{N}^S, \mathcal{L}^S)$, where $a_{w,ij} = 1$ if substrate node w is the tail of the directed substrate link ij, $a_{w,ij} = -1$ if substrate node w is the head of the directed substrate link ij, and $a_{w,ij} = 0$ otherwise.

A RAN slice request is also modeled by a weighed directed graph $\mathcal{G}^\mathcal{V} = (\mathcal{N}^\mathcal{V}, \mathcal{L}^\mathcal{V})$, where $\mathcal{N}^\mathcal{V}$ and $\mathcal{L}^\mathcal{V}$ denote the set of requested network functions and virtual links respectively. We assume that network functions used by a specific RAN slice are not shared by other RAN slices for isolation purpose [4]. We use $\mathcal{N}^\mathcal{V}_\mathcal{F}$ and $\mathcal{L}^\mathcal{V}_\mathcal{F}$ to denote the set of failed network functions and links respectively. Similarly, the set of available substrate nodes and links are denoted by $\mathcal{N}^\mathcal{V}_\mathcal{A}$ and $\mathcal{L}^\mathcal{V}_\mathcal{A}$ respectively, where $\mathcal{N}^\mathcal{V}_\mathcal{A} = \mathcal{N}^\mathcal{V} - \mathcal{N}^\mathcal{V}_\mathcal{F}$ and $\mathcal{L}^\mathcal{V}_\mathcal{A} = \mathcal{L}^\mathcal{V} - \mathcal{L}^\mathcal{V}_\mathcal{F}$. A request is represented by a binary vector $\boldsymbol{\alpha}$ and a bandwidth matrix $\mathbf{B}_v$, where binary variable $\alpha^k = 1$ indicates that the request requires network function k, and $\alpha^k = 0$ otherwise. For simplicity, we use kl to denote the virtual link between network function k and l. We introduce matrix $\mathbf{B}_v = \{b^\mathcal{V}_{kl}\}_{L\times L}$, where $b^\mathcal{V}_{kl}$ denotes the bandwidth requirement of the virtual link kl. Without loss of generality, we assume $1 \leq k < l \leq L$.

We use $b^{kl}_{(s,t),ij} \geq 0$ to denote the bandwidth allocated on substrate link ij, when virtual link kl is mapped onto a substrate path between source s and destination t. This means that substrate link ij is a segment link of the path (s, t), where $i, j \in \mathcal{L}^S_\mathcal{A}$. We introduce a binary variable m^k_s, where $m^k_s = 1$ stands for network function k mapped onto substrate node s, and $m^k_s = 0$ otherwise. Furthermore, we introduce a *pair-decision* variable $\gamma^{kl}_{(s,t)} = m^k_s \cdot m^l_t$, which equals 1 only when network functions k and l are mapped onto substrate nodes s and t respectively at the same time, otherwise $\gamma^{kl}_{(s,t)} = 0$. To avoid the quadratic constraint, such relationship among m^k_s, m^l_t, and $\gamma^{kl}_{(s,t)}$ can be represented by a triangle inequality [24] in (10.1).

$$m^k_s + m^l_t - \gamma^{kl}_{(s,t)} \leq 1, \forall s, t \in \mathcal{N}^S_\mathcal{A}, k, l \in \mathcal{N}^\mathcal{V}_\mathcal{F}. \tag{10.1}$$

Since one network function can be only mapped onto one substrate node and the fact that m^k_s and m^l_t are independent variables, we can obtain the value of m^k_s by summing up all the possible values of m^l_t in $\gamma^{kl}_{(s,t)}$. We express this relationship in (10.2). Similarly, we can also obtain the equation in (10.3).

$$\sum_{l\in\mathcal{N}^\mathcal{V}_\mathcal{F}} \sum_{t\in\mathcal{N}^S_\mathcal{A},} \gamma^{kl}_{(s,t)} = m^k_s, \forall s \in \mathcal{N}^S_\mathcal{A}, k \in \mathcal{N}^\mathcal{V}_\mathcal{F} \tag{10.2}$$

$$\sum_{k\in\mathcal{N}^\mathcal{V}_\mathcal{F}} \sum_{s\in\mathcal{N}^S_\mathcal{A},} \gamma^{kl}_{(s,t)} = m^l_t, \forall t \in \mathcal{N}^S_\mathcal{A}, l \in \mathcal{N}^\mathcal{V}_\mathcal{F}. \tag{10.3}$$

In our system model, the connectivity of virtual network functions is modeled as a multi-commodity flow problem [25]. The equations of (10.4)–(10.6) ensure flow conservations, i.e. the amount of flow going into substrate node w equals the amount of flows going out of it, except for $w = s$, which has only outgoing flows, or $w = t$, which has only incoming flows.

$$\sum_{ij\in\mathcal{L}^S_\mathcal{A}} a_{s,ij} \cdot b^{kl}_{(s,t),ij} = -\gamma^{kl}_{(s,t)} \cdot b^\mathcal{V}_{kl}, \forall s \neq t, s, t \in \mathcal{N}^S_\mathcal{A}, kl \in \mathcal{L}^\mathcal{V}_\mathcal{F} \tag{10.4}$$

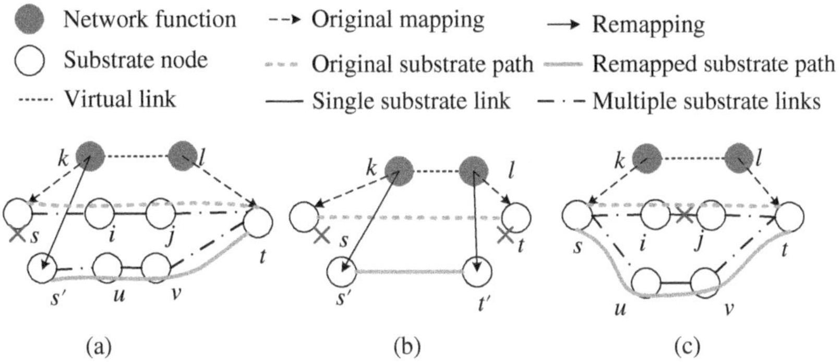

Figure 10.2 Cases of substrate node and link failures. (a) source (or destination) node failure, (b) source and destination nodes failures, (c) substrate link failure.

$$\sum_{ij \in \mathcal{L}^S_A} a_{t,ij} \cdot b^{kl}_{(s,t),ij} = \gamma^{kl}_{(s,t)} \cdot b^{\mathcal{V}}_{kl}, \forall s \neq t, s, t \in \mathcal{N}^S_A, kl \in \mathcal{L}^{\mathcal{V}}_F \tag{10.5}$$

$$\sum_{ij \in \mathcal{L}^S_A} a_{w,ij} \cdot b^{kl}_{(s,t),ij} = 0, \forall w \neq s, w \neq t, s \neq t, w, s, t \in \mathcal{N}^S_A, kl \in \mathcal{L}^{\mathcal{V}}_F. \tag{10.6}$$

Three kinds of failures may occur in the SN as shown in Figure 10.2: (i) source (or destination) node fails in a substrate node pair (s, t) as shown in Figure 10.2a; (ii) both source and destination nodes fail as shown in Figure 10.2b; (iii) only the substrate link fails as shown in Figure 10.2c. These three kinds of failures may coexist in the same RAN slice. Since migration and instantiation of virtual network functions may lead to a long time of service interruption [26, 27], we only remap the failed network functions and keep the unaffected ones on the original substrate nodes.

10.3 Robust RAN Slicing

We first formulate a recovery problem to solve remapping of a slice with deterministic traffic demands in Section 10.3.2. Based on that, we propose two robust RAN slicing algorithms for slice recovery under stochastic demands. The algorithm proposed in Section 10.3.2 is based on the robust optimization theory with a high computational complexity and slow convergence. To tackle the slow convergence problem, we propose a heuristic algorithm based on VNS in Section 10.3.3.

10.3.1 Failure Recovery Problem Formulation

We consider a robust RAN slicing algorithm that remaps the failed network functions and re-routes the failed virtual links. We assume that the bandwidth requirement in the request is deterministic. In other words, the bandwidth of each virtual link is unchanged after the creation of the corresponding RAN slices. The objective is to minimize the total bandwidth consumption. The failure recovery problem (FRP) is formulated as follows:

$$\min_{b^{kl}_{(s,t),ij}, m^k_s, m^l_t} \sum_{s,t \in \mathcal{N}^S_A} \sum_{ij \in \mathcal{L}^S - \mathcal{L}^S_F} \sum_{k,l \in \mathcal{N}^{\mathcal{V}}_F} b^{kl}_{(s,t),ij}, \tag{10.7}$$

$$\text{s.t.} m_s^k + m_t^l - \gamma_{(s,t)}^{kl} \leq 1, \forall s,t \in \mathcal{N}_{\mathcal{A}}^{S}, k,l \in \mathcal{N}_{\mathcal{F}}^{\mathcal{V}}, \tag{10.8}$$

$$\sum_{l \in \mathcal{N}_{\mathcal{F}}^{\mathcal{V}}} \sum_{t \in \mathcal{N}_{\mathcal{A}}^{S},} \gamma_{(s,t)}^{kl} = m_s^k, \forall s \in \mathcal{N}_{\mathcal{A}}^{S}, k \in \mathcal{N}_{\mathcal{F}}^{\mathcal{V}}, \tag{10.9}$$

$$\sum_{k \in \mathcal{N}_{\mathcal{F}}^{\mathcal{V}}} \sum_{s \in \mathcal{N}_{\mathcal{A}}^{S},} \gamma_{(s,t)}^{kl} = m_t^l, \forall t \in \mathcal{N}_{\mathcal{A}}^{S}, l \in \mathcal{N}_{\mathcal{F}}^{\mathcal{V}}, \tag{10.10}$$

$$\sum_{s,t \in \mathcal{N}_{\mathcal{A}}^{S}} \sum_{k,l \in \mathcal{N}_{\mathcal{F}}^{\mathcal{V}}} b_{(s,t),ij}^{kl} \leq b_{ij}^{S}, \forall ij \in \mathcal{L}_{\mathcal{A}}^{S}, \tag{10.11}$$

$$\sum_{ij \in \mathcal{L}_{\mathcal{A}}^{S}} a_{s,ij} \cdot b_{(s,t),ij}^{kl} = -\gamma_{(s,t)}^{kl} \cdot b_{kl}^{\mathcal{V}}, \forall s \neq t, s,t \in \mathcal{N}_{\mathcal{A}}^{S}, kl \in \mathcal{L}_{\mathcal{F}}^{\mathcal{V}}, \tag{10.12}$$

$$\sum_{ij \in \mathcal{L}_{\mathcal{A}}^{S}} a_{t,ij} \cdot b_{(s,t),ij}^{kl} = \gamma_{(s,t)}^{kl} \cdot b_{kl}^{\mathcal{V}}, \forall s \neq t, s,t \in \mathcal{N}_{\mathcal{A}}^{S}, kl \in \mathcal{L}_{\mathcal{F}}^{\mathcal{V}}, \tag{10.13}$$

$$\sum_{ij \in \mathcal{L}_{\mathcal{A}}^{S}} a_{w,ij} \cdot b_{(s,t),ij}^{kl} = 0, \forall w \neq s, w \neq t, s \neq t, w,s,t \in \mathcal{N}_{\mathcal{A}}^{S}, \tag{10.14}$$

$$\sum_{s \in \mathcal{N}_{\mathcal{A}}^{S}} m_s^k = \alpha^k, \sum_{t \in \mathcal{N}_{\mathcal{A}}^{S}} m_t^l = \alpha^l, \forall k,l \in \mathcal{N}_{\mathcal{F}}^{\mathcal{V}}, \tag{10.15}$$

$$m_s^k = \bar{m}_s^k, m_t^l = \bar{m}_t^l, \forall s,t \in \mathcal{N}_{\mathcal{A}}^{S}, k,l \in \mathcal{N}_{\mathcal{A}}^{\mathcal{V}}, \tag{10.16}$$

$$m_s^k \leq \beta_s^k, m_t^l \leq \beta_t^l, \forall s,t \in \mathcal{N}_{\mathcal{A}}^{S}, k,l \in \mathcal{N}_{\mathcal{F}}^{\mathcal{V}}, \tag{10.17}$$

$$m_s^k, m_t^l, \gamma_{(s,t)}^{kl} \in \{0,1\}, \forall s,t \in \mathcal{N}_{\mathcal{A}}^{S}, k,l \in \mathcal{N}_{\mathcal{F}}^{\mathcal{V}}, \tag{10.18}$$

$$b_{(s,t),ij}^{kl} \geq 0, \forall s,t \in \mathcal{N}_{\mathcal{A}}^{S}, \forall ij \in \mathcal{L}_{\mathcal{A}}^{S}, kl \in \mathcal{L}_{\mathcal{F}}^{\mathcal{V}},. \tag{10.19}$$

where constraint (10.11) ensures that the allocated bandwidth on a substrate link does not exceed its link capacity; constraint (10.15) ensures that all the failed VNFs are remapped onto the new ones; constraint (10.16) is to keep the unaffected VNFs; and constraint (10.17) ensures that the number of VNFs requested is within the capacity of the substrate node. (10.18) guarantees that the allocated VNFs are integers. (10.19) ensures that the allocated bandwidth is nonnegative.

Problem in (10.7) is a mixed integer program (MIP). To simplify the description, we reformulate (10.7) into an equivalent compact form, i.e. the matrix form. Let $\mathbf{H}$ represent the coefficient matrix of the constraints in FRP. $\mathbf{H}$ is a $I \times J$ dimensional matrix where $I = |\mathcal{L}_{\mathcal{F}}^{\mathcal{V}}| \cdot (|\mathcal{N}_{\mathcal{A}}^{S}|)^2 + 2 \cdot |\mathcal{N}_{\mathcal{A}}^{S}| \cdot |\mathcal{L}_{\mathcal{F}}^{\mathcal{V}}| + |\mathcal{L}_{\mathcal{A}}^{S}| + 2 \cdot |\mathcal{N}_{\mathcal{A}}^{S}|^3 \cdot |\mathcal{L}_{\mathcal{F}}^{\mathcal{V}}| \cdot |\mathcal{L}_{\mathcal{A}}^{S}| + 5 \cdot |\mathcal{L}_{\mathcal{F}}^{\mathcal{V}}|$ and $J = |\mathcal{L}_{\mathcal{F}}^{\mathcal{V}}| \cdot (|\mathcal{N}_{\mathcal{A}}^{S}|)^2 \cdot |\mathcal{L}_{\mathcal{A}}^{S}| + |\mathcal{L}_{\mathcal{F}}^{\mathcal{V}}| \cdot (|\mathcal{N}_{\mathcal{A}}^{S}|)^2 + |\mathcal{N}_{\mathcal{A}}^{S}| \cdot |\mathcal{L}_{\mathcal{F}}^{\mathcal{V}}|$. Let h_{ij} denote the component of $\mathbf{H}$. Let the variables of the compact form a problem denoted by a vector $\mathbf{x} = \begin{pmatrix}\mathbf{b} & \boldsymbol{\gamma} & \mathbf{m}\end{pmatrix}^T = \{x_j\} \in \mathbb{R}^J$, where vectors $\mathbf{b}$, $\boldsymbol{\gamma}$, and $\mathbf{m}$ are composed by $b_{(s,t),ij}^{kl}$, $\gamma_{(s,t)}^{kl}$, and m_s^k, respectively. $\mathbf{c}$ is a J dimensional constant vector, where the first $|\mathbf{b}|$ components of $\mathbf{c}$ are all ones and the rest of the components are all zeros. $\mathbf{d}$ is a I dimensional constant vector. Thus, the FRP can be rewritten as

$$\min_{\mathrm{x}} \quad \mathbf{c}^T \mathbf{x}, \tag{10.20}$$

$$\text{s.t.} \quad \mathbf{H}\mathbf{x} \leq \mathbf{d}, \tag{10.21}$$

$$\mathbf{x} \in \mathcal{X}, \tag{10.22}$$

where $\mathcal{X} = \{x_j | x_j \geq 0, \text{ for } j = 1, \ldots, |\mathbf{b}|, \text{and } x_j \in \{0,1\}, \text{for } j = |\mathbf{b}| + 1, \ldots, J\}$.

We use the Matlab solver to solve FRP. Since FRP is somewhat idealized, we take it as a benchmark for comparing with the robust RAN slicing algorithms. In Section 10.3.2, we extend FRP with more realistic assumptions.

10.3.2 Robust RAN Slicing Problem Formulation

The assumption that the bandwidth of virtual and substrate links is deterministic in (10.20) can be too ideal. In particular,

- the substrate link resources are likely to be under-utilized, since the request usually requires its link bandwidth according to the peak hour demand;
- if traffic shaping is not enabled, any increase in traffic demands can lead the mapped request to be infeasible;
- the availability of the substrate link is time-varying. The reasons can be, for instance, the time-varying switch port/interface failures in fiber connected network [15], or the time-varying interferences in microwave connected network [28].

Consequently, the bandwidth of both substrate and virtual links could be uncertain (stochastic). To capture this uncertainty, we use the stochastic variables $\tilde{b}^S_{ij}$ and $\tilde{b}^V_{kl}$ to replace the corresponding coefficients b^S_{ij} and b^V_{kl} in formulation (10.7) respectively. We assume that $\tilde{b}^S_{ij}$ follows a uniform distribution inside a symmetric interval $[b^S_{ij} - \hat{b}^S_{ij}, b^S_{ij} + \hat{b}^S_{ij}]$, which centers on b^S_{ij} with perturbation of $\hat{b}^S_{ij}$. Similarly, we also assume that $\tilde{b}^V_{kl}$ follows another uniform distribution inside a symmetric interval $[b^V_{kl} - \hat{b}^V_{kl}, b^V_{kl} + \hat{b}^V_{kl}]$, which centers on b^V_{kl} with perturbation of $\hat{b}^V_{kl}$. Note that we allow $\hat{b}^S_{ij} = 0$ or $\hat{b}^S_{kl} = 0$.

In what follows, we first formulate the MIP with stochastic variables. We aim to leverage the robust optimization technique [18] to solve it. Based on robust optimization, we then transform the MIP with stochastic variables into a computational tractable form.

We assume that virtual or substrate link uncertainties only affect the elements of the coefficient matrix. Let $\tilde{\mathbf{H}}$ represent the coefficient matrix of the optimization problem in (10.20) considering the existence of stochastic variables. $\tilde{\mathbf{H}}$ is a $I \times J$ dimensional matrix. We denote the component of $\tilde{\mathbf{H}}$ by $\tilde{h}_{ij}$, whose perturbation and mean value are $\hat{h}_{ij}$ and h_{ij} respectively. Remember that h_{ij} is the component of $\mathbf{H}$ in (10.20). $\tilde{h}_{ij}$ follows a uniform distribution inside a symmetric interval $[h_{ij} - \hat{h}_{ij}, h_{ij} + \hat{h}_{ij}]$. The FRP with stochastic variables is formulated as follows:

$$\min_{\mathrm{x}} \quad \mathbf{c}^T\mathbf{x}, \tag{10.23}$$

$$\text{s.t.} \quad \tilde{\mathbf{H}}\mathbf{x} \leq \tilde{\mathbf{d}}, \tag{10.24}$$

$$\mathbf{x} \in \mathcal{X}. \tag{10.25}$$

Unfortunately, we cannot solve problem (10.23) directly due to the stochastic variables. To cope with this problem, we need to transform (10.23) into its *robust counterpart* [18]. The robust counterpart is a solution method by creating a deterministic equivalent, which is computationally tractable, to deal with an uncertain problem like (10.23). It should be mentioned that in each row of $\tilde{\mathbf{H}}$, there will be at most one stochastic variable (either $\tilde{b}^S_{ij}$ or $\tilde{b}^V_{kl}$). In other words, for the constraint containing one stochastic variable of either $\tilde{b}^S_{ij}$ or $\tilde{b}^V_{kl}$, we only need to introduce one intermediate variable (denoted by y_{ij}) to transform such constraint into its robust counterpart. Consequently, we obtain two types of constraints: constraints (10.27)–(10.29) represent the robust counterpart of those rows in (10.24) containing one stochastic variable; constraint (10.30) are those rows in (10.24) without stochastic variables. The robust counterpart of (10.23) (or called ROBUST) is formulated as follows:

$$\min_{\mathbf{x}} \quad \mathbf{c}^T\mathbf{x}, \tag{10.26}$$

$$\text{s.t.} \quad \sum_{j=1}^{J^{\mathrm{RC}}} h_{ij}x_j + y_{ij} \le b_i, i = 1, \dots, I^{\mathrm{RC}}, \forall j \in \{j|\hat{h}_{ij} > 0, j = 1, \dots, J^{\mathrm{RC}}\}, \tag{10.27}$$

$$-y_{ij} \le \hat{h}_{ij}x_j \le y_{ij}, i = 1, \dots, I^{\mathrm{RC}}, \forall j \in \{j|\hat{h}_{ij} > 0, j = 1, \dots, J^{\mathrm{RC}}\}, \tag{10.28}$$

$$y_{ij} \ge 0, i = 1, \dots, I^{\mathrm{RC}}, \forall j \in \{j|\hat{h}_{ij} > 0, j = 1, \dots, J^{\mathrm{RC}}\}, \tag{10.29}$$

$$\sum_{j=1}^{J^{\mathrm{RC}}} h_{ij}x_j \le b_i, i = 1, \dots, I^{\mathrm{RC}}, \forall j \in \{j|\hat{h}_{ij} = 0, j = 1, \dots, J^{\mathrm{RC}}\}, \tag{10.30}$$

$$\mathbf{x} \in \mathcal{X}^{\mathrm{RC}}, \tag{10.31}$$

where $\mathcal{X}^{\mathrm{RC}} = \{\mathbf{x} \in \mathbb{R}^{J^{\mathrm{RC}}} | x_j \ge 0, \forall j \in C, x_j \in \{0,1\}, \forall j \in \mathcal{B}, \text{and } x_j \in \mathbb{R} \text{ otherwise}\}$, in which

$$C = \{1, 2, \dots, |\mathcal{L}_{\mathcal{F}}^{\mathcal{V}}| \cdot (|\mathcal{N}_{\mathcal{A}}^{S}|)^2\} \bigcup \{J + |\mathrm{L}_{\mathcal{A}}^{S}| + 1, J + |\mathcal{L}_{\mathcal{A}}^{S}| + 2, \dots, J^{\mathrm{RC}}\},$$

and

$\mathcal{B} = \{|\mathcal{N}_{\mathcal{A}}^{S}|^3 \cdot |\mathcal{L}_{\mathcal{F}}^{\mathcal{V}}| \cdot |\mathcal{L}_{\mathcal{A}}^{S}| + 1, |\mathcal{N}_{\mathcal{A}}^{S}|^3 \cdot |\mathcal{L}_{\mathcal{F}}^{\mathcal{V}}| \cdot |\mathcal{L}_{\mathcal{A}}^{S}| + 2, \dots, (J - |\mathcal{L}_{\mathcal{A}}^{S}| - |\mathcal{L}_{\mathcal{F}}^{\mathcal{V}}|)\}$ are sets of the subscripts of continuous and binary variables respectively. We use a $I^{\mathrm{RC}} \times J^{\mathrm{RC}}$ dimensional matrix, denoted by $\tilde{\mathbf{H}}^{\mathrm{RC}}$, to represent the coefficients of (P), where $I^{\mathrm{RC}} = |\mathcal{L}_{\mathcal{F}}^{\mathcal{V}}| \cdot (|\mathcal{N}_{\mathcal{A}}^{S}|)^2 + 2 \cdot |\mathcal{N}_{\mathcal{A}}^{S}| \cdot |\mathcal{L}_{\mathcal{F}}^{\mathcal{V}}| + 2 \cdot |\mathcal{L}_{\mathcal{A}}^{S}| + 2 \cdot |\mathcal{N}_{\mathcal{A}}^{S}|^3 \cdot |\mathcal{L}_{\mathcal{F}}^{\mathcal{V}}| \cdot |\mathcal{L}_{\mathcal{A}}^{S}| + 6 \cdot |\mathcal{L}_{\mathcal{F}}^{\mathcal{V}}|$ and $J^{\mathrm{RC}} = J + 2 \cdot |\mathcal{L}_{\mathcal{A}}^{S}| + |\mathcal{L}_{\mathcal{F}}^{\mathcal{V}}|$. We can see that problem (P) is also an MIP. However, we need to introduce $(2 \cdot |\mathcal{L}_{\mathcal{A}}^{S}| + |\mathcal{L}_{\mathcal{F}}^{\mathcal{V}}|)$ new variables and $(|\mathcal{L}_{\mathcal{A}}^{S}| + |\mathcal{L}_{\mathcal{F}}^{\mathcal{V}}|)$ new constraints to transform (10.23) into (P), which means that the robust counterpart transformation scales linearly with the number of substrate and virtual links. Consequently, utilizing Matlab solver as a black box for solving (P) may lead to a slow convergence and scalability problem [29]. With the aim of finding better solutions faster than the MIP solver alone, in Section 10.3.3, we propose a heuristic algorithm to solve problem (P).

10.3.3 Variable Neighborhood Search Based Heuristic for Robust RAN Slicing

Problem (P) is a 0–1 mixed integer programming, which is proved to be NP-hard [19] and time-consuming when solved by some exact methods, such as branch-and-bound, branch-and-cut, and Lagrangian relaxation [30]. To address this problem, we develop a heuristic algorithm to solve problem (P), whose basic idea is as follows:

- First, we give an initial feasible solution utilizing Matlab solver as a black box so that we can add an objective cut to (P). Then we solve the linear relaxation of P, denoted by LP(P), which is formulated by relaxing the integer variables of (P) from $\{0,1\}$ to $[0,1]$. The *distance* between the feasible solution and the solution of LP(P) is calculated. By rounding the relaxed variables according to the distance, we can solve a *reduced problem* (which will be elaborated later) to obtain an initial searching point.

- Second, a local search procedure is applied around the neighborhood of the initial searching point to find the local optimal solution, where the neighborhood structure is defined in [29]. The local search procedure follows the idea of VNS, which changes the neighborhood of the initial searching point, both in descent toward local optimal and in the escape from the valleys that contain them. While the local optimal is found, problem (P) is then updated by adding a new objective cut to avoid revisiting the same local search area.

Finite iterations of the aforementioned steps can lead to an optimal solution of problem (P). Otherwise, (P) is proved to be infeasible in a finite number of iterations [30].

Next, we define the distance between two points in the solution space of (P), saying $\mathbf{x}^1$ and $\mathbf{x}^2$, as $D(\mathbf{x}^1, \mathbf{x}^2) = \sum_{j \in \mathcal{B}} |x_j^1 - x_j^2|$, where $\mathcal{B}$ is defined in Section 10.3.2. Similarly, the distance between the jth element of $\mathbf{x}^1$ and $\mathbf{x}^2$, which are x_j^1 and x_j^2 respectively, is defined as $D_j(\mathbf{x}^1, \mathbf{x}^2) = |x_j^1 - x_j^2|$. Then we define the partial distance related to $\mathcal{B}_k$ between $\mathbf{x}^1$ and $\mathbf{x}^2$ as $D(\mathcal{B}_k, \mathbf{x}^1, \mathbf{x}^2) = \sum_{j \in \mathcal{B}_k} |x_j^1 - x_j^2|$, where $|\mathcal{B}_k| = k$ and $\mathcal{B}_k \subseteq \mathcal{B}$.

Now, we introduce the reduced problem of P specifically. The reduced problem is derived from P except that those variables whose subscripts fall in $\mathcal{B}_k$ are fixed at $\bar{\mathbf{x}}$, where $\bar{\mathbf{x}} = \{\bar{x}_j\}_{1 \times J^{\mathrm{RC}}}$ is a constant vector. The reduced problem denoted by $P(\bar{\mathbf{x}}, \mathcal{B}_k)$ is given as follows.

$$\min_{\mathrm{x}} \quad \mathbf{c}^T \mathbf{x}, \tag{10.32}$$

$$\text{s.t.} \quad \sum_{j=1}^{J^{\mathrm{RC}}} h_{ij} x_j + y_{ij} \leq b_i, i = 1, \ldots, I^{\mathrm{RC}}, \forall j \in \{j | \hat{h}_{ij} > 0, j = 1, \ldots, J^{\mathrm{RC}}\}, \tag{10.33}$$

$$-y_{ij} \leq \hat{h}_{ij} x_j \leq y_{ij}, i = 1, \ldots, I^{\mathrm{RC}}, \forall j \in \{j | \hat{h}_{ij} > 0, j = 1, \ldots, J^{\mathrm{RC}}\}, \tag{10.34}$$

$$y_{ij} \geq 0, i = 1, \ldots, I^{\mathrm{RC}}, \forall j \in \{j | \hat{h}_{ij} = 0, j = 1, \ldots, J^{\mathrm{RC}}\}, \tag{10.35}$$

$$\sum_{j=1}^{J^{\mathrm{RC}}} h_{ij} x_j \leq b_i, i = 1, \ldots, I^{\mathrm{RC}}, \forall j \in \{j | \hat{h}_{ij} = 0, j = 1, \ldots, J^{\mathrm{RC}}\}, \tag{10.36}$$

$$\mathbf{x} \in \mathcal{X}^k., \tag{10.37}$$

where $\mathcal{X}^k = \{\mathbf{x} \in^{J^{\mathrm{RC}}} | x_j \geq 0, \forall j \in \mathcal{C}, x_j = \lceil \bar{x}_j \rceil, \forall j \in \mathcal{B}_k, x_j \in \{0, 1\}, \forall j \in \mathcal{B} - \mathcal{B}_k\}$. After fixing one or several variables, the resulting reduced problem, which has the same structure as (P) but a smaller solution space, is solved by Matlab ILP solver.

The kth neighborhood of $\mathbf{x}^1$ is defined as $\mathrm{N}_k(\mathbf{x}^1) = \{\mathbf{x}^1 \in \mathbb{R}^{J^{\mathrm{RC}}} | D_j(\mathbf{x}^1, \mathbf{x}) \leq k\}$, where k is the size of $\mathcal{B}_k$. According to [30], we can linearize $\mathrm{N}_k(\mathbf{x}^1)$ as

$$\sum_{j \in \mathcal{B}} (x_j(1 - x_j^1) + x_j^1(1 - x_j)) \leq k. \tag{10.38}$$

Performing local search in $\mathrm{N}_k(\mathbf{x}^1)$ means to add (10.38) as a new constraint to reduce the solution space of P. We use $(P|(10.12))$ to denote the problem of adding constraint (10.38) to P.

Algorithm 10.1: VNS-ROBUST

Input: P, an initial feasible solution $\mathbf{x}^*$;

Output: The optimal solution $\mathbf{x}^*$;

1: Add an objective cut: $P = \left(P|\mathbf{c}^T\mathbf{x} \leqslant \mathbf{c}^T\mathbf{x}^*\right)$;

2: Solve the LP(P) to obtain an optimal solution $\tilde{\mathbf{x}}$.

3: **while** $\left\lfloor \sum_{j\in\mathcal{B}} \left(c_j x_j^* - c_j\tilde{x}_j\right) \right\rfloor > 1$ **do**

4:   **if** $\tilde{x}_j$ is binary for all $j \in \mathcal{B}$ **then**

5:     **break**;

6:   **end if**

7:   Set $D_j(\mathbf{x}^*, \tilde{\mathbf{x}}) = |x_j^* - \tilde{x}_j|, j \in \mathcal{B}$;

8:   Index $\tilde{x}_j$ to ensure $D_j(\mathbf{x}^*, \tilde{\mathbf{x}}) \leq D_{j+1}(\mathbf{x}^*, \tilde{\mathbf{x}}), j \in \mathcal{B}$;

9:   Set $q = |\left\{j \in \mathcal{B}|D_j(\mathbf{x}^*, \tilde{\mathbf{x}}) \neq 0\right\}|$;

10:   Set $k_{\min} = |\mathcal{B}| - q$, $k_{\max} = |\mathcal{B}|$, $k_{step} = \lfloor q/2 \rfloor$, $k = k_{\max}$;

11:   **while** $\left\lfloor \sum_{j\in\mathcal{B}} \left(c_j x_j^* - c_j\tilde{x}_j\right) \right\rfloor > 1$ and $k \leqslant k_{\max}$ **do**

12:     Solve the reduced problem $P(\tilde{\mathbf{x}}, \mathcal{B}_k)$ to obtain $\mathbf{x}'$;

13:     **if** $\mathbf{c}^T\mathbf{x}' < \mathbf{c}^T\mathbf{x}^*$ **then**

14:       Local search $\mathbf{x}'$ in the neighborhood of $\mathbf{x}'$ to update $\mathbf{x}^*$;

15:       Update P with an objective cut $P = \left(P|\mathbf{c}^T\mathbf{x} \leqslant \mathbf{c}^T\mathbf{x}^*\right)$;

16:       Solve the LP(P) to obtain an optimal solution $\tilde{\mathbf{x}}$;

17:       **break**;

18:     **else**

19:       $k = k - k_{step}$;

20:     **end if**

21:   **end while**

22: **end while**

23: **return** the best solution $\mathbf{x}^*$ of P if one is generated.

We illustrate our VNS based heuristic algorithm (VNS-ROBUST). There are two loops in VNS-ROBUST: in the outer loop (between lines 3 and 22); we calculate the distance between the feasible solution $\mathbf{x}^*$ of P and the solution $\tilde{\mathbf{x}}$ of LP(P) to decide the set $\mathcal{B}_k$, which represents those integer variables to be fixed in the inner loop. In line 10, we do not fix k, which denotes the size of $\mathcal{B}_k$. Instead, we change k from $k_{\max}$ to $k_{\min}$ with a given step size k_{step}. In the inner loop of VNS-ROBUST (between lines 11 and 21), the initial searching point $\mathbf{x}'$ is obtained by solving the reduced problem $P(\tilde{\mathbf{x}}, \mathcal{B}_k)$, using the Matlab solver (i.e. intlinprog). If the solution $\mathbf{x}'$ of the reduced problem is better than the incumbent $\mathbf{x}^*$, a local search (which will be elaborated in the local search algorithm within the neighborhood of $\mathbf{x}'$) is performed to update $\mathbf{x}^*$. After updating $\mathbf{x}^*$, an objective cut $\mathbf{c}^T\mathbf{x} \leq \mathbf{c}^T\mathbf{x}^*$ is added to prevent the local search process from revisiting solution $\mathbf{x}^*$. The solution $\tilde{\mathbf{x}}$ of LP(P) is solved to decide whether the stopping criteria $\left\lfloor \sum_{j\in\mathcal{B}} (c_j x_j^* - c_j\tilde{x}_j) \right\rfloor > 1$ is fulfilled to terminate both the inner and outer loops. Otherwise, the solution $\mathbf{x}'$ is not better than the incumbent $\mathbf{x}^*$; we set $k = k - k_{\text{step}}$ to reduce the size of $\mathcal{B}_k$, which means less integer variables will be fixed in the next iteration of the inner loop. The inner and outer loops proceed until the stopping criterion is reached.

Algorithm 10.2: Local search

Input: the solution of the reduced problem $\mathbf{x}'$;the neighborhood size k;
Output: the local minimal solution $\mathbf{x}^*$;

1: Set $\mathbf{x}^* = \mathbf{x}', i = 1$;
2: **while** $i \leq k$ and $iter \leqslant MAX_{ITER}$ **do**
3: $\mathbf{x}'' = \text{intlinprog}\left(P|\text{N}_i(\mathbf{x}')\right)$;
4: **if** $\mathbf{c}^T\mathbf{x}'' \leq \mathbf{c}^T\mathbf{x}'$ **then**
5: Set $i = 1, \mathbf{x}^* = LocalSearch(P, \mathbf{x}'', k)$;
6: **else**
7: Set $i = i + 1$;
8: **end if**
9: $iter = iter + 1$.
10: **end while**

Local search is the same as the general VNS process in [29] to find the local optimal $\mathbf{x}^*$. VNS systematically exploits the solution space based on the following observation [29]: a local minimum with respect to one neighborhood structure is necessarily but insufficient condition for the local minimum of another neighbourhood, i.e. a local minimum within the ith neighborhood $\text{N}_i(\mathbf{x}')$ is not necessarily within the $(i+1)$th neighborhood $\text{N}_{i+1}(\mathbf{x}')$. In VNS, the neighborhood size, which is denoted by i, increases from 1 to k. Given a solution $\mathbf{x}'$ of the reduced problem, we explore the ith neighborhood $\text{N}_i(\mathbf{x}')$ to obtain the best solution $\mathbf{x}''$. The loop of local search continues if $\mathbf{x}''$ is found to be better than the incumbent solution $\mathbf{x}'$ (line 4); then another local search re-centers around $\mathbf{x}''$ and begins again with the first neighborhood $N_i(\mathbf{x}'')$; otherwise, if there is no improvement on the objective function value, the neighborhood size is increased by 1. The local search process proceeds until the total iterations, which is denoted by *iter*, exceed the $MAX - ITER$, i.e. the maximum allowed iterations.

10.4 Numerical Results

10.4.1 Performance Metrics

Performance metrics include the failure rate of requests (R_F, defined in (10.39)), the average load of substrate links (U_L, defined in (10.40)), and the average time of mapping a request. The total number of substrate links is denoted by $|\mathcal{L}^S|$. The initialized substrate link bandwidth within the substrate node pair (i,j) is denoted by $\bar{b}^S_{ij}$.

$$R_F = \frac{\text{the number of failed requests}}{\text{the number of arrived requests}}. \tag{10.39}$$

$$U_L = 1 - \frac{1}{|\mathcal{L}^S|} \sum_{ij \in \mathcal{L}^S} \frac{b^S_{ij}}{\bar{b}^S_{ij}}. \tag{10.40}$$

10.4.2 Simulation Scenarios and Settings

We model the arrivals of slice requests, occurrence of VNF failures, and substrate link failures as Poisson processes with arrival rates of 1, 0.001, and 0.001 per second, respectively. The duration of slice requests follows an exponential distribution with 100 seconds on average. In real scenarios, there can be permanent and/or temporary slices. We are interested in investigating whether our proposed algorithm can ensure the robustness of slices during their lifecycles. Consequently, only temporary slices are generated in our simulation. To evaluate the impact of the uncertainties of substrate and virtual links, we assume that both the substrate and virtual links' arrival uncertainties follow Poisson processes, and two different scenarios are considered:

- High uncertainty scenario, where the uncertainties' arrival rate is 0.2 per second, for both substrate and virtual links
- Medium uncertainty scenario, where the arrival rate is set to 0.1 per second, for both substrate and virtual links' uncertainties

The following simulation settings are the same for both the high and medium uncertainty scenarios. The duration of the uncertainties of substrate and virtual links follows an exponential distribution with 20 seconds on average. We consider an SN that is randomly generated with 15 nodes and the average substrate node connectivity is 0.5. Let the substrate node capacity be 30 units (number of VNFs). The number of the l type VNF is in proportion to q_l, which is uniformly distributed within $(0, 1)$, where $\sum_{1 \leq l \leq L} q_l = 1$ and $l \in \mathcal{N}^S$. In the simulation, we consider a RAN slice to be the data-plane protocol of LTE layer-2. We have $L = 3$ kinds of VNFs, which are packet data convergence protocol (PDCP), radio link control (RLC), and media access control (MAC) layer protocol functionalities, respectively. Note that the time data of processing of each VNF, such as PDCP, RLC, and MAC, is from real tests of Huawei Technologies Co., Ltd [31]. Let the substrate link bandwidth (i.e. b_{ij}^S) be a number uniformly distributed between 50 and 100 units. The virtual link bandwidth (i.e. b_{kl}^V) is also a number uniformly distributed between 1 and 20 units. The indicator α_l, where $1 \leq l \leq L$, is no more than 1 because each VNF appears in an RAN slice at most once. Three types of requests are randomly generated to represent RAN slices of mMTC, eMBB, and URLLC [4], respectively, which are constituted by MAC sublayer, MAC+RLC sublayers, and PDCP+RLC+MAC sublayers, respectively.

We assume that the maximum value of virtual link perturbation (i.e. $\hat{b}_{kl}^V$) is 30% of the nominal value b_{kl}^V. Similarly, we assume that the maximum value of substrate link perturbation (i.e. $\hat{b}_{ij}^S$) is 30% of the nominal value b_{ij}^S. We can adjust the tolerance of virtual link and substrate link uncertainties of ROBUST by tuning the $\hat{b}_{kl}^V$ and $\hat{b}_{ij}^S$ at the same time from 10% to 30% of the nominal value of b_{kl}^V and b_{ij}^S, respectively. In other words, if we need ROBUST to tolerate at most 10% bandwidth uncertainties of virtual and substrate links, $\hat{b}_{kl}^V$ and $\hat{b}_{ij}^S$ are set to $10\% b_{kl}^V$ and $10\% b_{ij}^S$ respectively at the same time. If $\hat{b}_{kl}^V$ and $\hat{b}_{ij}^S$ exceed 10% of their nominal values, the remapping results of requests solved by ROBUST become invalid. In contrast, FRP cannot tolerate any perturbations on bandwidth of substrate and virtual links.

10.4.3 Results

In Figure 10.3a,b, we compare the failure rate of requests (R_F) over time for FRP, ROBUST, and VNS under the high and medium uncertainty scenarios respectively. The tolerance of link uncertainties (including both virtual and substrate links) can be adjusted from 10% to 30%. We notice that in high uncertainty scenario, R_F of FRP, ROBUST, and VNS are roughly three times higher than the corresponding R_F of FRP, ROBUST, and VNS in

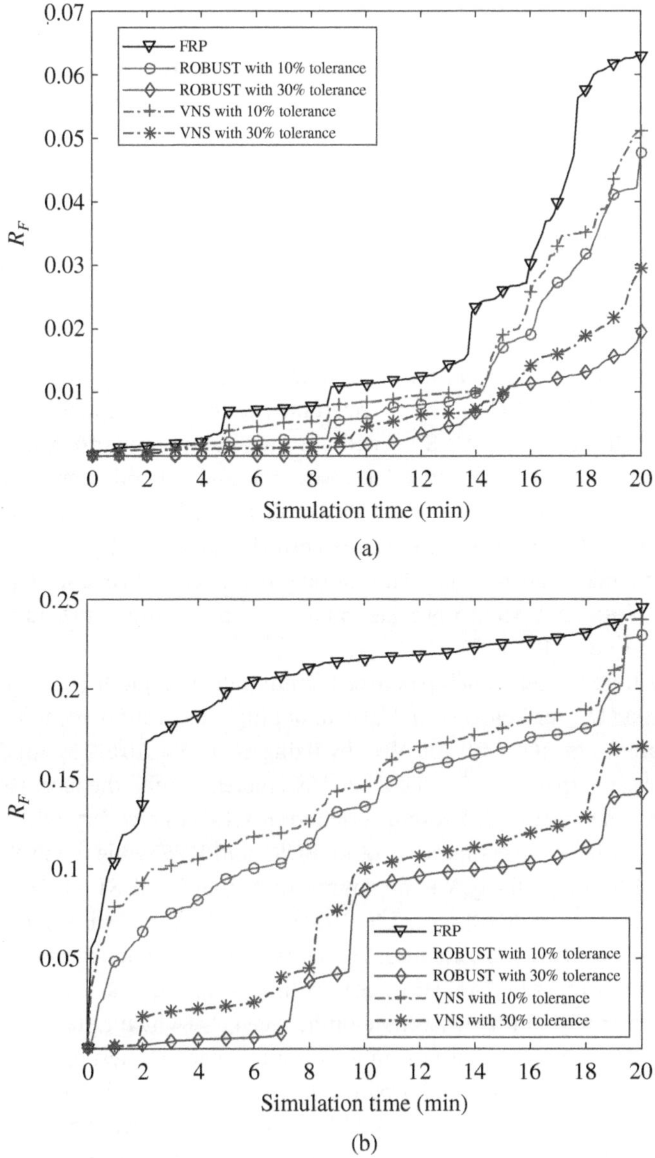

Figure 10.3 Failure rate of requests over time. (a) Medium uncertainty scenario and (b) high uncertainty scenario.

medium uncertainty scenario. Moreover, we can observe that in both scenarios, R_F of FRP, ROBUST, and VNS rise over time due to the increasing failures of substrate nodes and links as well as the increasing uncertainties of links (including both substrate and virtual links). We can also observe that in both scenarios, R_F of FRP is higher than that of ROBUST and VNS due to a lack of redundant bandwidth to protect requests from link uncertainties. By adjusting the tolerance of link uncertainties from 10% to 30%, R_F of ROBUST and VNS decreases. R_F of FRP increases faster than those of ROBUST and VNS, due to which FRP is zero-tolerant to uncertainties; in other words, uncertainties of substrate links will affect all the incumbent requests mapped on those substrate links. In contrast, ROBUST and VNS are uncertainties tolerant algorithms, which reserve link resources beforehand in case of link uncertainties. The performance of ROBUST is slightly better than that of VNS, due to which VNS is a heuristic algorithm that provides near optimal solutions while ROBUST always provides optimal solutions. Comparing the robust RAN slicing algorithms, i.e. ROBUST and VNS, R_F of ROBUST with 30% link uncertainty tolerance has the lowest failure rate. The reason is that 30% tolerance of link uncertainties equals our assumed maximum bandwidth perturbation of virtual and substrate links, and in this case, all failed requests are incurred by substrate node and link failures.

In Figure 10.4a,b, we compare the average load of links requests (U_L) over time for FRP, ROBUST, and VNS under the high and medium uncertainty scenarios respectively. We notice that U_L of FRP, ROBUST, and VNS in high uncertainty scenario are roughly 5% higher than the U_L of FRP, ROBUST, and VNS in medium uncertainty scenario. Moreover, we can see that U_L of FRP, ROBUST, and VNS increases with time. In both scenarios, U_L of ROBUST and VNS increase more rapidly than FRP. This is because ROBUST and VNS introduce redundancies while recovering against link perturbations. Specifically, VNS brings slightly higher load on links than ROBUST. The reason is that in order to enable VNS to perform fast recovery of requests, VNS just provides a near-optimal solution instead of the optimal solution as produced by ROBUST.

In Figure 10.5a,b, we illustrate the relationship between failure rate of requests (R_F) and average load of substrate links (U_L) for ROBUST and VNS under high uncertainty scenario. In Figure 10.5a, we compare R_F of ROBUST and VNS by fixing U_L at 5%, 10%, 15%, and 20%, respectively. We can observe that R_F of ROBUST and VNS increases with the increase in U_L. Larger U_L results in fewer substrate link resources left for RAN slices recovery, which in turn leads to higher R_F. Similarly, we compare U_L of ROBUST and VNS while fixing R_F at the value of 5%, 10%, and 15%, respectively. We can observe that U_L of ROBUST and VNS increases with the increase of R_F. R_F of ROBUST (resp. VNS) with 30% tolerance is smaller than that of ROBUST (resp. VNS) with 10% tolerance. Therefore, more redundancies introduced can result in higher robustness of RAN slices against link perturbations. From the results obtained so far, we can see that there is a trade-off between R_F (which reflects the robustness of a request) and U_L (which reflects how much resource that a request recovery needs to consume). For ease of comparison, we introduce the same redundancies in RAN slices of mMTC, eMBB, and URLLC. However, it should be noted that for slices with critical requirements on reliability (e.g. URLLC slices), we can further promote its tolerance of link uncertainties by slightly sacrificing the link resource utilization, to ensure that the RAN slices are service-specific.

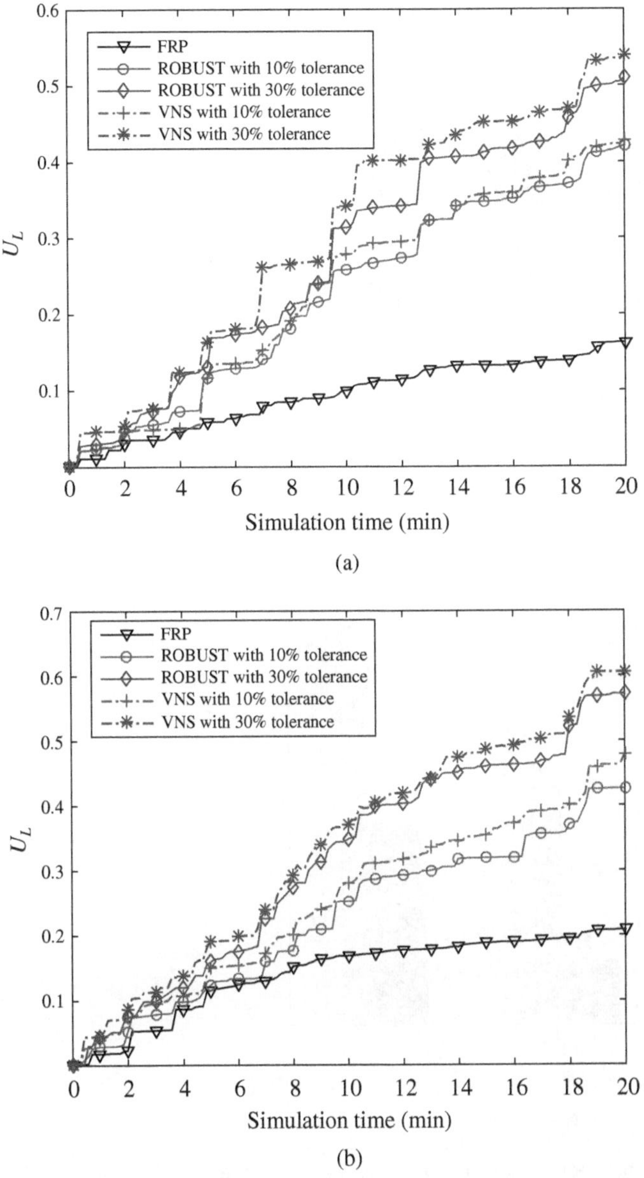

Figure 10.4 Average load of substrate links over time. (a) Medium uncertainty scenario and (b) high uncertainty scenario.

In Figure 10.6a,b, we compare the average time of recovering a request by adopting the robust recovery algorithms under the high and medium uncertainty scenarios respectively. The curves of the average time of recovering a request in both scenarios have the same trend. In the beginning of simulation, the average time of recovering a request of all the algorithms increases slowly. However, when simulation time approaches 18 minutes, the average time of recovering a request of all the algorithms increases drastically. This is because in the

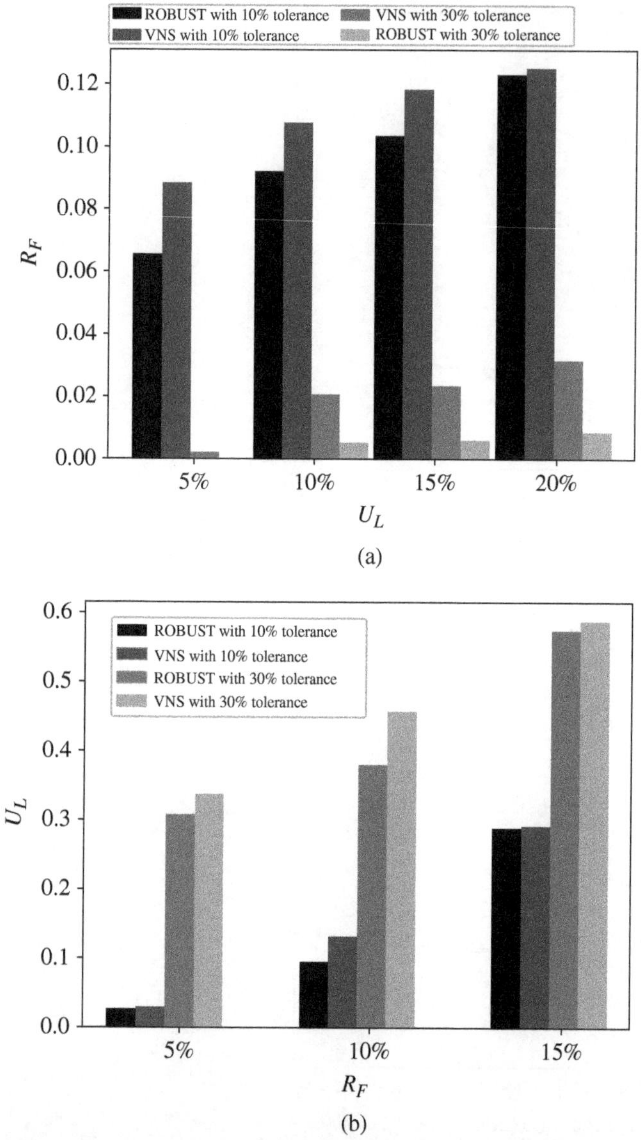

Figure 10.5 The relationship between failure rate of requests and average load of substrate links. (a) Variation of R_F with fixed U_L and (b) variation of U_L with fixed R_F.

beginning of simulation, small amounts of failure occur in substrate nodes and links, which results in enough residual substrate resources to enable successful recoveries by ROBUST and VNS, while near the end of the simulation, in which almost all the substrate nodes and links are failed, it becomes difficult for either ROBUST or VNS to find a feasible solution. As expected, we can observe that ROBUST with 30% tolerance of link uncertainty requires the longest time compared with the rest of the curves, while VNS with 10% tolerance of link uncertainty requires the least time compared with the rest of the curves. Owing to

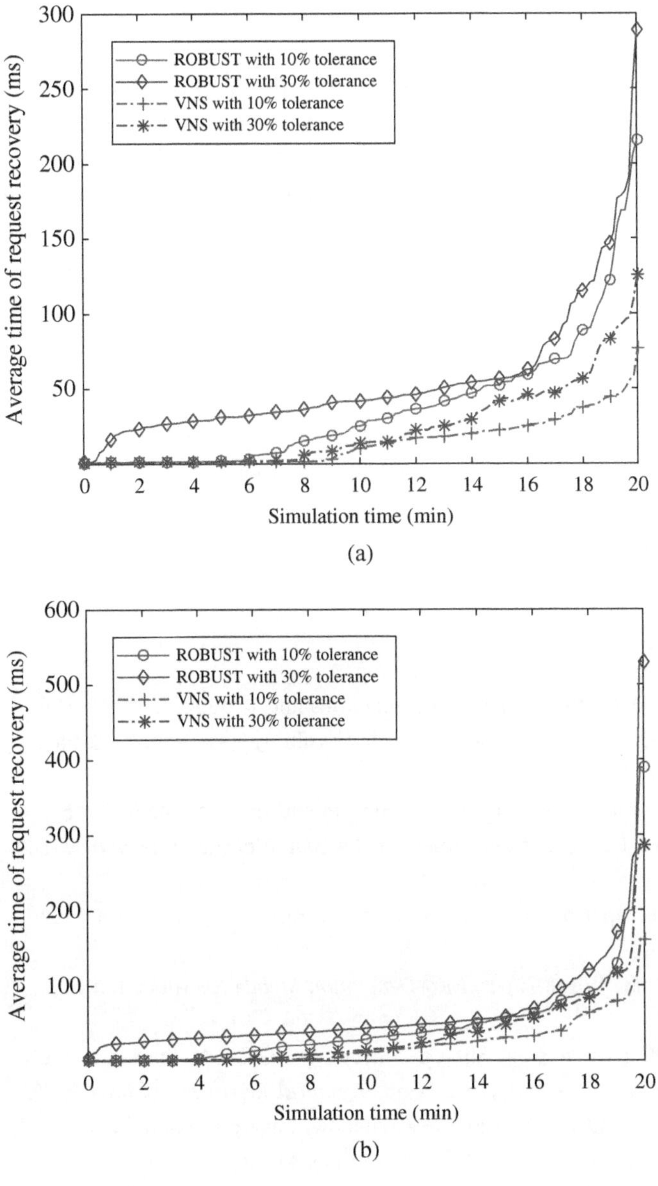

Figure 10.6 Average time of recovering a request over time. (a) Medium uncertainty scenario and (b) high uncertainty scenario.

this, ROBUST with 30% tolerance needs more computation time for an optimal solution compared with ROBUST with 10% tolerance and VNS. In contrast, VNS with 10% tolerance needs much less time to converge for a sub-optimal solution. This fact reveals that there is also a trade-off between the robustness of a request and the time consumption for request recovery.

10.5 Conclusions and Future Work

Network slicing has become one of the most discussed capabilities of 5G to provide multiple end-to-end networks with logical isolation for the intended usage scenarios. The robustness of a RAN slice is essential throughout its lifecycle. In this chapter, we have formulated and solved two problems, i.e. a non-robust (i.e. the baseline) and a robust problem, to address the failures of SNs and the uncertainties of traffic demands in a RAN slice. Furthermore, we developed a heuristic algorithm to tackle the robust problem, which produces the near-optimal solution. Numerical results have revealed that by solving the proposed robust RAN slicing algorithms, we can provide adjustable tolerance of link uncertainties.

In this chapter, we assume that network functions are dedicated to one RAN slice. In the next stage, we will consider how to ensure isolation of RAN slices when some of the network functions are shared by multiple RAN slices. Moreover, we can combine RAN slicing with machine learning to develop an automatic RAN slicing model, since the monitoring data collected from the softwarized functions can provide effective features to refine slices' management decisions in return.

References

1 del Río, J. and Gandía, D. (2016). D1.1 Refined Scenarios and Requirements, Consolidated Use Cases, and Qualitative Techno-Economic Feasibility Assessment, Europe, METIS. *Technical Report No. D1.1.*

2 An, X., Zhou, C., Trivisonno, R. et al. (2016). On end to end network slicing for 5G communication systems. *Transactions on Emerging Telecommunications Technologies* 28: 1–13.

3 IMT-2020 (2016). White Paper on 5G Network Architecture Design, China, IMT-2020. *Technical Report No. 2016-05.*

4 NGMN Alliance (2015). 5G White Paper. *Next-Generation Mobile Networks, White Paper*, pp. 1–125.

5 NGMN (2016). Description of Network Slicing Concept by NGMN Alliance, Frankfurt, Germany, NGMN P1 WS1 E2E Architecture Team. *Technical Report No. NGMN 5G P1.*

6 Pentikousis, K., Wang, Y., and Hu, W. (2013). Mobileflow: toward software-defined mobile networks. *IEEE Communications Magazine* 51 (7): 44–53.

7 Mijumbi, R., Serrat, J., Gorricho, J.-L. et al. (2016). Network function virtualization: state-of-the-art and research challenges. *IEEE Communications Surveys & Tutorials* 18 (1): 236–262.

8 Zhang, H. and Canada, H.T. (2016). Future Wireless Network: MyNET Platform and End-to-End Network Slicing. https://arxiv.org/ftp/arxiv/papers/1611/1611.07601.pdf (accessed 01 June 2020).

9 Nguyen, T.-T. and Bonnet, C. (2016). SDN-based distributed mobility management for 5G networks. *IEEE Wireless Communications and Networking Conference*, Doha, Qatar, pp. 1–7.

10 Ramirez-Perez, C. and Ramos, V. (2016). SDN meets SDR in self-organizing networks: fitting the pieces of network management. *IEEE Communications Magazine* 54 (1): 48–57.

11 Wen, R., Feng, G., Member, S. et al. (2017). Protocol function block mapping of software defined protocol for 5G mobile networks. *IEEE Transactions on Mobile Computing* 2018 (7): 1651–1665.

12 Taleb, T., Member, S., Ksentini, A. et al. (2016). On service resilience in cloud-native 5G mobile systems. *IEEE Journal on Selected Areas in Communications* 34 (3): 483–496.

13 Basta, A., Kellerer, W., Hoffmann, M. et al. (2014). Applying NFV and SDN to LTE mobile core gateways, the functions placement problem. *Proceedings of the 4th Workshop on All Things Cellular: Operations, Applications, & Challenges*, Chicago, Illinois, USA, August 2014, pp. 33–38.

14 Hawilo, H., Shami, A., Mirahmadi, M., and Asal, R. (2014). NFV: state of the art, challenges, and implementation in next generation mobile networks (vEPC). *IEEE Network* 28 (6): 18–26.

15 Smith, W.E., Trivedi, K.S., Tomek, L.A., and Ackaret, J. (2008). Availability analysis of blade server systems. *IBM Systems Journal* 47 (4): 621–640.

16 Gill, P., Jain, N., and Nagappan, N. (2011). Understanding network failures in data centers: measurement, analysis, and implications. *ACM SIGCOMM*, Toronto, Canada, August 2011, pp. 350–361.

17 BusinessWire (2003). Blade Logic Sets Standard for Data Center Automation and Provides Foundation for Utility Computing with Operations Manager Version 5. http://findarticles.com/p/articles/mi_m0EIN/is_2003_Sept_15/ai_107753392/pg_2 (accessed 01 June 2020).

18 Ben-Tal, A., Ghaoui, L.E., and Nemirovski, A. (2009). *Robust optimization.* Princeton University Press.

19 Garey, M.R. and Johnson, D.S. (1990). *Computers and intractability: A guide to the theory of NP-completeness.* New York, USA: W.H. Freemann, pp. 1–338.

20 Klotz, E., Newman, A.M., and Tyson, M. (2012). Practical guidelines for solving difficult mixed integer linear programs. *Surveys in Operations Research and Management Science* 18 (1): 18–32.

21 5G-PPP (2016). View on 5G Architecture, Heidelberg, Germany, 5G PPP. *Report number: Version 1.0.*

22 OASIS TOSCA (2015). TOSCA Simple Profile for Network Functions Virtualization (NFV) Version 1.0. *Technical Report.* http://docs.oasis-open.org/tosca/tosca-nfv/v1.0/tosca-nfv-v1.0.html (accessed 01 June 2020).

23 Jain, R. and Paul, S. (2013). Network virtualization and software defined networking for cloud computing: a survey. *IEEE Communications Magazine* 51 (11): 24–31.

24 Tversky, A. and Gati, I. (1982). Similarity, separability, and the triangle inequality. *Psychological Review* 82 (2): 123–154.

25 Hu, T.C. (1963). Multi-commodity network flows. *Operations Research* 11 (3): 344–360.

26 Abid, H. and Samaan, N. (2013). A novel scheme for node failure recovery in virtualized networks. *2013 IFIP/IEEE International Symposium on Integrated Network Management (IM 2013)*, Ghent, Belgium, May 2013, pp. 1154–1160.

27 Raza, M., Samineni, V., and Robertson, W. (2016). Physical and logical topology slicing through SDN. *2016 IEEE Canadian Conference on Electrical and Computer Engineering (CCECE)*, Vancouver, Canada, May 2016, pp. 1–4.

28 Hong, S., Brand, J., Choi, J. et al. (2014). Applications of self-interference cancellation in 5G and beyond. *IEEE Communications Magazine* 52 (2): 114–121.

29 Hansen, P., Mladenović, N., and Urošević, D. (2006). Variable neighborhood search and local branching. *Computers & Operations Research* 33 (10): 3034–3045.

30 Hanafi, S., Laz, J., Mladenov, N. et al. (2015). New variable neighbourhood search based 0-1 Mip heuristics. *Yugoslav Journal of Operations Research* 25 (3): 343–360.

31 Huawei (2015). 5G Software Defined Protocol Runtime Test Report. *Technical Report No. 1.0*. Shenzhen, China: Huawei Technologies Co. Ltd.

11

Flexible Function Split Over Ethernet Enabling RAN Slicing

Ghizlane Mountaser and Toktam Mahmoodi

Centre for Telecommunications Research, Department of Engineering, King's College London, UK

5G network aims at providing an infrastructure that is flexible to support a wide range of services and applications and satisfy their communication needs. Several factors, such as very high data rate, ultra-low latency, and high reliability, play a defining role in designing 5G infrastructure. Additionally, 5G network needs to be future proofing infrastructure, and hence, it needs to be scalable to support new features over the next several years. In addition, it needs to be flexible to adapt to a wide range of different service requirements and needs to be open. Above all, 5G network needs to be cost-effective for operators. In this context, Cloud-radio access network (RAN) and RAN slicing are the key technology components for 5G by enabling isolation [1], cloudification, and softwarization [2] of RAN functionalities.

Cloud-RAN is envisaged as a promising solution for wireless systems to support the flexibility and high scalability of the network to keep up with traffic growth. Besides, RAN slicing has emerged as an efficient way to deploy diverse requirements for heterogeneous 5G services at the same time. RAN slicing aims at guaranteeing service experience requirements and maintaining performance. This could happen by slicing RAN and transport network resources.

11.1 Flexible Functional Split Toward RAN Slicing

11.1.1 Full Centralization and CPRI

Cloud-RAN, as one of the enablers of 5G, consists of two components, namely, remote radio head (RRH), also known as distributed unit (DU), that holds analog and radio frequency (RF) functions and baseband unit (BBU), also known as central unit (CU), holding baseband functionalities.

The traditional Cloud-RAN corresponds to the so-called *full centralization*. Full centralization approach provides several advantages. It facilitates upgrade, enabling new features to be added without change in the radio equipment. It is easy to maintain because of the simpler radio equipment, which further lowers the cost on capital expenditure (CAPEX) and operational expenditures (OPEX). A recent trial from China Mobile has shown that OPEX and CAPEX can be lowered by 53% and 30% respectively using Cloud-RAN [3]. In

Radio Access Network Slicing and Virtualization for 5G Vertical Industries, First Edition.
Edited by Lei Zhang, Arman Farhang, Gang Feng, and Oluwakayode Onireti.

addition, the centralized BBUs can be shared and turned off when necessary, potentially reducing energy. It also facilitates the implementations of schemes that mitigate inter-cell interference and ensure effective coordination among adjacent cells, i.e. enhanced inter-cell interference coordination and coordinated multi-point, which significantly improve system capacity and coverage. In full centralization, fronthaul (FH), which is an interface between BBU and RRH, transports data in the form of in-phase and quadrature (IQ) samples over typically common public radio interface (CPRI), Open Base Station Architecture Initiative, and Open Radio Interface protocols, which are open BBU–RRH interfaces. CPRI, created in 2004, was the most adopted for Cloud-RAN. Despite the attractive advantages of conventional Cloud-RAN, the architecture faces several challenges. The first challenge is the constant data rate of CPRI that is independent of user activity. In fact, it requires continuous transport of CPRI stream even if no user traffic is present. The second challenge is IQ data transmission, which requires large bandwidth. The data of one antenna for one carrier is mapped to one IQ data flow that is carried by one CPRI link, which makes the FH data rate to relatively scale up with the number of antennas and system bandwidth. Consequently, the FH bandwidth requirement substantially increases with the use of Multiple Inputs Multiple Outputs. Thereby, the high throughput requirement poses challenges for the FH interface and the system may encounter capacity bottlenecks in 5G. Finally, CPRI requires stringent latency and jitter requirements. Such critical requirements make CPRI very challenging.

11.1.2 RAN Functional Split

To relax the excessive bandwidth and latency requirements, as well as to enhance the flexibility of the FH, functional split is introduced in Cloud-RAN whereby a more flexible placement of baseband functionality between the DU and the CU is considered. Third Generation Partnership Project (3GPP) [4] has defined multiple possible function splits as shown in Figure 11.1 taking Evolved Universal Terrestrial Radio Access (E-UTRA) protocol stack as a reference stack for discussion.

Option 1: Only Radio Resource Control (RRC) is in CU; Packet Data Convergence Protocol (PDCP), radio link control (RLC), medium access control (MAC), physical layer (PHY), and RF are in DU. The main benefits of this split are that the FH data rate scales flexibly according to the user plane traffic and the interface typically copes with relatively larger latency. Moreover, having all the user plane protocol in DU, i.e. closer to the edge, this split is beneficial for low-latency use cases. However, because PDCP, which performs security, is in DU, this split requires distribution of security key.

Option 2: For this split, RRC and PDCP are centralized whereas RLC, MAC, PHY, and RF are distributed. This split point is intensively considered by standard bodies and researchers and it is like split 3C, which has been standardized in long-term evolution (LTE) dual connectivity [5]. Having PDCP in CU, the split is effectively suitable for aggregation at PDCP level because it does not necessarily require strict lower layer synchronization. It is also suitable for mobility and handover as it enables reducing handover failure probability. As in the case of split option 1, FH data rate scales relative to user traffic.

Option 3: The split is performed within RLC sublayer. From 3GPP point of view, two options are defined in this split. The first option centralizes RLC automatic repeat request (ARQ), which can help recover from FH interface failure using ARQ recovery

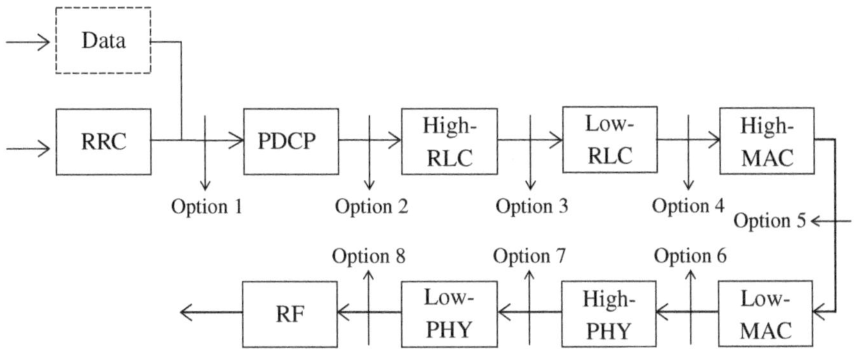

Figure 11.1 Options for functional splits for Cloud-RAN [4].

mechanism. However, since ARQ is responsible for end-to-end re-transmission, this option is more sensible to latency than split option 2. The second option separates transmit and receive RLC where transmit RLC is in DU whereas receive RLC is in CU. There is, thus, no constraint on downlink (DL) data transmission. However, placing transmit RLC in DU increases processing and buffer requirements in DU.

Option 4: RRC, PDCP, and RLC are in CU. MAC, PHY, and RF are located in DU. The FH interface transports RLC packet data units (PDUs). Thus, the data rate of the interface is dependent on user activity. The drawback is that the split is not simple to implement and might be impractical because of the tight interaction between MAC and RLC. For example in DL, RLC considers its buffer size along with the scheduling decisions of MAC in terms of resource blocks to generate the RLC PDUs. This mechanism should not take a long time.

Option 5: The split is within MAC, where part of the MAC functionality, for instance scheduling decision, is in CU whereas the time critical MAC processing is located in DU.

Option 6: For this option, the split is between MAC and PHY wherein only PHY and RF are in DU. The split offers a high level of centralization and pooling gain compared to the options above. Transport blocks are transmitted over the FH interface, and hence the data rate of the FH scales with user traffic. The latency requirement is relaxed compared to CPRI. In [6], the latency is found to be, on average, within 60 μs.

Option 7: This is an intra-PHY split whereby parts of the PHY functions are in CU. The other parts of PHY and RF are in DU. The key advantage of this option is the high degree of centralization with a significant reduction on the FH data rate requirement compared to CPRI by moving antenna-related operations to the CU (e.g. DL antenna mapping, fast Fourier transform). However, the DU in this option is more complex than the one in option 8.

Option 8: This corresponds to a fully centralized RAN architecture. FH transports IQ data in time domain. While this option benefits from the advantages of full centralization, it has very high data rate requirement.

Nevertheless, each of these envisioned splits has its own requirement. Depending on the split point, the latency and bandwidth requirements change as shown in Table 11.1 [7]. In general, the lower the split point, the greater the level of centralization, the higher is the required interface data rate, and the more stringent is the latency requirement.

Table 11.1 Bandwidth and latency requirements for different split points.

Split point	One-way latency	DL bandwidth (Mbps)	UL bandwidth (Mbps)
RRC-PDCP	30 ms	151	48
PDCP-RLC	30 ms	151	48
Split MAC	6 ms	151	49
MAC-PHY	2 ms	152	49
PHY Split III	250 μs to 2 ms	1075	922

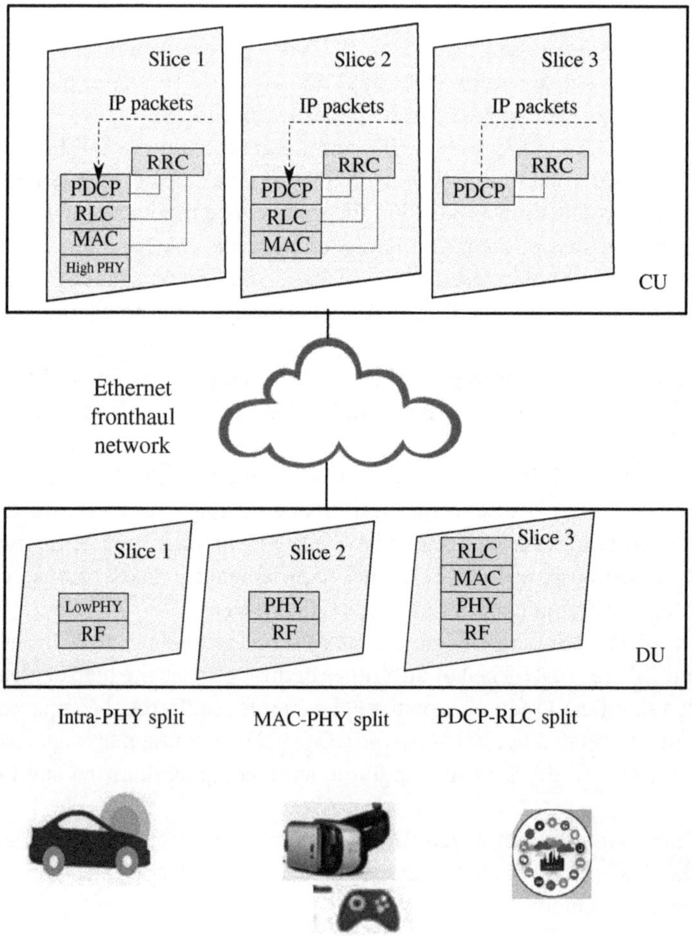

Figure 11.2 Example of the envisioned architecture of RAN functional split supporting RAN slicing.

11.1.3 Flexible Functional Split as RAN Slicing Enabler

The 5G network aims at supporting a wide range of services and applications with diverse key performance indicator (KPI) requirements. As such, there is no optimal split point that can fulfill the diverse requirements.

On this basis, to flexibly meet the different requirements of 5G services, different splits between CU and DU can be configured. In this context, each functional split may belong to a slice. Accordingly, the requirements provided by functional split can be guaranteed by the slice, and thus the features provided differ from slice to slice. A specific type of service is dedicated to an appropriate slice according to the application need. Resource allocations can be made in different ways. For instance, each slice might be assigned a set of resources statically to guarantee quality of service (QoS), such as latency, data rate, and reliability, for each of them. An alternative way is to dynamically allocate resources to slices according to the traffic load. Nevertheless, slicing should allow efficient resource utilization. In general, there is a trade-off between system performance and effective radio resource utilization.

Figure 11.2 shows an example of functional split deploying RAN slicing. For instance, smart cars will require low-latency communication while smart cities would typically require high bandwidth and as such different slices may be configured to offer different requirements for latency and FH throughput helping to serve different types of services. This architecture provides Cloud-RAN the flexibility needed to support the diverse requirements of 5G services.

11.2 Fronthaul Reliability and Slicing by Deploying Multipath at the Fronthaul

11.2.1 Packet-based Fronthaul

In traditional Cloud-RAN architecture, fiber has been defined as an ideal attractive solution to meet the strict requirements of high bandwidth and low latency of CPRI. However, there are situations where deployment of fiber is not possible or is not a good choice due to cost. To this end, packet-based FH can be considered as a promising alternative transport. This shifts the paradigm of fronthaul network design.

This highly cost-effective solution allows sharing and convergence with Ethernet-based fixed networks and offers great flexibility. However, packet-based FH imposes many challenges such as high latency and high jitter. Nevertheless, it can be used in functional split where the latency and jitter requirements are relaxed. For example, the feasibility of splitting between MAC and PHY was demonstrated in [8]. In standard deployments of packet-based FH, enhanced CPRI [9] and IEEE 1914 Radio over Ethernet provide a framework for transport of radio data over Ethernet.

11.2.2 Multipath Packet-based Fronthaul for Enhancing Reliability

In addition to low latency and excessive data rate, reliability is also an important metric in 5G [10], [11] as many use cases will rely on reliability. Under the Cloud-RAN architecture, FH needs to provide comparative reliability to enable adoption of Cloud-RAN. The most well-used methods to improve reliability are retransmission, multipath with packet duplication (MPD), and multipath with coding (MPC). Retransmission is a straightforward way to achieve reliability. However, retransmission can have significant impact on increasing the latency, making it a non-viable solution on the FH where delays cannot be afforded. By contrast, path diversity with duplication offers better latency in the expense of significant transmission overhead by duplicating packets over multiple interfaces increasing FH network congestion. Two important considerations in such approaches are latency and FH overhead. An alternative solution that provides trade-off between latency and FH overhead is channel coding, which can add redundancy to achieve the desired reliability and splits the total amount of information to transmit across different paths. For example, fountain coding is used in a number of different domains; such domains include broadcasting system [12], [13], content download [14], and over air interface [15], and has been proved to enhance reliability.

11.2.3 Slicing Within Multipath Fronthaul

FH resource sharing can be categorized into two primary categories, non-orthogonal sharing and orthogonal sharing. In the formal, all FH resources are shared among the services.

In the case of orthogonal sharing, a dedicated amount of resources are allocated to services. For instance, a percentage of the bandwidth or a percentage of the available path is allocated to a service to guarantee the required QoS. This type of orthogonality enables service slicing in the coexistence of heterogeneous service to offer the requirement according to the specific service.

Traffic is steered to the appropriate slice using, for example, the concept of identification. The packets, hence, are enqueued into the specific FH path associated to a slice identity. Each slice offers a set of reserved KPIs such as delay and packet loss.

11.3 Experimentation Results Evaluation of Flexible Functional Split for RAN Slicing

11.3.1 Experimental Setup

In this section, we study the flexibility in RAN configuration in terms of splitting the baseband functionalities between CU and DU entities, and the impact of this flexibility on communication latency and jitter is examined. Higher layer and lower layer splits have been implemented, i.e. PDCP-RLC, MAC-PHY, and intra-PHY, using the real-time platform Open Air Interface (OAI).

The experimental setup for the evaluation of the FH functional split is depicted in Figure 11.3 [6]. The overall experimental testbed consists of an end-to-end LTE system from eNodeB to user equipment (UE), with full implementation of functionalities of the protocol stack of the eNodeB and the UE.

The OAI UE is based on LTE UE developed by the OAI community [16]. The software runs on Intel Core i5-650 @ 3.20 GHz with 2 cores and it is connected via Universal Serial Bus 3 to a USRP (Universal Software Radio Peripheral) used to transmit and receive data.

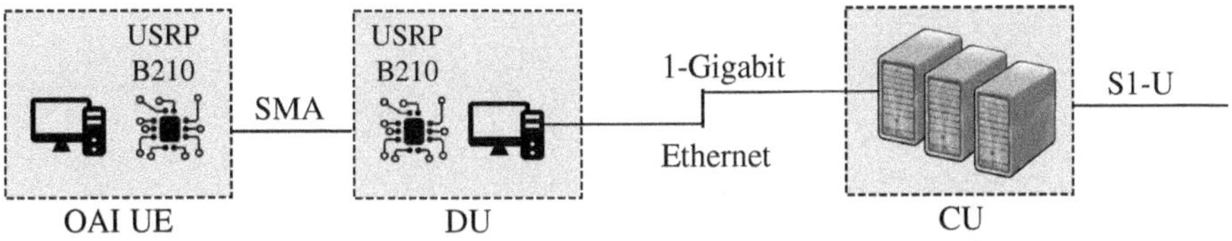

Figure 11.3 Setup of the testbed platform. Source: © 2017, IEEE Reprinted, with permission Mountaser et al. 2017 [6].

Table 11.2 Experimentation parameters.

Parameters	Values
Carrier frequency	2.68 GHz
System bandwidth	5 MHz (25 RBs)
Frame type	FDD
Tx/Rx antennas	1 Tx antenna/1 Rx antenna
Modulation schemes	16 QAM and 64 QAM
IP packet size (bytes)	1500 eMBB, 500 URLLC, and 300 mMTC

The OAI eNodeB is modified and divided into two entities, the DU and the CU, to support Cloud-RAN. They run on separate servers, each with an 8 GB RAM with Xeon 1220 and 4 cores. The DU and CU are connected with Ethernet link with a capacity of 1 gigabit. Focusing on the DL direction, the IP packets are injected in the eNodeB PDCP layer and are then processed by the whole protocol stack in CU. Afterwards, the PDU in the CU is packetized. The Ethernet packet is then transmitted to the DU via Ethernet. Upon arrival, the DU depacketizes the Ethernet packet into PDU, which undergoes further baseband processing. The configuration parameters are listed in Table 11.2.

11.3.2 Evaluation and Discussion of the Results

Figure 11.4 evaluates the total latency from PDCP to PHY for user plane packets. The aim is to compare the latency added by the different functionality splits in addition to the legacy protocol stack processing latency. We can note from this analysis that enhanced mobile broadband (eMBB) is the most affected service. This is due to the large packet size, which introduces higher computation load and hence delay. The massive machine type communications (mMTC) and ultra-reliable low-latency communications (URLLC) services are impacted by the splits in almost the same way. The only difference is that URLLC has a latency of 10 μs higher than mMTC due to the larger packet size. From this analysis we can observe that intra-PHY split adds a larger latency than PDCP-RLC and MAC-PHY splits.

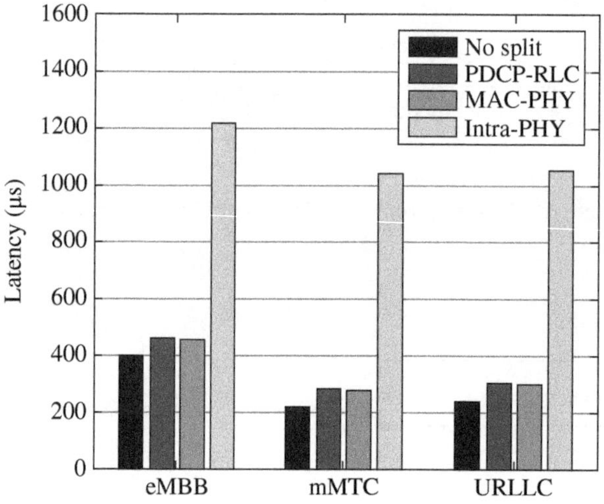

Figure 11.4 Latency for an IP Packet from when it is injected to PDCP layer to when it is transmitted to the UE. Source: © 2017, IEEE Reprinted, with permission Mountaser et al. 2017 [6].

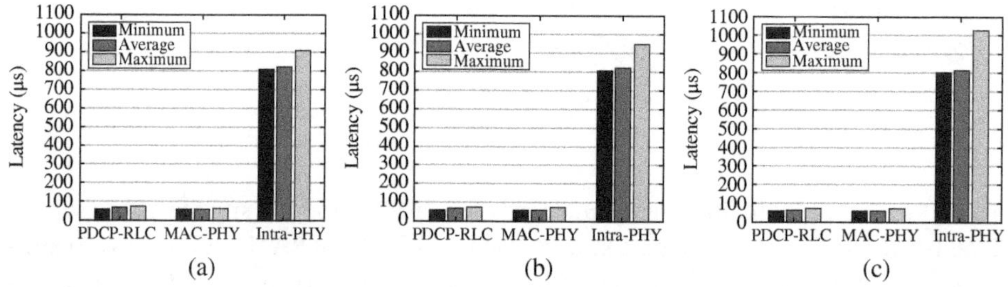

Figure 11.5 Latency from the upper to the lower layer for different splits for 5G services. (a) eMBB; (b) mMTC; (c) URLLC. Source: © 2017, IEEE Reprinted, with permission Mountaser et al. 2017 [6].

This is because, for the intra-PHY split, all 14 symbols, and thus the 14 packets, need to be transmitted by the DU for each transmission time.[1]

In Figure 11.5 the focus is on the latency introduced by each split (i.e. the time interval from when a packet is transmitted by the CU to when the packet is successfully received by the DU). From Figure 11.5, we can observe that the average latency of PDCP-RLC and MAC-PHY splits is almost constant for all 5G services, which equals ~65 μs for PDCP-RLC and ~60 μs for MAC-PHY. Hence, the PDCP-RLC and MAC-PHY splits work in a more consistent way compared to intra-PHY in terms of average latency. While the average latency of the intra-PHY split is high and equal to ~810 μs this is due to congestion caused by many packets to be sent in each transmission time. From latency point of view, PDCP-RLC and MAC-PHY are both suitable split options for URLLC and eMBB as being able to guarantee low latency.

1 This behavior is due to current OAI implementation; improvements could be achieved by employing, for example, compression and multiplexing.

For mMTC, intra-PHY split may be a candidate solution for the following reason. All the traffic received by the CU (i.e. both user- and control-plane traffic) will be translated on the FH with a fixed data rate as it depends only on the channel bandwidth. This is beneficial especially for new technologies expected to be used for mMTC such as Narrow-Band Internet of Things (NB-IoT) [17].

11.4 Simulation Results Analysis of Multipath Packet-Based Fronthaul for RAN Slicing

11.4.1 Simulation System Model

The aim of this section is to investigate how FH resources should be shared within a slice that is composed of MAC-PHY split to increase the performance of the system. In this section, we first introduce the system model for the Cloud-RAN system with multipath FH and then compare the solution with single-path (SP) FH transport and MPD under both orthogonal and non-orthogonal sharing of FH resources in the coexistence of eMBB and URLLC services.

As discussed, in order to improve the reliability of FH transfer, a solution that applies multipath transmission with coding is used [18] (Figure 11.6). The simulation model consists of Cloud-RAN with a single CU and a single DU connected with multiple FH paths (n different paths), where each path i has a capacity ψ_i. Packets of size B bits arrive at the system with exponential inter-arrival periods with average $1/\lambda$ seconds (s). We assume that the FH links are identical. Each link is modeled as a single queue. We suppose that the service time of each queue follows an exponential distribution. The mean service time to transmit a packet of size B bits from CU to DU is $1/\mu = B/\psi$ s. The packets within each queue are served in a first-in first-out manner and the buffer length is assumed to be infinite. The focus of our model is on DL direction. However, all arguments are valid for uplink (UL). The system model parameters are shown in Table 11.3.

To elaborate, we assume that eMBB and URLLC traffics are independent and characterized by arrival rates and packet sizes as shown in Table 11.3. For each traffic, PDUs are tagged independently and with equal probability as intended for either UE.

We compare three different FH resource allocation strategies for the eMBB and URLLC services:

- *FH bandwidth orthogonal allocation*: A fraction of the capacity of each path is exclusively given to each service.
- *FH path orthogonal allocation*: Each path is allocated exclusively to either one or the other service.
- *Shared FH*: All FH paths are shared between traffic types.

Figure 11.7 shows the probability of error as a function of the latency under orthogonal bandwidth allocation with four-fifth of the bandwidth allocated to URLLC and one-fifth to eMBB. MPC with $k = 2$ can reduce latency by 0.3 ms in eMBB and 0.0075 ms in URLLC, with respect to MPD at the error probability of 10^{-5}. Furthermore, dedicating a large bandwidth to URLLC can significantly enhance the probability of error as compared to shared FH, but this happens at the cost of a larger latency for the eMBB service.

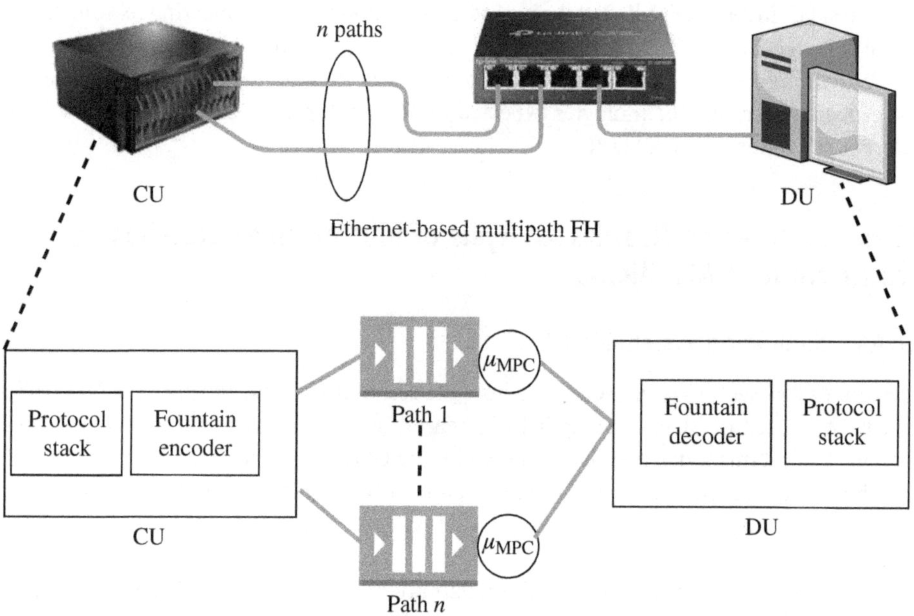

Figure 11.6 Multipath FH with erasure coding (MPC) for downlink communication with MAC-PHY split [18].

Table 11.3 System model parameter.

Parameters	**Values**
Ethernet capacity	100 Mbps
n	10
Packet arrival rates (packet/ms)	8 eMBB, 24 URLLC
IP packet size (bytes)	1500 eMBB, 500 URLLC

Figure 11.8 shows the probability of error obtained with path split where $n_u = 8$ is the number of paths allocated to URLLC of the available path, $n = 10$. For URLLC, MPC can manage to reduce latency by 0.023 ms as compared to MPD at the error probability of 10^{-5}. Moreover, the probability of error of path split is improved by 60% as compared to shared FH.

To summarize, the error probability is improved as compared to MPD by proposing a solution based on erasure coding and multipath transmission on the FH network. Furthermore, by slicing the FH resources adequately via orthogonal allocation transmission, the error probability can be reduced compared to non-orthogonal FH allocation.

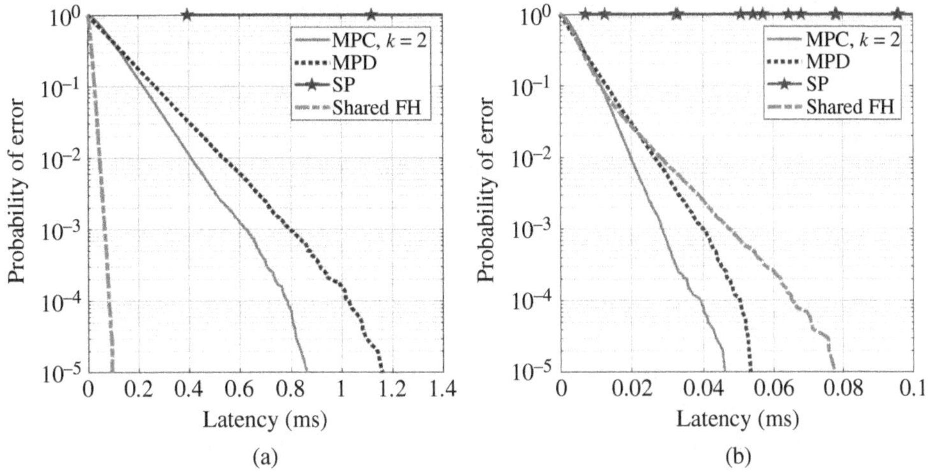

Figure 11.7 Probability of error vs. latency functions for SP, MPD, and MPC under orthogonal FH bandwidth split with eMBB bandwidth fraction 1/5: (a) eMBB; (b) URLLC. Source: © 2018, IEEE Reprinted, with permission Mountaser et al. 2018 [18].

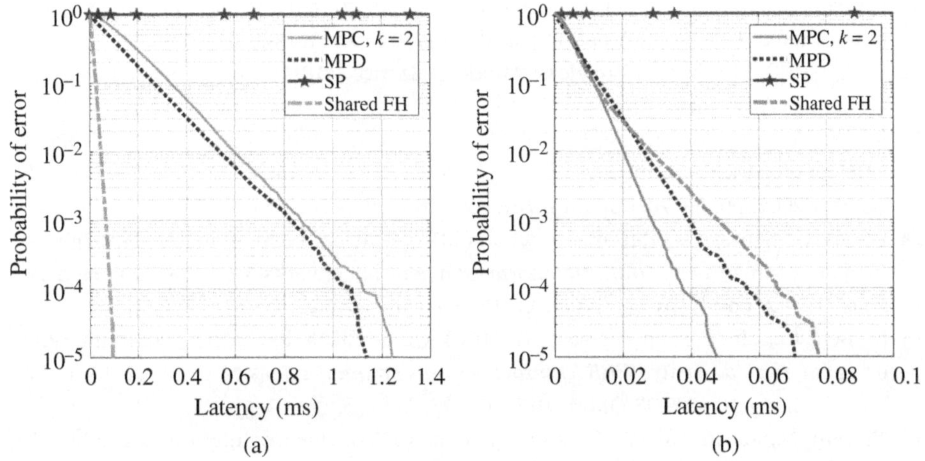

Figure 11.8 Probability of error vs. latency functions for SP, MPD, and MPC under orthogonal FH bandwidth split with eMBB path fraction 1/5 ($n_e = 2$ and $n_u = 8$): (a) eMBB; (b) URLLC. Source: © 2018, IEEE Reprinted, with permission Mountaser et al. 2018 [18].

References

1 Marabissi, D. and Fantacci, R. (2019). Highly flexible RAN slicing approach to manage isolation, priority, efficiency. *IEEE Access* 7: 97130–97142. doi: https://doi.org/10.1109/ACCESS.2019.2929732.

2 Rodriguez, V.Q. and Guillemin, F. (2017). Towards the deployment of a fully centralized cloud-RAN architecture. *2017 13th International Wireless Communications and Mobile Computing Conference (IWCMC)*, June 2017, pp. 1055–1060. doi: https://doi.org/10.1109/IWCMC.2017.7986431.

3 China Mobile (2011). C-RAN: the road towards green RAN. *White Paper*, 2, 2011.

4 3GPP (2017). Study on New Radio Access Technology: Radio Access Architecture and Interface (Release 14). *Technical Report 38.801.*

5 3GPP (2013). Study on Small Cell Enhancements for E-UTRA and E-UTRAN - Higher layer aspects. *Technical Report 36.842.*

6 Mountaser, G., Condoluci, M., Mahmoodi, T. et al. (2017). Cloud-RAN in support of URLLC. *2017 IEEE GC Wkshps*, December 2017. doi: https://doi.org/10.1109/GLOCOMW.2017.8269135.

7 Small Cell Forum (2016). Small cell virtualization functional splits and use cases. Small Cell Forum Document 159.07.02.

8 Mountaser, G., Rosas, M.L., Mahmoodi, T., and Dohler, M. (2017). On the feasibility of MAC and PHY split in cloud RAN. *IEEE WCNC*, March 2017. doi: https://doi.org/10.1109/WCNC.2017.7925770.

9 CPRI (2017). eCPRI Specification V1.0, Common Public Radio Interface.

10 Sachs, J., Wikstrom, G., Dudda, T. et al. (2018). 5G radio network design for ultra-reliable low-latency communication. *IEEE Network* 32 (2): 24–31. doi: https://doi.org/10.1109/MNET.2018.1700232.

11 Sulieman, N.I., Balevi, E., Davaslioglu, K., and Gitlin, R.D. (2017). Diversity and network coded 5G fronthaul wireless networks for ultra reliable and low latency communications. *IEEE PIMRC*. doi: https://doi.org/10.1109/PIMRC.2017.8292238.

12 3GPP (2019). Multimedia Broadcast/Multicast Service (MBMS); Protocols and codecs. *Technical Report 26.3462.*

13 Ma, Y., Yuan, D., and Zhang, H. (2006). Fountain codes and applications to reliable wireless broadcast system. *2006 IEEE Information Theory Workshop - ITW '06 Chengdu*, October 2006, pp. 66–70. doi: https://doi.org/10.1109/ITW2.2006.323758.

14 Parisis, G., Sourlas, V., Katsaros, K. et al. (2016). Efficient content delivery through fountain coding in opportunistic information-centric networks. *Computer Communications* 100: 12. doi: https://doi.org/10.1016/j.comcom.2016.12.005.

15 Nielsen, J.J., Liu, R., and Popovski, P. (2018). Ultra-reliable low latency communication using interface diversity. *IEEE Transactions on Communications* 66 (3): 1322–1334. doi: https://doi.org/10.1109/TCOMM.2017.2771478.

16 Nikaein, N., Marina, M.K., Manickam, S. et al. (2014). OpenAirInterface: a flexible platform for 5G research. *ACM SIGCOMM Computer Communication Review* 44 (5): 33–38.

17 Wang, Y.P.E., Lin, X., Adhikary, A. et al. (2017). A primer on 3GPP narrowband internet of things. *IEEE Communications Magazine* 55 (3): 117–123. doi: https://doi.org/10.1109/MCOM.2017.1600510CM.

18 Mountaser, G., Mahmoodi, T., and Simeone, O. (2018). Reliable and low-latency fronthaul for tactile internet applications. *IEEE Journal on Selected Areas in Communications* 36 (11): 2455–2463. doi: https://doi.org/10.1109/JSAC.2018.2872299.

12

Service-Oriented RAN Support of Network Slicing

Wei Tan, Feng Han, Yinghao Jin and Chenchen Yang

Huawei Technologies Co., Ltd.

12.1 Introduction

The era of the fifth generation (5G) of cellular networks is approaching rapidly. The services foreseen in 5G fall into three typical scenarios: enhanced mobile broadband (eMBB), ultra-reliable low-latency communications (URLLC), and massive machine type communications (mMTC) [1]. eMBB focuses on services characterized by high data rates, such as high definition (HD) videos, virtual reality (VR), augmented reality (AR), and fixed mobile convergence (FMC). URLLC focuses on latency-sensitive services, such as self-driving cars, remote surgery, or drone control. Lastly, mMTC focuses on services that have high requirements for connection density, such as those typical for smart city and smart agriculture use cases. Moreover, 5G is expected to support diverse and unprecedented vertical segments, including factories, oil, gas and energy, utilities and public safety, etc. These vertical services present quite different requirements in terms of reliability, data rate, latency, deployment density, and mobility [2]. All these vertical services and requirements require the 5G system to be flexible and agile [3]. The diverse services onto operators' networks expand the reach of operators' business and open new sources of revenue to service providers and mobile operators.

Network slicing, as one of the key technologies in 5G, allows network operators to define logical networks on top of the single infrastructure to adapt to specific requirements and service level agreement (SLA) [4]. It can facilitate the support of vertical segments with lower operating expense (OpEx), greater capital expenditure (CapEx) efficiency, fast deployment, and short time-to-market [5]. As an integral part of end-end network slice, radio access network (RAN) support of network slice is essential to empower the diverse vertical services, and can provide customized differentiation toward different slice requirements. But due to scarcity of spectrum, RAN slicing shares spectrum which is significantly different from core network (CN) slicing in which to decouple the user from the control plane, places the control plane on a logical centralized location, and exposes state and abstract resources to applications and tenants. In RAN slicing, several of the complex functions executed in base stations are tightly coupled with the hardware so as to achieve the desired performance. Therefore, designing new mechanisms to improve spectrum efficiency, meet different key

Radio Access Network Slicing and Virtualization for 5G Vertical Industries, First Edition.
Edited by Lei Zhang, Arman Farhang, Gang Feng, and Oluwakayode Onireti.

performance indicators (KPIs) for different vertical services, and bring sufficient incentives toward service providers to use slicing, becomes an essential and challenging issue.

The rest of the chapter is organized as follows. The general concept and principles of network slicing are presented in Section 12.2. In Section 12.3, we describe RAN subsystem deployment scenarios. Then, we propose key technologies to enable service-oriented RAN slicing in Section 12.4. Finally, Section 12.5 concludes this chapter.

12.2 General Concept and Principles

Before we introduce the RAN subsystem concept, a whole picture of end-to-end slicing is shown. Network slicing is an end-to-end concept, which always includes RAN slicing and CN slicing [6], although from a larger perspective end-to-end slicing includes terminal, RAN, transport network (TN) and CN, as well as a data center (DC) domain that hosts third-party applications from vertical industries (see Figure 12.1).

This section aims to clarify the definition of slicing concepts, such as network slicing, slice instance, slice type, and their corresponding relationships. Then, we will present the overall RAN subsystem and focus on key principles that apply for support of network slicing in RAN subsystems.

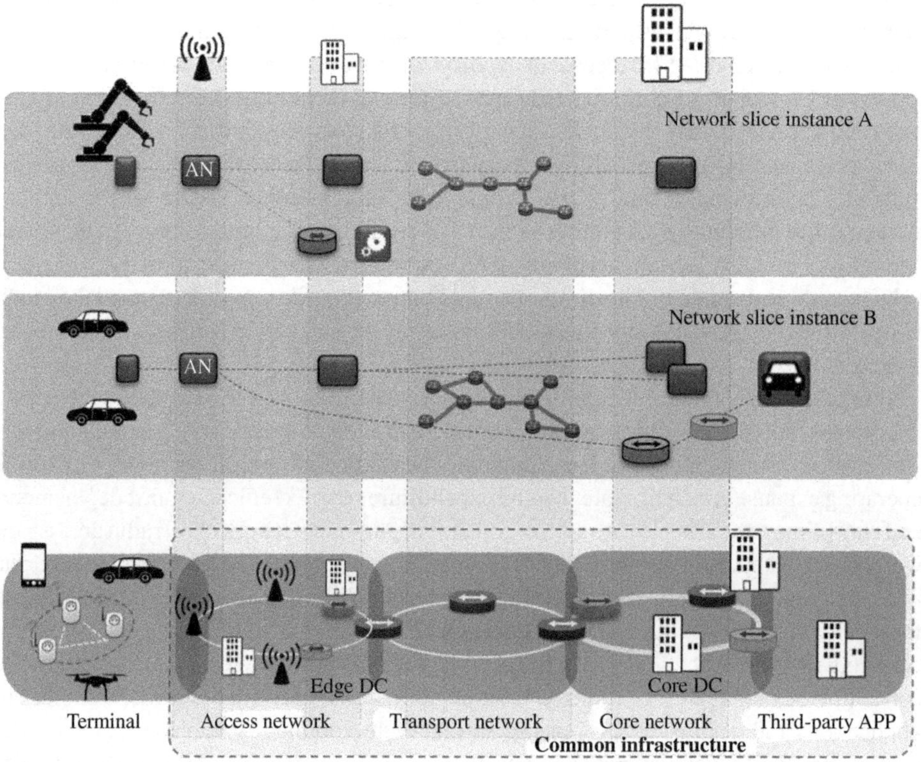

Figure 12.1 End-to-end network slicing.

12.2.1 Network Slicing Concepts

In this section, we will introduce the basic concepts related to network slicing as follows [6, 7]:

- *Network slicing*: this is the collection of a set of technologies to create specialized, dedicated logical networks as a service (NaaS) in order to support different network services and meet the diversified requirements from vertical industries. Through flexible and customized design of functions, isolation mechanisms, and operation and maintenance (O&M) tools, network slicing is capable of providing logical dedicated networks upon a common infrastructure.
- *Network slice instance (NSI)*: an NSI is the realization of a network slice. It is an end-to-end logical network, which comprises of a group of network functions, resources, and connection relationships. An NSI typically covers multiple technical domains, which include terminal, RAN, TN and CN as well as DC domain. Different NSIs may have different network functions and resources. They may also share some of the network functions and resources such as computing, storage, and transmission.
- *Network slice types*: these are high-level categories for NSIs, which reflect the distinct demands for network solutions. Three fundamental network slice types have been identified for 5G: eMBB, mMTC, and URLLC. These could be further extended, e.g. according to the operator's policies or with the development of 5G.
- *Network slice template*: this is the output of the slice design phase used to create NSIs.
- *Tenants*: these are the operators' customers (for example, customers from vertical industries) or the operators themselves. They utilize the NSIs to provide services to their users. Tenants typically will have independent O&M requirements, which are uniquely applicable to the NSIs.

The following descriptions of relationship between the aforementioned key concepts help understand them better before proposing deployment scenarios and key technologies:

- *Network slice types and tenants are important references for creating an NSI*: an NSI is instantiated from one network slice template with a specific network slice type. A tenant that provides different service types may use multiple NSIs with different network slice types. Tenants who may provide services of the same service type can still use differentiated NSIs via customization of the network slice template with the same network slice types.
- *Template design is separated from the NSI operation*: in the design phase, the network slice template is generated based on the network capability of each technical domain and a tenant's particular requirements. In the operation phase, an NSI is instantiated based on the network slice template, which includes the deployment and configuration of related network functions and related resources in different technical domains. The network slice design is separated from the operation to enable repeated use of a network slice template.
- *NSIs require multidimensional management*: an NSI usually includes multiple technical domains such as RAN domain, CN domain, and TN domain. An NSI may also include multiple administrative domains that belong to different operators. To guarantee NSI's

fast deployment, it is essential to use efficient multidimensional management via coordination and cooperation across such different domains.

- *NSIs ensure SLA compliance*: tenants will sign SLA with operators, which may include requirement agreements related to security/confidentiality, visibility/manageability, specific service characteristics (service type, air interface standard, and customized functions), and the corresponding performance indicators (latency, throughput, packet loss rate, call drop rate, and reliability/availability).
- *Terminals may be involved in the selection of NSIs*: terminals can access one or multiple NSIs. Terminals could assist NSI selection based on, for instance, network slice type, while the network performs the final selection decision. Simple terminals, such as sensors, are usually in a static and one-to-one relationship with NSIs, because the costs and power consumption requirements limit the terminal capability. Therefore, the network solely performs NSI selection.

12.2.2 Overall RAN Subsystem

The RAN subsystem, i.e. the RAN part of network slice, holds the following features:

- A service device accesses the Third Generation Partnership Project (3GPP) network via a RAN subsystem as if it were connected to the service provider's network.
- The RAN subsystem is characterized by service-specific network topology, protocol configuration, connection control, and mobility management.

The key incentive is to allow operators to provide tailored solutions to vertical service providers over the same network infrastructure [1, 8]. An example is given in Figure 12.2. More specifically:

- For devices supporting factory automation slice, which is a combination of mMTC and URLLC, they would treat their accessed RAN subsystem as a factory automation "private" network with subsystem2. The low latency, high reliability, and even the deterministic data transmission would be through this "private" network.
- For cars supporting V2X slice, which is a combination of eMBB and URLLC, they may access the V2X "private" network to have both high-quality video service and

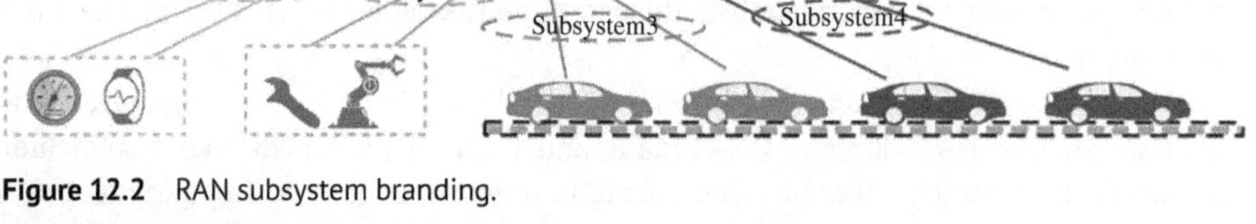

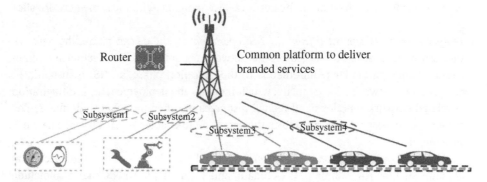

Figure 12.2 RAN subsystem branding.

fully automated driving. Different tenants used for different car companies may be treated with different RAN subsystems for different subsystems such as subsystem3 and subsystem4.

- Those sensors supporting metering slice, a kind of mMTC service, in the number of millions, can perceive to be accessing their own "private" network with subsystem1. These low-cost devices characterized by low burst data rate, long battery life, and large coverage requirements communicate via their own RAN subsystem network [9].

12.2.3 Key Principles of Network Slicing in RAN

In this section, the general principles related to the realization of network slicing in the next generation radio access network (NG-RAN) for new radio (NR) are given. The support of network slicing relies on the principle that traffic for different slices is handled by different packet data unit (PDU) sessions [10]. A network can realize the different network slices by scheduling and providing different physical layer (PHY) and medium access control (MAC) layer configurations [11].

Each network slice is uniquely identified by a Single Network Slice Selection Assistance Information (S-NSSAI), as defined in [11]. The NSSAI includes one or a list of where an S-NSSAI is a combination of:

- Mandatory SST (slice/service type) field, which identifies the slice type and consists of 8 bits (with range 0–255);
- Optional SD (slice differentiator) field, which differentiates among slices with the same SST field and consists of 24 bits.

The UE provides NSSAI for network slice selection by radio resource control (RRC), if it has been provided by non-access stratum (NAS). While the network can support a large number of slices, the UE does not need to support more than eight slices simultaneously.

The following key principles in [11] apply for the support of network slicing in NG-RAN:

- **RAN awareness of slices:**
  - NG-RAN supports a differentiated handling of traffic for different network slices that have been preconfigured. How NG-RAN supports slice enabling in terms of NG-RAN functions is implementation dependent.
- **Selection of RAN of the network slice:**
  - NG-RAN supports the selection of the RAN part of the network slice, by NSSAI provided by the UE or the 5G core network (5GC), which unambiguously identifies one or more of the preconfigured network slices in public land mobile network (PLMN).
- **Resource management between slices:**
  - NG-RAN supports policy enforcement between slices as per SLAs. It should be possible for a single NG-RAN node to support multiple slices. The NG-RAN should be free to apply the best radio resource management (RRM) policy for the SLA in place to each supported slice.
- **Support of quality of service (QoS)**:
  - NG-RAN supports QoS differentiation within a slice.

- **RAN selection of CN entity:**
  - For initial attachment, the UE may provide NSSAI to support the selection of an access management function (AMF). If available, NG-RAN uses this information for routing the initial NAS to an AMF. If the NG-RAN is unable to select an AMF using this information or the UE does not provide any such information, the NG-RAN sends the NAS signaling to one of the default AMFs.
  - For subsequent accesses, the UE provides a temporary identifier (Temp ID), which is assigned to the UE by the 5GC, to enable the NG-RAN to route the NAS message to the appropriate AMF as long as the Temp ID is valid (NG-RAN is aware of and can reach the AMF that is associated with the Temp ID). Otherwise, the methods for initial attachment apply.
- **Resource isolation between slices:**
  - The NG-RAN supports resource isolation between slices. NG-RAN resource isolation may be achieved by means of RRM policies and protection mechanisms that should prevent the shortage of shared resources in one slice from breaking the SLA for another slice. It should be possible to fully dedicate NG-RAN resources to a certain slice.
- **Access control:**
  - By means of the unified access control (see clause 12.4), operator-defined access categories can be used to enable differentiated handling for different slices. NG-RAN may broadcast barring control information (i.e. a list of barring parameters associated with operator-defined access categories) to minimize the impact of congested slices.
- **Slice availability:**
  - Some slices may be available only in part of the network. The NG-RAN supported S-NSSAI(s) is configured by O&M. Awareness in the NG-RAN of the slices supported in the cells of its neighbors may be beneficial for inter-frequency mobility in connected mode. It is assumed that the slice availability does not change within the UE's registration area.
  - The NG-RAN and the 5GC are responsible for handling a service request for a slice that may or may not be available in a given area. Admission or rejection of access to a slice may depend on factors such as support for the slice, availability of resources, and support of the requested service by NG-RAN.
- **Support for UE associating with multiple network slices simultaneously:**
  - In case a UE is associated with multiple slices simultaneously, only one signaling connection is maintained and for intra-frequency cell reselection, the UE always tries to camp on the best cell. For inter-frequency cell reselection, dedicated priorities can be used to control the frequency on which the UE camps.
- **Granularity of slice awareness:**
  - Slice awareness in NG-RAN is introduced at PDU session level, by indicating the S-NSSAI corresponding to the PDU session, in all signaling containing PDU session resource information.
- **Validation of the UE rights to access a network slice:**
  - It is the responsibility of the 5GC to validate that the UE has the rights to access a network slice. Prior to receiving the initial context setup request message, the NG-RAN may be allowed to apply some provisional/local policies, based on the

awareness of which slice the UE is requesting access to. During the initial context setup, the NG-RAN is informed of the slice for which resources are being requested.

12.3 RAN Subsystem Deployment Scenarios

The support of diverse vertical services with distinctly different requirements requires flexible RAN subsystem deployment. In particular, for shopping mall or factory automation, the deployment coverage could be limited with a few base stations, while for V2X services they may need larger area because of high mobility [8]. Flexible deployment alternatives are needed to adapt to the specific requirements while maintaining overall cost efficiency. Several examples are given as follows:

- *Scenario 1*: deployments on a single frequency band (cell/tracking area granularity):
  The example on a single frequency band is given in Figure 12.3, where subsystem1 is supported by all RAN nodes. RAN subsystem2 and subsystem3, with limited coverage requirements, are supported by different RAN nodes.
- *Scenario 2*: deployment over multifrequency bands:
  For multifrequency bands, the example given in Figure 12.4 shows that RAN subsystem1 is supported on the first frequency band and subsystem2 on the second band. This could be applicable to RAN subsystems with large service areas [10].

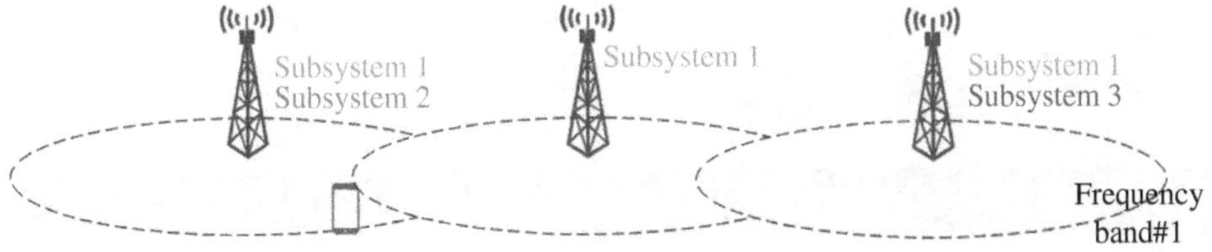

Figure 12.3 RAN subsystem deployment example on a frequency band.

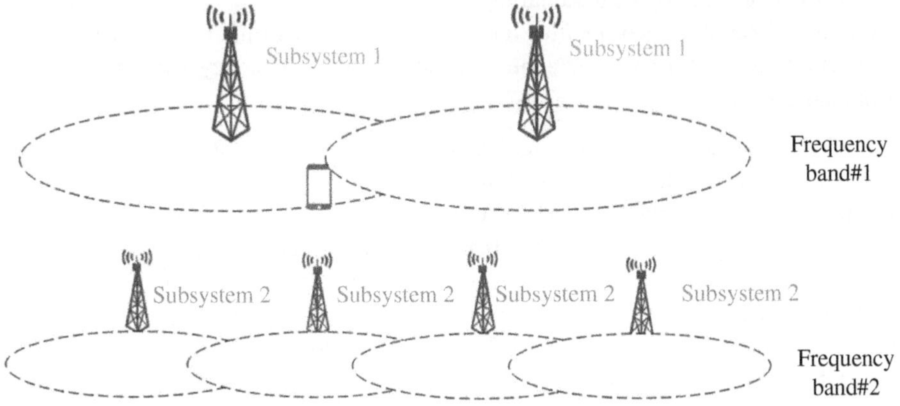

Figure 12.4 RAN subsystem deployment example over multifrequency bands.

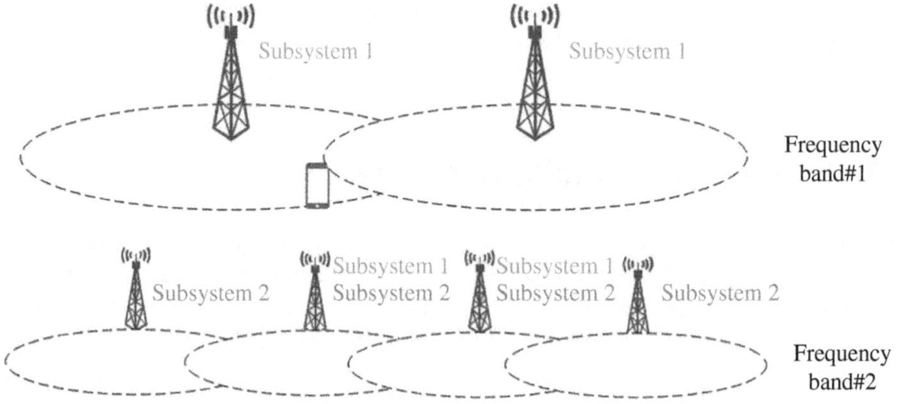

Figure 12.5 Hybrid deployment example over multifrequency bands.

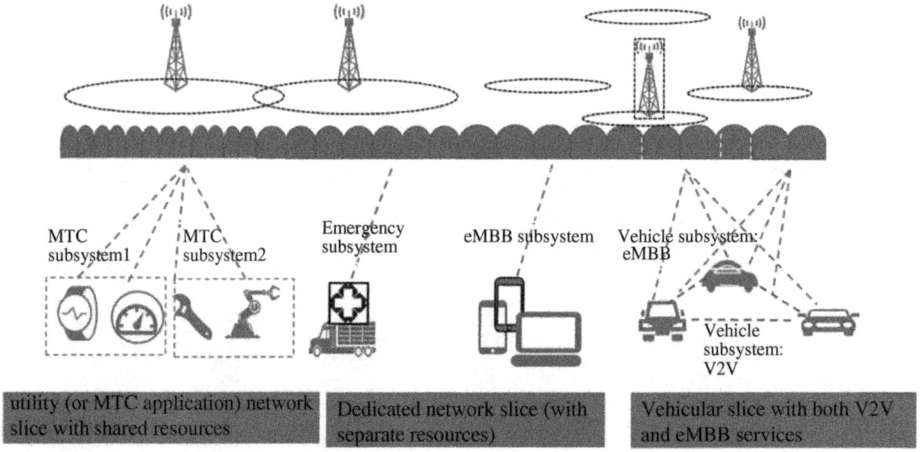

Figure 12.6 Deployment scenarios of RAN parts of network slice.

- *Scenario 3*: hybrid deployment scenarios:
  In this hybrid example, incorporating the above two scenarios, different RAN subsystems can be supported within a single frequency band with multifrequency bands deployment, as in Figure 12.5.

In 3GPP Release 15, it is assumed that UE supported network slices are homogeneously available in a single registration area including a tracking area list. However, for the above deployments, the RAN subsystem could be deployed even in a cell granularity. There is no strict relationship of registration area assignment and UE supported slices. Hence, this can highly increase the RAN deployment flexibility, and allow mobile operators to have lower OpEx and greater CapEx efficiency [8]. Overall, Figure 12.6 gives the deployment scenario. Specifically, some RAN subsystems could be deployed on lower frequency bands below 6 GHz, while other RAN subsystems could be supported on high frequency bands, e.g. above 24 GHz [12]. Also, it can be observed that some RAN subsystems may share common radio resources while others would require dedicated and separate resources.

12.4 Key Technologies to Enable Service-Oriented RAN Slicing

To support a branded subsystem over a NG-RAN from the branded subsystem provided by the NG-RAN, the UE can obtain its "private" and specific services, e.g. potential vertical services. To achieve this, the 3GPP network needs to inform the UE of the correlations or mappings between the supported slices and the supported branded subsystems, e.g. via NAS or access stratum (AS) signaling.

To allow customized architecture and protocols in connecting devices of different service providers, different logical dedicated networks should be provided upon a common 3GPP infrastructure via network slicing for the efficient connection of different services. To achieve this, the architecture and protocols for different slices should be allowed to be customized and designed differently upon a common 3GPP infrastructure, e.g. the slice-specific protocol layer, and protocol functionalities are customized to adapt to the slice-oriented services.

Detailed objectives of the work items are as follows:

- Device awareness of RAN part of network slice:
  - Detection of the availability of the corresponding service: it aims to solve the problems of (i) how NG-RAN/CN notifies UEs that services related to the RAN part of network slice are supported in related NG-RAN cells, and (ii) how UEs notify NG-RAN/CN that services related to the RAN part of network slice are requested by the UEs.
  - Support of service-specific connectivity control and service continuity in mobility: it aims to solve the problem of how the service continuity can be guaranteed when the UE moves to areas that do not support the connected RAN subsystem, and it can be guaranteed more easily when the UE moves within areas where the connected RAN subsystem is supported.
- Slice-specific RAN part of network slice, e.g.:
  - PHY/MAC configuration, access control, and mobility: it aims to achieve the goal that PHY/MAC parameters and resources can be configured and scheduled differently and flexibly in UE or NG-RAN for different slice-specific services during the UE access and mobility procedures, for example, different datalink protocols or PHY parameters such as transmission time interval (TTI) numerology can be configured differently for different services supported by eMBB, URLLC, and mMTC slices.
  - Network topology and air interface protocol adaptation: it aims to achieve the goals that (i) the NG-RAN may configure and support service-oriented network topologies dynamically and flexibly for different slices based on the requested slice types of the connected UEs; (ii) dynamic and flexible air interface adaptation and scheduling should be supported by the NG-RAN based on the service-oriented network topologies, for example, the NG-RAN should provide different multi-layer air connections or communication modes to different UEs supporting multiple slices based on the service-oriented network topologies to maximize the network performance.
- Mission-driven resource utilization, sharing, and aggregation, e.g. by specifying the following:

- Mechanisms to support multi-layer connection hierarchy to UE with service subscriptions to multiple providers: it aims to achieve the goal that the RAN may provide slice-specific multi-layer connection to one UE supporting multiple network slices simultaneously, for example, the multi-layer connection may include side-link V2X communication, UE-to-UE communication, and broadcast/multicast mode communication supported by different slices, respectively.
- Mechanisms to allow UE to have multiple data paths for various services based on SLA requirements: it aims to achieve the goal that the NG-RAN can enable UE to have various data paths for various services toward RAN subsystems related to different slices, for example, semi-static radio resources and data paths for services with different SLA requirements can be allocated to the UE.
- The support of dual connectivity across RAN parts of different network slices: it aims to achieve the goal that Master Node (MN) and Secondary Node (SN) PHY can be flexibly configured and released for the UE supporting multiple branded subsystems related to multiple slices, for example, dual connection can be configured to the UE if both the subsystem supported by MN and the subsystem supported by SN are supported by the UE, and the UE falls back to single connection when only one of the subsystems supported by MN and SN is supported by the UE.

12.4.1 Device Awareness of RAN Part of Network Slice

To support the branded RAN subsystem, the UE should access the 3GPP network as if it were connected to its supported RAN subsystem directly. As Figure 12.7 shows, when the UE registers and accesses the network, it may perform the selection and reselection of the RAN subsystem, and treat the RAN subsystem as its own "private" network. In order to achieve this, the RAN node may broadcast its supported RAN subsystems, to assist the UE to perform RAN subsystem (re)selection. It may also broadcast those neighbor frequency bands and their supported RAN subsystems for inter-frequency subsystem (re)selection.

We envision two potential solutions for the UE to be aware of RAN subsystem. An example is given in Figure 12.7.

- *Option 1*: Each cell broadcasts its supported S-NSSAIs:
  The UE tries to access the 3GPP network based on its supported S-NSSAIs and cell broadcasted S-NSSAIs. As the RAN may support hundreds of slices and should ensure scalability with more slices in the future, it is essential to reduce the broadcasting overhead.

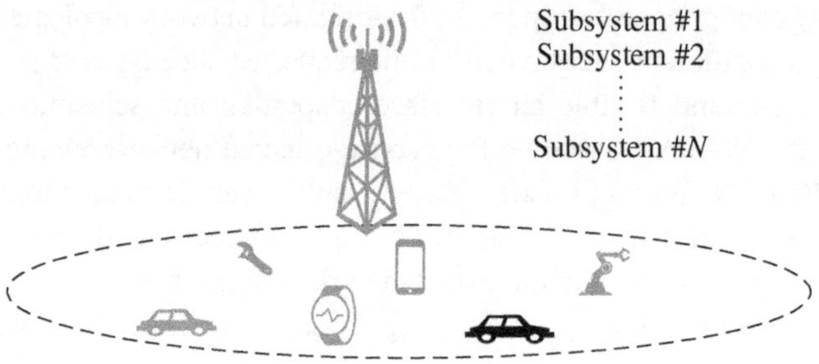

Figure 12.7 Device awareness of RAN subsystem.

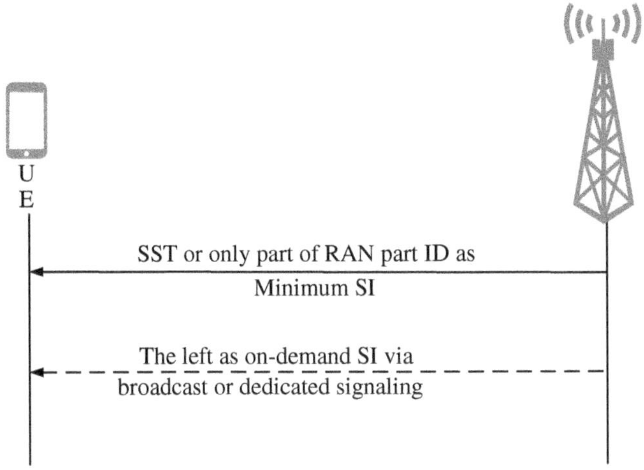

Figure 12.8 UE acquisition of system information.

Hence, the RAN node may broadcast the SST as minimum system information (SI), and treat optional SD as other SI either to be broadcast or provisioned in a dedicated manner upon UE request. The UE acquisition of SI is provided in Figure 12.8.

- *Option 2*: Each cell broadcasts its supported RAN part IDs:
  In this option, the UE may map its supported S-NSSAIs to RAN part IDs, and try to camp on those cells that support its supported RAN part IDs. Similar to option 1, this option may treat the essential information of RAN part IDs as minimum SI, and the optional SD as other SI to reduce the broadcast overhead. This option requires the UE to maintain the mapping relationship between S-NSSAIs and RAN part IDs, which could be provided in case registration update request by the RAN node.

The principles of RAN subsystem selection are as follows:

- The UE NAS layer identifies the network slices the UE supports and transfers to UE AS.
- The UE AS searches the NR frequency bands, and for each carrier frequency identifies the strongest cell that broadcasts S-NSSAIs or RAN part IDs, and tries to camp on the cell.
- If the UE is not able to identify a suitable cell that supports its associated slices, it seeks to identify an acceptable cell:
  - a suitable cell is the one for which the measured cell attributes satisfy the cell selection criteria, and the UE's associated RAN subsystem is supported;
  - an acceptable cell is the one for which the measured cell attributes satisfy the cell selection criteria but the cell does not support the UE associated cells.

The principles of RAN subsystem reselection for idle and inactive UEs are given as follows:

- The UE makes measurements of attributes of the serving and neighbor cells to enable the reselection process.
- The UE will prioritize the frequency band that supports its associated RAN subsystems with the corresponding S-NSSAIs or RAN part IDs.
- The UE will prioritize the cell that supports its associated RAN subsystems from those cells satisfying the cell reselection criteria.

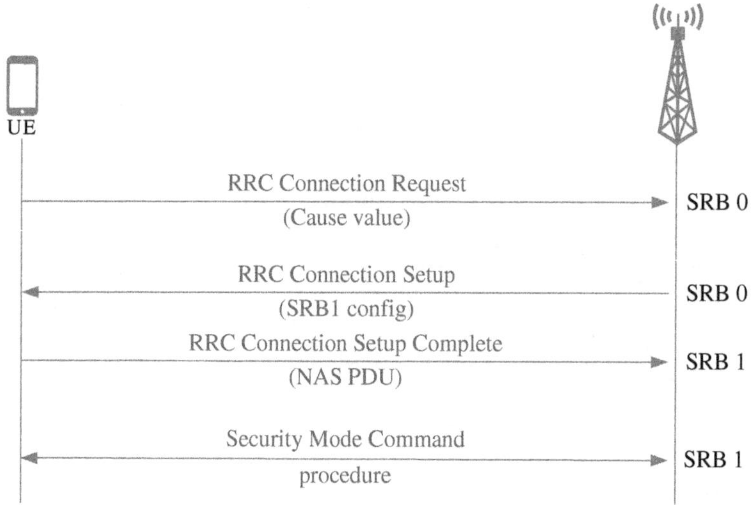

Figure 12.9 Transitions from RRC_IDLE to RRC_CONNECTED.

12.4.2 Slice-Specific RAN Part of Network Slice

When the UE successfully accesses the network, the RAN may provide slice-specific configuration and differentiated handlings to meet the slice requirements and SLA.

Figure 12.9 gives the UE transition procedure, which involves the following steps:

- After the random access procedure with the network, in Message3, the UE may carry the cause value to indicate the RAN subsystem the UE tries to access. Based on this, the RAN may perform early admission control, for example, to reject some access request in case of congestion situations.
- In Message5, the UE may include slice information during the registration procedure for AMF selection.

Figure 12.10 gives the PDU session setup/modify procedure, in which the RAN may identify the specific slice requirements based on the S-NSSAI per PDU session and perform the slice-specific RAN part of network slice configurations.

The detailed slice level configuration includes the following aspects.

- Slice-specific network topology:
  Based on the requested slice type, the RAN may configure and support service-oriented network topologies for different slices. The following gives some examples:
  - If the requested slice is a V2X slice, the RAN may support both side-link V2X communication and UE-to-UE communication. Also, it can support both unicast and broadcast communication mode for this UE.
  - If the requested slice is for media delivery, the broadcast/multicast mode communication could be supported.
  - For wearables and Internet of Things (IoT) slice, the RAN may use multi-hop and relaying like integrated access and backhaul (IAB) topology.

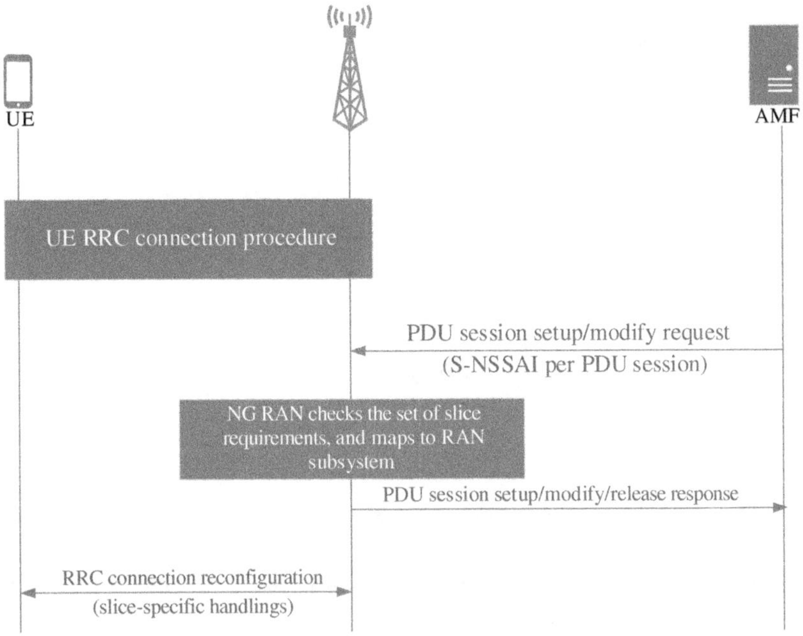

Figure 12.10 RAN subsystem-aware PDU session setup/modify.

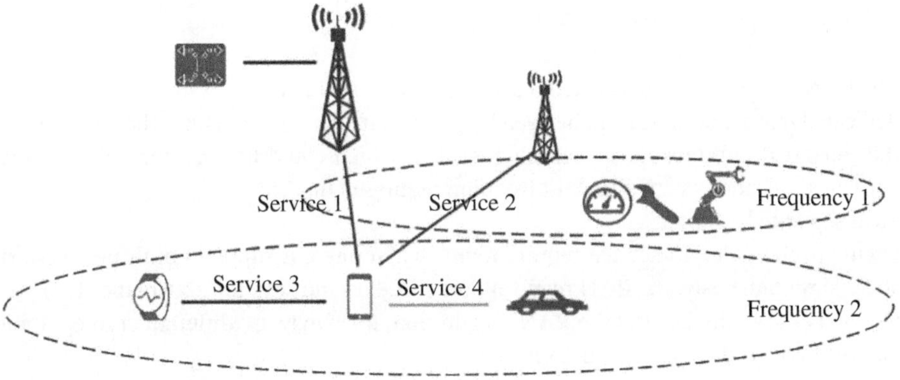

Figure 12.11 Service-oriented RAN topology.

- For UEs supporting multiple slices simultaneously, multi-layer connection hierarchy should be supported.

Figure 12.11 gives an example in which the UE communicates with one RAN node via frequency F1 in RAN sub-system 2 and communicates with another RAN node via frequency F2 in another RAN sub-system 1.

- Slice-customized protocol adaptation:
  The RAN can tailor the slice-specific protocol layer and protocol functionalities to adapt to slice-specific requirements. The following gives several illustrative examples.

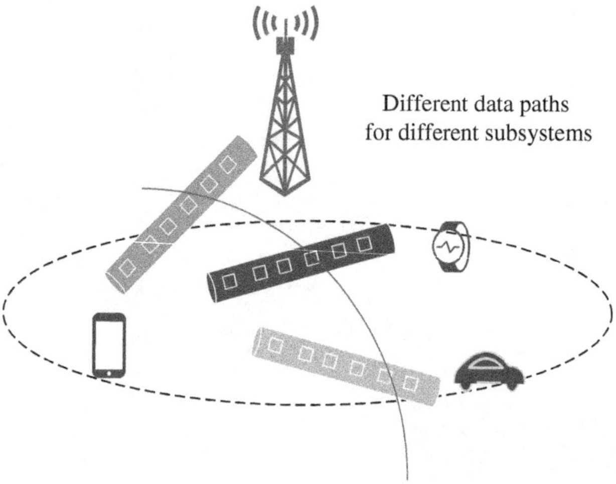

Figure 12.12 Single-connectivity with different RAN subsystems.

- The UE is configured with full data link layer protocol stack and all functionalities for eMBB slices.
- To support URLLC slices, the UE in carrier aggregation/dual connectivity (CA/DC) case is configured with data link layer and short TTI numerology mapping.
- mMTC slices may not be configured with ciphering function or even without packet data convergence protocol (PDCP) sublayer.

- Slice-separated resource allocation:
  The RAN can also allocate isolated resources for different RAN subsystems. For example, dedicated frequency bands can be used for particular slices. Even within the single band, the separated radio resources can be allocated, which could be used for public safety, vertical applications with stringent isolation requirements, etc.
- Slice-separated security:
  Owing to the different service requirements, when one UE supports multiple network slices simultaneously, the RAN may support slice-specific keys for user plane (UP) protection between the UE and the RAN. In addition, RAN may use different cryptographic algorithms or key length for different slices.

12.4.3 Mission-Driven Resource Utilization, Sharing, and Aggregation

When the UE simultaneously supports multiple slices via different RAN subsystems, the RAN can provide efficient resource utilization to meet the requirements of these slices with high spectrum efficiency. The following analyzes single connectivity and dual connectivity respectively.

- Resource allocation in single-connectivity:
  When the RAN node identifies UE-supported slices, it can enable the UE to have multiple data paths for various services toward RAN subsystems based SLA requirements. One example is given in Figure 12.12, where the UE has three communication paths

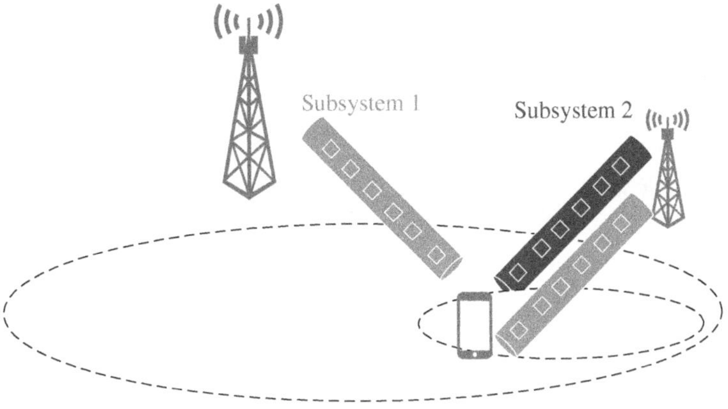

Figure 12.13 Dual-connectivity with different RAN subsystems.

toward the different RAN subsystems that it supports. The RAN may allocate semi-static radio resources for which the UE performs autonomous resource selection, or scheduled resources to support the data transmission for different RAN subsystems.

- Resource sharing and aggregation in dual connectivity:
  Dual connectivity can be used to provide aggregated resources for some slices while it is also allowed for the RAN node to support some slices via single link. One example is given in Figure 12.13, where the RAN subsystem1 can be supported by both MN and SN, while subsystem2 is supported only by SN. In order to allow multiple connectivity across different subsystems to achieve system efficiency, the following should be considered:
  - The MN should be aware of UE-supported RAN subsystems and SN-supported subsystems for SN selection.
  - When the UE is within the SN coverage, it may access the RAN subsystem1 and RAN subsystem2 simultaneously.
  - When the UE moves out of the RAN subsystem coverage, service continuity can be ensured if the traffic of the unsupported slice is allowed to be switched to another slice. In addition, inter-slicing data forwarding is needed.

12.4.4 Slice-Aware Connected UE Mobility

Network controlled mobility should take slice information into account. Specifically, the RAN will take RAN subsystems supported by neighbor cells as one factor to select the proper target cell for the connected UE. The RAN node and its neighboring nodes may exchange their supported RAN subsystems with each other. Based on this, the RAN can endeavor to support the service connectivity for this UE.

When the UE moves within areas where the RAN subsystem is supported, service continuity could be ensured. However, when the UE moves to areas that do not support the RAN subsystem, as Figure 12.14 shows, the RAN can try to ensure service continuity. For example, some slicing mapping policies could be predefined, and the traffics could be switched to another slice for data continuity.

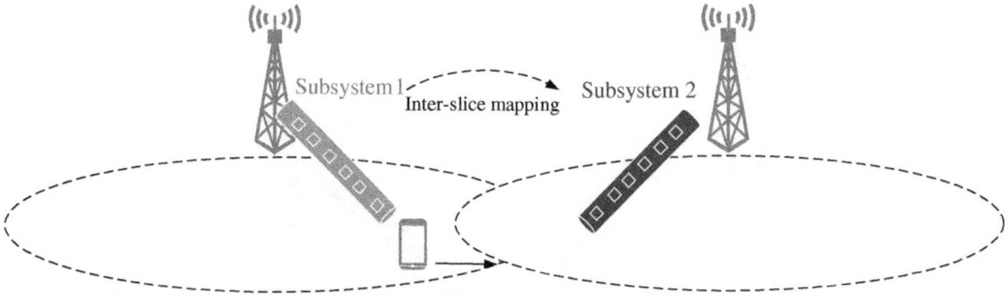

Figure 12.14 Slice service continuity during inter-cell handover.

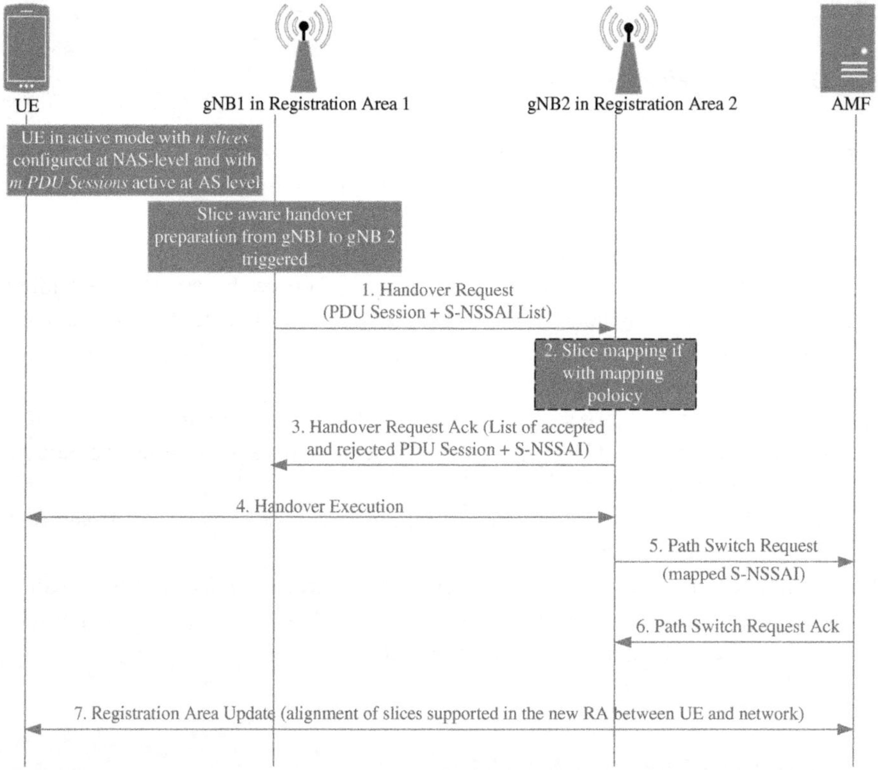

Figure 12.15 Xn based inter-cell handover.

The detailed network interface between gNBs (next generation NodeB) based handover is given in Figure 12.15. When gNB2 finds that the active PDU sessions associated with S-NSSAI#1 are not supported, based on the mapping policy as one of the UE contexts, it could map the S-NSSAI#1 to S-NSSAI#2, so that the service continuity of S-NSSAI#1 is ensured.

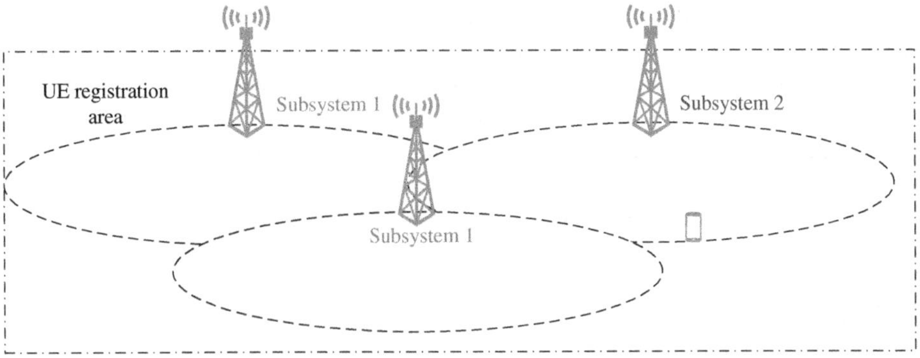

Figure 12.16 Illustration of UE RA with nonuniform RAN subsystems.

12.4.5 Slice-Level Handlings for Idle/Inactive UEs

- Paging procedure:
  As discussed above, the UE's registration area is not coupled with the UE-supported network slices. Hence, within the UE's routing area (RA), different RAN nodes support different RAN subsystems. When the CN initiates a paging message, it may happen that the RAN node does not support the CN's initiated PDU sessions of network slices. One example is given in Figure 12.16, where the paging procedure is initiated by S-NSSAI corresponding with RAN subsystem1 while the UE is located in the coverage of subsystem2. We envision several options as follows:
  - The paging message may carry the slice information to indicate which network slice triggers the paging request. Then the UE can respond to the request if it camps on the cell that supports the corresponding RAN subsystem, or stores the remapping policies that allow the RAN subsystem to support the initiated network slice.
  - The paging message may not carry the slice information. After the UE accesses the network, the CN may not initiate the PDU sessions for which the serving RAN node does not support the RAN subsystem. It may signal the RAN so that the RAN may perform the redirection procedure toward this UE to the cell/frequency.
- Slice-aware access control:
  In order to ensure that the congestion of one slice does not adversely affect other slices, the RAN may perform slice-specific access control. Moreover, different RAN subsystems could be treated differently using the unified access barring mechanism applicable for all RRC states. Each UE can perform access control based on its associated RAN subsystems accordingly.

12.5 Summary

In this chapter, we present the general concept, principles, and deployment scenarios of service-oriented RAN slicing. The primary objective of service-oriented RAN slicing is a major enabler for mobile operators to support all potential of vertical services. Service-oriented RAN slicing provides the essential incentives to allow operators to provide tailored solutions to vertical service providers over the same network infrastructure, thus bringing new business models for mobile operators and vendors. We identify key enabling technologies to improve spectrum efficiency and meet different KPIs. We anticipate with well-developed standards, tight collaboration with operators, vendors, and vertical industries to provide a healthy 5G system via the introduced service-oriented RAN slicing.

References

1 Popovski, P., Trillinsgaard, K.F., Simeone, O., and Durisi, G. (2018). 5G wireless network slicing for eMBB, URLLC, and mMTC: a communication-theoretic view. *IEEE Access* 6: 55765–55779.

2 Kaloxylos, A. (2018). A survey and an analysis of network slicing in 5G networks. *IEEE Communications Standards Magazine* 2 (1): 60–65.

3 Zhang, H., Liu, N., Chu, X. et al. (2017). Network slicing based 5G and future mobile networks: mobility, resource management, and challenges. *IEEE Communications Magazine* 55 (8): 138–145.

4 Yan, M., Feng, G., Zhou, J. et al. (2019). Intelligent resource scheduling for 5G radio access network slicing. *IEEE Transactions on Vehicular Technology* 68 (8): 7691–7703.

5 Afolabi, I., Taleb, T., Samdanis, K. et al. (2018). Network slicing and softwarization: a survey on principles, enabling technologies and solutions. *IEEE Communication Surveys and Tutorials* 20 (3): 2429–2453.

6 3GPP TS 23.501 (December 2019). System Architecture for the 5G System; Stage 2. Release 16, v. 16.3.0. 127-139.

7 Samdanis, K., Costa-Perez, X., and Sciancalepore, V. (2016). From network sharing to multi-tenancy: the 5G network slice broker. *IEEE Communications Magazine* 54 (7): 32–39.

8 NGMN Alliance (2017). Description of network slicing concept. https://www.ngmn.org/uploads/media/160113_Network_ Slicing_v1_0.pdf (accessed January 2017). Page(s): 94-100

9 Foukas, X., Patounas, G., Elmokashfi, A., and Marina, M.K. (2017). Network slicing in 5G: survey and challenges. *IEEE Communications Magazine* 55 (5).

10 3GPP TS 36.300 (January 2020). Evolved Universal Terrestrial Radio Access (E-UTRA) and Evolved Universal Terrestrial Radio Access Network (E-UTRAN); Overall Description. Release 16, v15.8.0.

11 3GPP TS 38.300 (January 2020). Group Radio Access Network; NR; NR and NG-RAN Overall Description; Stage 2. Release 16, v. 15.8.0.

12 Ghatak, G., De Domenico, A., and Coupechoux, M. (2019). Small cell deployment along roads: coverage analysis and slice-aware RAT selection. *IEEE Transactions on Communications* 67 (8): 5875–5891.

13

5G Network Slicing for V2X Communications: Technologies and Enablers

Claudia Campolo[1], Antonella Molinaro[1] and Vincenzo Sciancalepore[2]

[1] *Dipartimento DIIES, Università Mediterranea di Reggio Calabria, Via Graziella, Italy*
[2] *EC Laboratories Europe GmbH, Heidelberg, Germany*

13.1 Introduction

Fifth generation (5G) systems are envisioned to support enhanced mobile broadband communication services as well as a flurry of new use cases from vertical industries. Among them, the automotive vertical is considered an undoubted key driver of 5G systems because of the ongoing technological transformations in the field of communication, automation, sensing, and positioning, which are expected to revolutionize future transportation and quality of life.

Vehicle-to-everything (V2X) communications play a crucial role in enabling the exchange of information to support different applications, which range from cooperative autonomous and tele-operated driving to traffic efficiency and infotainment services, including advertisements as well as software updates and diagnostics [1]. Owing to their heterogeneity, vehicular applications bring about a plethora of connectivity requirements. Most of them are particularly challenging, i.e. ultralow latency (below 10 ms), ultrahigh reliability (near 100%), and high data rate (in the order of Gbps), and their fulfillment is further exacerbated by the high mobility and density of vehicles.

Thus, the *one-size-fits-all* design philosophy applied in existing networks is not flexible and scalable enough for the simultaneous support of a diversified set of V2X applications, in terms of performance, security, availability, and cost. *Network slicing* has been recently proposed by academia and industry in the 5G research arena as a promising approach to fulfill divergent service requirements in a flexible way, as analyzed in [2]. Network and computing resources can be tailored to specific service needs on a common *programmable end-to-end* network infrastructure, by logically isolating control plane (CP) and user plane (UP) network functions.

UP and CP functionalities of different slices can be virtualized through network function virtualization (NFV) and independently scaled and displaced in convenient locations to flexibly support various services based on changing user demands. For instance, UP functions can be located close to the user, in edge cloud facilities, to reduce service access latency. By abstracting network resources, programmable and software-defined networking

Radio Access Network Slicing and Virtualization for 5G Vertical Industries, First Edition.
Edited by Lei Zhang, Arman Farhang, Gang Feng, and Oluwakayode Onireti.

can dynamically steer data traffic for network slices through the setting up of paths that can be automatically reconfigured either to handle traffic engineering (TE) requirements (e.g. load balancing, traffic prioritization) or to react to possible network failures and changing conditions (e.g. mobility).

Thanks to these notable features, network slicing has been initially advocated as a key solution for the automotive vertical in [3], followed by flourishing literature in the last couple of years. Notwithstanding, the topic is still in its infancy and deserves further research to properly understand the V2X ecosystem, its evolution in terms of connectivity and computing demands as the level of driving automation increases, and the ongoing progress in standardization bodies toward 5G (and beyond) systems. In this context, the contributions of this chapter are as follows:

- to dissect the V2X applications and their demands, by accounting for the latest taxonomies proposed by involved standardization bodies and stakeholders (Section 13.2);
- to discuss the main V2X communication technologies, with special focus on the latest progress in the Third Generation Partnership Project (3GPP) for what concerns the radio interfaces as well as architectural enhancements to the cellular network to support V2X connectivity (Section 13.3);
- to analyze the crucial role of multi-access edge computing (MEC) technologies for V2X applications (Section 13.4);
- to scrutinize other prominent enabling technologies for V2X network slicing such as segment routing (Section 13.5);
- to present the network slicing concept and its instantiation in the vehicular ecosystem, by scanning 3GPP efforts and the literature (Section 13.6);
- to emphasize the main lessons learnt and provide design guidelines, while also sharing perspectives about future research directions and open issues (Section 13.7).

13.2 Vehicular Applications

V2X applications cover a wide range of use cases, which are typically grouped together based on their purpose and minimum requirements. The 5G Automotive Association (5GAA)[1], formed by major telecommunications operators, automotive industries, and chip manufacturers, groups them in four categories: *(i) safety* use cases are aimed at reducing the frequency and severity of vehicle collisions, e.g. through the exchange of warnings; *(ii) convenience* use cases provide services to manage the health of vehicles, such as diagnostics and software updates; *(iii)* vulnerable road user (VRU) use cases target safe interactions between vehicles and other non-vehicle road users; *(iv) advanced driving assistance* use cases share similar objectives with safety use cases, but they aim to support (semi-)autonomous vehicle operation. Hence, they exhibit more demanding requirements and are typically treated separately. Indeed, they have specifically catalyzed the interest of 3GPP, which further classifies them as follows (3GPP TR 22.186, 2018): *(i) vehicles platooning* enables vehicles to dynamically form a group travelling together with short

1 http://5gaa.org/.

Table 13.1 Connectivity requirements of V2X autonomous driving applications (3GPP TR 22.186)

Application	Payload (bytes)	Latency (ms)	Reliability	Data rate (Mbps)
Vehicles platooning	50–6500 B	10–20	90–99.999	0.012–65
Advanced driving	300–12 000	3–100	90–99.999	0.096–53
Extended sensors	1600	3–100	90–99.999	10–1000
Remote driving	–	5	99.999	25 (UL) 1 (DL)

inter-vehicle distances; *(ii) advanced driving* foresees vehicles to share data obtained from local sensors and their driving intention with other vehicles in proximity, thus allowing to coordinate their trajectories or maneuvers; *(iii) extended sensors* enable the exchange of raw/processed sensor data or live video data among vehicles, road-side units (RSUs), and VRUs, *(iv) remote driving* allows a remote driver or a cloud application to tele-operate a (private or public) vehicle; this is useful for those passengers who cannot drive themselves (e.g. impaired people) or when the vehicle is located in dangerous or uncomfortable environments (e.g. earthquake-affected regions, road construction work zones).

Requirements for such services get stricter as the degree of automation increases: e.g. latency requirements pass from 100 ms for information sharing in case of advanced driving to 5 ms in case of remote driving, as illustrated in Table 13.1. Although the objective of 3GPP in targeting connectivity requirements is neat, the aforementioned applications also need computing resources, e.g. for developing remote vehicle control systems and for the analysis and aggregation of sensor/video data retrieved from vehicles and other sources, as clarified in Section 13.4.

13.3 V2X Communication Technologies

V2X communication is a key enabler for a safer, greener, more connected and autonomous driving experience. After more than a decade of extensive industrial and academic research, standardization activities, and field trials, the choice of the enabling V2X communication technology is still under debate.

IEEE 802.11p (ITS-G5 in Europe) was considered as the "de facto" access technology for vehicular networking until a few years ago, with worldwide field trials demonstrating its feasibility for *day-one* V2X applications (e.g. emergency brake light, stationary vehicle warning) [4]. However, this technology has been slow to be developed on a large scale, and in recent years its supremacy has been undermined by the rising role of 3GPP and cellular networks. Starting with the support of telematics and infotainment services for connected cars, cellular networks are now noticeably involved at a much wider scope.

The Cellular V2X (C-V2X) technology is part of the completed 3GPP Long-Term Evolution (LTE) Releases 14 and 15. It is intended to support ultrahigh reliable and ultralow latency V2X applications in the 5G Release 16 (e.g. autonomous driving, car platooning), with specifications to be frozen by the first half of 2020. C-V2X is meant to be a unified technology that will allow vehicles to communicate with other vehicles, pedestrians, infrastructure nodes, and remote servers, to guarantee full coverage and service continuity.

In 3GPP documents (3GPP TR 22.185, 2018) V2X communication types are classified as follows: (i) vehicle-to-vehicle (V2V) for direct communications between vehicular user equipments (VUEs) in close proximity; (ii) vehicle-to-infrastructure (V2I) for communications between vehicles and the RSU in radio range, which can be implemented either in an eNodeB or in a standalone device (e.g. a traffic light); (iii) vehicle-to-pedestrian (V2P) between vehicles and VRUs (e.g. pedestrians, bikers); and (iv) vehicle-to-network (V2N) for communications with remote servers and cloud-based services reachable through the cellular infrastructure.

13.3.1 The C-V2X Technology

C-V2X spans both the PC5 and LTE-Uu radio interfaces as well as core network (CN) functionalities, as discussed in the following.

13.3.1.1 The PC5 Radio Interface

V2V, V2P, and V2I communications occur over the PC5 radio interface (sidelink) and bypass the cellular infrastructure. The 5.9 GHz band is used, independently or even in the absence of a cellular network, in order to ensure ultrahigh availability under all geographies, regardless of the specific mobile network operator (MNO). 3GPP efforts resulted in the specifications (3GPP TR 36.213, 2018) of two direct communications modes for broadcast interactions over the sidelink, namely *Mode 3* or scheduled, and *Mode 4* or autonomous. In Mode 3, operating only in-coverage of an eNodeB, the allocation of radio resources is supervised by the network, but a specific resource allocation design is left to the operators. In Mode 4, instead, preconfigured resources can be accessed by vehicles in an autonomous manner without the network control, both in- and out-of coverage of an eNodeB (e.g. in urban canyons, tunnels).

13.3.1.2 The LTE-Uu Interface

V2N communications occur over the conventional cellular LTE-Uu interface operating in the licensed spectrum. Less disruptive, although not negligible, modifications have been applied to this interface in order to support both unicast and multicast V2X communications. The multicast primitive has especially catalyzed the interest of 3GPP and of the scientific community, because it enables efficient V2X information sharing (e.g. high definition maps, infotainment services) with many vehicles. Some enhancements to the evolved multimedia broadcast multicast service (eMBMS) architecture have been specified to meet the latency requirements of V2X applications and the localized nature of their data traffic (3GPP TS 23.785, 2016; 3GPP TR 36.885, 2016; 3GPP TS 36.331, 2018), with promising results preliminarily shown in [5].

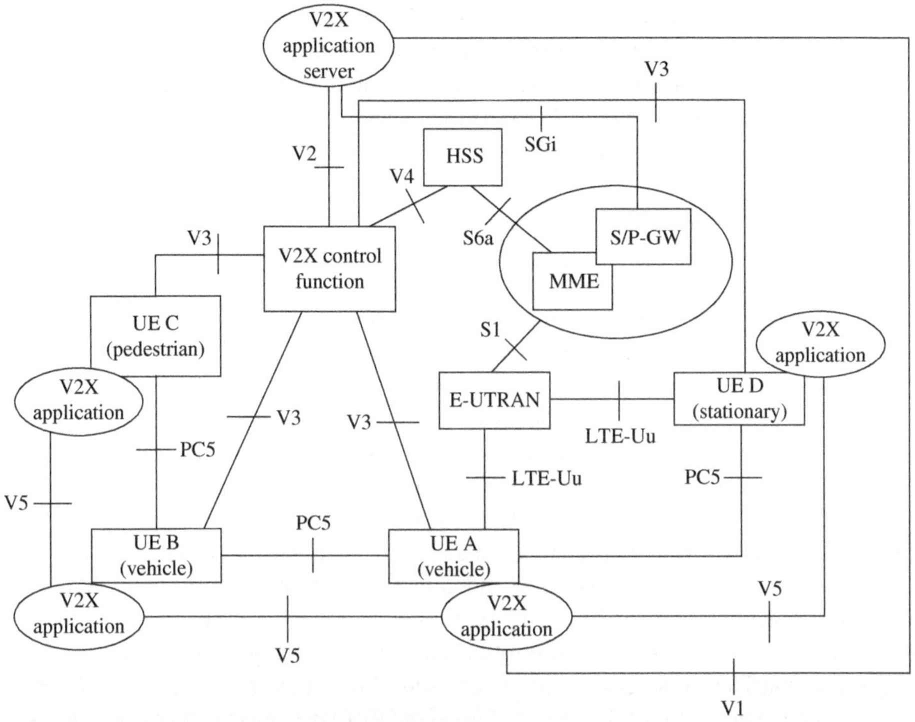

Figure 13.1 Reference architecture for PC5 and LTE-Uu based V2X communications (3GPP TS 23.285, 2019).

13.3.1.3 Core Network

Architectural enhancements have been specified in the LTE Evolved Packet System (EPS) to support the operations of VUEs (3GPP TS 23.285, 2019), as depicted in Figure 13.1. Two new entities have been added to manage V2X communications. The *V2X control function* is the logical function for V2X network-related actions; it is responsible for the provisioning of V2X policy and parameter configuration over the PC5 (in- and out-of-coverage) and the LTE-Uu interfaces. The *V2X application server* (AS) has a wide range of functionalities including the reception of uplink unicast data from the VUEs, VRUs, and RSUs, and the delivery of data to the VUEs in a target area using unicast and/or multicast interfaces.

13.3.2 C-V2X Toward 5G

3GPP is enhancing C-V2X in several ways toward 5G specifications, for what concerns both the radio interface and the architectural functionalities.

13.3.2.1 Radio Interface

New radio (NR) V2X specifications in Release 16 (3GPP TR 38.885, 2019) are expected by early 2020 to align C-V2X connectivity with the NR standardization launched, independently of V2X, for the first phase of 5G in Release 15. The envisioned modifications are meant (i) to introduce more disruptive technologies such as new/flexible waveforms and numerologies and a larger (mmWave) spectrum to support the most challenging V2X applications' demands (ultralow latency and ultrahigh reliability), and (ii) to improve modes 3 and 4 in order to support peculiar V2X traffic patterns, such as asynchronous traffic, unicast/multicast/groupcast, as discussed in [6].

13.3.2.2 Core Network

As the 5G CN design is progressing in 3GPP Release 16, it is under discussion whether a V2X architecture, simpler than the one conceived in Release 14, can be developed by using some of the available features without any component from the EPS. The 5G CN architecture is natively designed with modularization and softwarization in mind; it is composed of network functions (NFs) and reference points connecting them.

In the 5G architecture, the UE is connected to either the radio access network (RAN) or a non-3GPP access network (AN) such as Wi-Fi[2], as well as with the Access and Mobility Management Function (AMF). The RAN represents a base station using new radio access technologies (RATs) and evolved LTE. The CN consists of various NFs, with a clear separation of CP and UP functions (UPFs), as shown in Figure 13.2: (i) the AMF provides UE-based authentication, authorization, mobility management, etc.; (ii) the Authentication Server Function (AUSF) stores data for authentication of UEs; (iii) the Unified Data Management (UDM) stores UEs' subscription data; (iv) the Session Management Function (SMF) is responsible for session management and also selects and controls the UPF for data transfer; (v) the Network Repository Function (NRF) is in charge of maintaining and providing information on the NF instances when deploying/updating/removing them; it also supports the service discovery function; (vi) the Network Exposure Function (NEF) is responsible for providing means to collect, store, and securely expose the services and

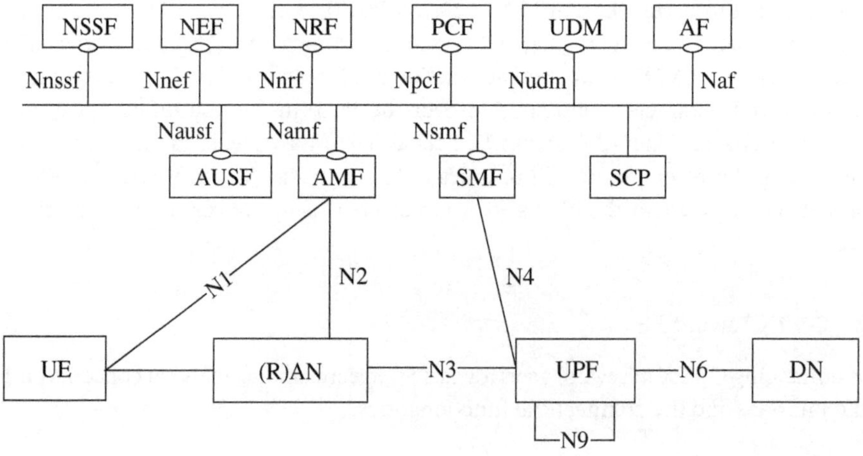

Figure 13.2 5G system architecture (3GPP TS 23.501, 2019).

2 For those non-3GPP RATs, access to the 5G CN is provided by the Non-3GPP InterWorking Function (N3WIF) (3GPP TS 24.502, 2019).

capabilities provided by 3GPP NFs (e.g. to third parties or among NFs themselves); (vii) the Application Function (AF) represents any additional CP functions that might be required, e.g. to implement network slicing.

Some alternatives are under evaluation (3GPP TR 23.786, 2019) to deploy V2X functionalities within the 5G CN, e.g. to reuse the 5G Policy Control Function (PCF), or to reuse the EPS-based V2X architecture, or to deploy the V2X Control Function as a new CP entity in the 5G CN.

13.4 Cloudification in V2X Environments

The 5GAA[3] considers the automotive sector, among the vertical segments in the 5G ecosystem, as a key driver toward the adoption of (edge) cloudification solutions. Indeed, V2X requires a very low-latency environment for a significant part of the related use cases (about 10 ms). Autonomous driving is the most relevant and critical example in this respect. Such a capability is only offered by an edge computing solution, which facilitates the exchange of V2X related information between (mobile) nodes via the underlying communication network as well as their fast processing, as discussed in [7].

The infotainment experience on board might also open up considerable business opportunities and be significantly impacted by (edge) cloudification system deployments; about US$7 trillion-worth passenger economy will emerge when drivers become passengers,[4] thereby freeing up to 300 hours per year they typically spend behind the wheel. Wearable devices, in-vehicle and roadside sensors, and vehicles themselves will be fully connected and exchange information related to traffic conditions, surrounding environment, and geographical maps, while generating, sharing, and retrieving rich media contents that enable advanced V2X applications [8].

Within this ecosystem, the role of edge computing is essential to provide fast access and distribution of information of local interest. Examples of such information range from simple safety warnings to contextual data advertising, such as the proximity of other vehicles (e.g. see-through service) or obstacles (e.g. real-time situational awareness), to real-time video streaming and augmented reality/virtual reality (AR/VR) services. Further evidence of the synergy between V2X and edge computing comes from several trials on automotive use cases resulting from the collaboration among different stakeholders, such as car original equipment manufacturers (OEMs), service providers, and telecommunication operators.

13.4.1 The Role of MEC

In the described context, the European Telecommunications Standards Institute (ETSI) specified the MEC paradigm. It is aimed to guarantee a preferred environment for locally preprocessing data and to save backhaul bandwidth, as well as for hosting – in very close

3 http://5gaa.org/wp-content/uploads/2017/12/5GAA_T-170219-whitepaper-EdgeComputing_5GAA.pdf
4 https://newsroom.intel.com/news-releases/intel-predicts-autonomous-driving-will-spur-new-passenger-economy-worth-7-trillion/#gs.vkg80p

proximity to the end user devices (or vehicles) – intelligent transport system (ITS) applications, such as the V2X AS in Figure 13.1. The MEC system facilitates IT cloud capability at the edge of the network, and its access-agnostic characteristics guarantee smoother deployment independently of the underlying communication network.

In a typical MEC-based deployment, cars communicate with RSUs and base stations (BSs), whereas MEC hosts are co-located with RAN elements to allow the consumption of ITS applications running on the MEC platform. The MEC-based system exposes a set of standardized interfaces to such applications, enabling developers and car OEM suppliers to implement ITS services in an interoperable manner across heterogeneous access networks, mobile operators, and vendors.

The engagement of the ETSI MEC Industry Specification Group (ISG) with V2X networking involved two stages. The first stage entailed the collection of V2X use cases and the definition of challenging requirements for the MEC-based system. The results of such activity are listed in the Group Report (GR) ETSI MEC ISG-MEC 022. Interestingly, it emerged that some of the collected requirements, commonly identified by automotive stakeholders, stem from the multi-MNO and multi-OEM scenarios. Specifically, an ITS operator might be interested in offering an ITS service to its own users across different communication networks, thereby involving a set of vehicles from different manufacturers. The second stage is envisaged to include in the future new normative activities to close the gap identified by the study, e.g. by defining standardized interfaces to overcome potential interoperability issues in the multi-MNO multi-OEM scenario.

13.4.2 ETSI MEC-Based Programmable Interfaces

The ETSI MEC ISG specified a list of programmable interfaces, available within published working documents. Such interfaces, dubbed as application programming interfaces (APIs), are designed to serve a wide set of 5G use cases also encompassing automotive scenarios. The definition of such APIs represents a key activity within the MEC ISG.

One of the first detailed services available on the MEC platform is the *Radio Network Information Service*, specified in ETSI MEC ISG-MEC 012, which provides up-to-date RAN information relating to UEs in the network. This might be useful while collecting channel state information of VUEs so as to perform, e.g. user profiling operations or localization procedures. Future API refinements will provide V2X applications with predictions on the quality of service (QoS) performance of the V2I link: this is a fundamental feature for car OEMs not only to improve reliability but also to support dynamic and fast decisions in the next road segment, where vehicles may switch to a safe off-line mode if the QoS forecast is poor.

In combination with such service, the *MEC Location Service* (ETSI MEC ISG-MEC 013) is a powerful tool enabling applications to exploit user proximity information, e.g. to retrieve and monitor the list of users connected to a particular cell or access point. This service appears essential for V2X applications to estimate the location of pedestrian/vehicles so as to improve the overall safety. Additionally, the *UE Identity Service* (ETSI MEC ISG-MEC 014) allows applications to trigger user-specific traffic rules on the MEC platform by e.g. steering traffic to a local network. Finally, the *Bandwidth Management Service* (ETSI MEC ISG-MEC 015) enables applications to reserve networking resources to ensure specific quality of experience (QoE) requirements. Currently, the ETSI MEC ISG is pursuing additional service APIs, one targeting specifically the V2X use cases, with the aim of facilitating V2X interoperability in a multi-vendor, multi-network, and multi-access environment.

13.4.3 MEC-Based Support for V2X Applications

Given the variety of differentiated services, 5G (and beyond) network design will come with the reasonable coexistence of multiple virtualized networks, namely *network slices*, tailored to automotive use cases. Such a design forces a compound upgrade of business plans as well as RAN and CN architectural blocks, as described in Section 13.6. A gradual transition is also expected toward a fully capable 5G network that might be expedited by the deployment of ETSI MEC platforms, being able to fulfill many of the automotive use case requirements.

MEC-based solutions should incorporate the following main sets of functionalities within a V2X communications system: (i) baseband units (BBUs) of a Cloud-RAN (C-RAN) architecture for access link termination, (ii) core gateway functions for session management, routing of local traffic, etc. and (iii) an interoperable and programmable computing environment for cloud-based vehicular applications [7].

In particular, to enhance hardware reconfigurability and network agility, especially in view of high mobility scenarios, the recent advent of software-defined networking (SDN) and NFV boosted the potential of a C-RAN deployment by means of virtualizing NFs, thus enabling general purpose processors to be exploited for optimized network operations. Nevertheless, the strict end-to-end latency requirements of the automotive use cases are arduous to achieve by means of a highly virtualized albeit remote cloud. The need to process data in a proximity-based manner can be hardly satisfied even by means of fractional RAN centralization procedures. Toward this end, the proximity-oriented deployment of MEC hosts at RSUs, or co-located with the BSs, can be useful for efficiently allocating all resources needed to run C-RAN equipment.

Architecturally renewed by the 5G service types and related key features, the CN shall support new highly demanding services with increased data rate, reduced end-to-end latency, massive connectivity, guaranteed QoS/QoE, and higher availability and efficiency. Therefore, MEC is expected to facilitate the transition to 5G deployments, as it accommodates the co-location of CN functions at edge cloud facilities with the MEC applications, hence having 5G use cases fulfilled using the fourth generation (4G) technology. For example, the serving gateway (SGW) function can be distributed at the edge and enhanced with an IP-based interface, very similarly to the SGi interface of the packet data network gateway, which selectively breaks out the traffic that needs to remain local, i.e. associated to MEC applications as well as to the V2X AS. The described solution is named serving gateway-local breakout (SGW-LBO) [9]. Similarly to the uplink classifier, it also allows the operator to steer the traffic based on different parameters such as the access point name, user identities, the packet header's 5-tuple, and the differentiated services code point (DSCP), while supporting user mobility, charging, lawful interception, and session management operations, without requiring changes in the 3GPP standards.

13.5 Transport and Tunneling Protocol for V2X

Tunneling protocol design is of paramount importance to properly support network slices and provide V2X-based services. We detail the 3GPP architecture and illustrate the main interfaces (Figure 13.3) that have been disclosed by the recently published standard guidelines. We then summarize the current General Packet Radio Service (GPRS) Tunneling Protocol (GTP) properties, while shedding light on the main novelties introduced by the Segment Routing IPv6 (SRv6) protocol in view of the supporting V2X slices.

13.5.1 GTP-U Encapsulation

GTP is the current encapsulation process used in mobile networks, being also part of the 5G network specifications. In particular, as described in Section 13.3.2, the UE establishes a data session and gets assigned the corresponding UPF that is in charge of routing and forwarding traffic between UE and the data network (DN). User data traffic arrives from the gNB (or eNB) to the UPF by means of the N3 interface, while the UPF is connected to the DN via the N6 interface. The N9 interface might be activated to enable a direct communication between UPFs. Both N3 and N6 interfaces implement GTP to identify and forward encapsulated traffic in the CN, while the N6 interface assumes routing plain IP protocol data units (PDUs).

Encapsulation consists of adding a GTP-user (GTP-U) header to the IP datagram (i.e. the user data packet T-PDU) and using User Datagram Protocol (UDP)/IP as transport layer. Both IPv4 and IPv6 traffic types are supported, as explained in (3GPP TS 29.281, 2019). The GTP-U header has a variable length that depends on the number of optional fields. The minimum length is 8 bytes and includes the Tunnel Endpoint Identifier (TEID), which is the key element to identify the packet data protocol (PDP) context of the T-PDU of any received packet. However, this protocol does not provide native path differentiation and it is already showing its limitations for some of the 5G use cases, including V2X scenarios [10].

13.5.2 Segment Routing v6

Recently envisioned as the successor to the GTP protocol, SRv6 uses an IPv6 routing header extension, the Segment Routing Header (SRH), as defined in [11]. Its structure consists of *segment lists* (SLs), wherein each segment is identified with an IPv6 address, i.e. the SegmentID (SID). The current active segment is the destination address of the packet, while the next address is indicated by the SL field of the SRH.

The use of segment routing fully enables network programmability, as one can associate a given SID with given processing function behaviors, which may be tailored onto specific applications, such as V2X services. Segment routing enables service providers to support network slicing without any additional protocol [12]. Such functions can be classified into two main categories: *endpoint behaviors*, which have a local SID to the destination address received, being associated to a processing function in the local node; and *transit behaviors*, wherein there is no local SID associated to the destination address of the packets received,

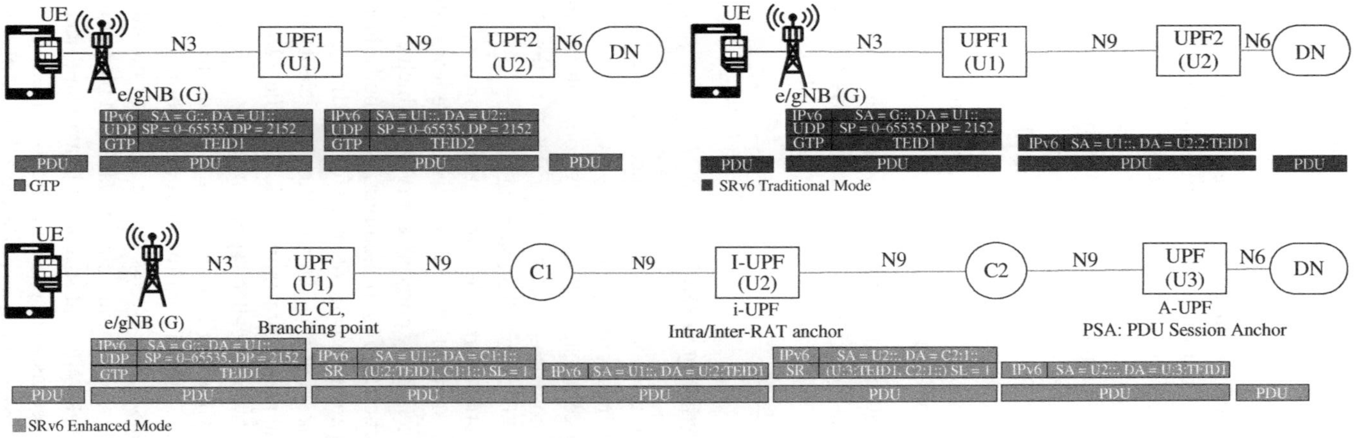

Figure 13.3 Uplink packet flow using Segment Routing v6 (SRv6). It shows the application of classical GTP, SRv6 in traditional mode, and SRv6 in enhanced mode.

so they are not bound to an SID, corresponding to either SRv6 source nodes that introduce the SRH in the packets or SRv6 transit nodes steering traffic.

We distinguish two modes of SRv6 operation in a mobile network, depending on whether it is considered as a replacement of GTP-U or a step toward the adoption of network programmability:

- *Traditional mode*: There is a direct mapping between GTP-U and SRv6 encapsulations. It does not introduce changes to the network architecture and the impact is, thus, kept minimum. This mode would be used, for instance, in the topology as per Figure 13.3, where the user packet traverses four hops to arrive at the DN.
- *Enhanced mode*: By introducing more than one SID in the SRH (i.e. a packet can traverse more nodes with respect to the traditional mode), the use of more advanced functions in the network path of the traffic user flows is allowed, enabling to steer traffic in this way by using different paths and supporting more sophisticated behaviors, e.g. service programming. Therefore, it would be possible to associate non-3GPP services, such as TE in nodes along the path, as illustrated in the bottom of Figure 13.3.

In specific V2X scenarios, it might be needed to apply the SRv6 enhanced mode that may enable handover scenarios (or in some more specific cases, cross-bordering scenarios), where multiple administrative domains (and, in turn, multiple UPFs) are involved while providing V2X services to the same vehicles.

13.5.3 Scalability and Flexibility in SRv6

SRv6 better supports source routing for TE without incurring the heavy workload of states and signaling in the nodes by means of IDs, such as Multiprotocol Label Switching (MPLS) labels or IPv6 addresses. In contrast, GTP is mainly coupling the mobile infrastructure to the underlying transport network, representing a threat to the status quo as it brings single-vendor monolithic mobile network implementations.

Adopting SRv6 in V2X-supported mobile infrastructures may bring an added value, as the operator may employ the same Layer-3 virtual private network (VPN) techniques that are used for managing and delivering enterprise services, which can be further improved in 5G fixed-mobile convergence (FMC). In addition, SRv6 boils down the complexity by reducing protocol overheads and removing the need for the UPF to keep states. This brings a big advantage in terms of scalability. SRv6 also allows efficient service level agreement (SLA)-enabled multicast content injection (suitable for specific V2X use cases) by means of the standard unicast core with the so-called "spray" method.

On the other hand, SRv6 will show its drawbacks when dealing with legacy suppliers, as a packet processing engine should continually dig deeply into the IPv6 header, especially when a big flow of data traffic arrives due to the high mobility in V2X scenarios. Essentially, SRv6 may require a programmable software data plane to be effectively implemented, UPF as fully virtualized NF, and specialized data plane acceleration solutions to continuously meet deep packet inspection requirements. Given the extent of advantages, the main technology limitations, and the recently started study item from 3GPP (3GPP TR 29.892, 2019), SRv6 might become the right candidate for updating the UP protocol while supporting upcoming stringent V2X services.

13.6 Network Slicing for V2X

Network slicing has gained momentum as a prominent technology to flexibly support the diverse and even conflicting demands of 5G use cases that can run on top of a common physical infrastructure, as described in [2]. It enables a new paradigm for flexible multi-tenancy among MNOs/infrastructure providers and mobile virtual network operators (MVNOs).

A network slice is intended as a composition of adequately configured NFs, network applications, specific radio access settings, and underlying computing resources bundled together to meet the requirements of a specific use case.

Slices are expected to be built in order to meet the demands of three reference service categories identified by the International Telecommunication Union (ITU): *enhanced mobile broadband* (eMBB), which encompasses all the applications that require very high throughput to ensure access to multimedia contents, services, and data, such as ultrahigh definition video streaming and AR/VR; *massive machine type communications* (mMTC) (a.k.a. massive IoT, mIoT), which include the non-delay-sensitive traffic generated by massively connected constrained devices; *ultra-reliable and low-latency communications* (URLLC), which are extremely sensitive to latency and require high reliability, such as emergency services and tactile Internet (ITU-R Rec. M2083, 2015).

Owing to the peculiarities of their requirements, V2X services cannot be mapped into the reference slices for the aforementioned ITU service categories (see Ref. [3] and 3GPP TR 23.786, 2019). Besides the proper mapping into service slice, the V2X ecosystem also raises daunting challenges to conveniently engineer the slicing design. Such issues motivated specific 3GPP amendments and encouraged literature works, which will be reviewed in the Sections 13.6.1 and 13.6.2.

13.6.1 3GPP Specifications

In order to meet the requirements of a specific slice, the relevant traffic needs to be properly marked, identified, and treated accordingly in the network element. While the specific slice traffic treatment is left to operators, the selection and identification functionalities are specified by 3GPP.

A VUE, wishing to attach to a slice, has to provide the network with the *Network Slice Selection Assistance Information* (NSSAI) parameters, identifying a slice, to enable the selection of a slice instance for it (3GPP TS 23.501, 2019). The Network Slice Selection Function (NSSF), specifically added in the 5G system architecture (see Figure 13.2), is then responsible for slice selection.

In particular, NSSAI includes *Slice Service Type* (SST) and, possibly, *Slice Differentiator* (SD). The SST refers to an expected network behavior. 3GPP proposes to define a specific standardized SST for V2X services (3GPP TR 23.786, 2019). A *customized* SST value facilitates the deployment of dedicated network slices for use of, for example, automotive industry. A *standardized* value could facilitate roaming support. Indeed, it can be expected that vehicles move fast and often cross over different countries/MNOs while being connected to the network. The standardized SST values (3GPP TR 23.501, 2019) are (i) 1 for the eMBB slice, (ii) 2 for the URLLC slice, (iii) 3 for the mIoT slice, and (iv) 4 for the V2X slice.

The SD is an optional information that complements the SST and can help select among several network slice instances of the same type, e.g. to isolate the traffic related to different services provided over different slice instances. An SD could identify the *Tenant ID*, e.g. the car manufacturer, the road authority.

13.6.2 Literature Overview

On a parallel and also faster lane than 3GPP efforts, literature works have been published on network slicing for V2X. Although the entire end-to-end chain of 5G radio, networks, applications, and services should be tailored to meet V2X requirements through network slicing, existing literature works target specific domains and network segments, as summarized in Table 13.2. The majority of works focus on allocating resources over the RAN, by neglecting the allocation of network resources over the CN segment and the assignment of computation resources. Analytical and/or simulation studies have been mainly performed, while a few works have conducted more practical experimental results.

In [3] a set of guidelines is dissected for the design of a network slicing V2X framework. The work advocates the vision of a dedicated slice for V2X services, embraced only recently in 3GPP specifications, as detailed before. A set of four slices are proposed to accommodate the needs of V2X services, i.e. autonomous driving, tele-operated driving, vehicular infotainment, and vehicle remote diagnostics and management.

In [20] a reference network slicing architecture is presented that builds upon the three-layers model, as in [21], illustrated in Figure 13.4. It includes the following layers: the *Infrastructure Layer*, encompassing user devices, RAN nodes, edge and remote computing, storage and networking resources; the *Service Layer*, including the Virtual Network Functions (VNFs) and logical network behaviors that realize a slice when chained together; the *Business Layer*, encompassing the mechanisms and tools to describe the slice behavior at a high level, and to capture the requirements of a given SLA for a vertical segment, whose assurance is tracked by the Operation/Business Support System (OSS/BSS); and the *Management and Orchestration* (MANO) layer, including all the tools that allow an operator to monitor, orchestrate, and adapt the slice's components to fulfill the negotiated SLAs.

Table 13.2 Overview of the main literature works

References	RAN	CN	MEC	Non-V2X slices
Campolo et al. [3]	✓	✓	✓	×
Zhang et al. [13]	✓	×	×	×
Campolo et al. [14]	×	✓	✓	×
Khan et al. [15]	✓	×	×	×
Albonda and Pérez-Romero [16]	✓	×	×	✓
Ge [17]	✓	×	×	×
Chekired et al. [18]	✓	✓	✓	×
Sanchez-Iborra et al. [19]	✓	✓	✓	×

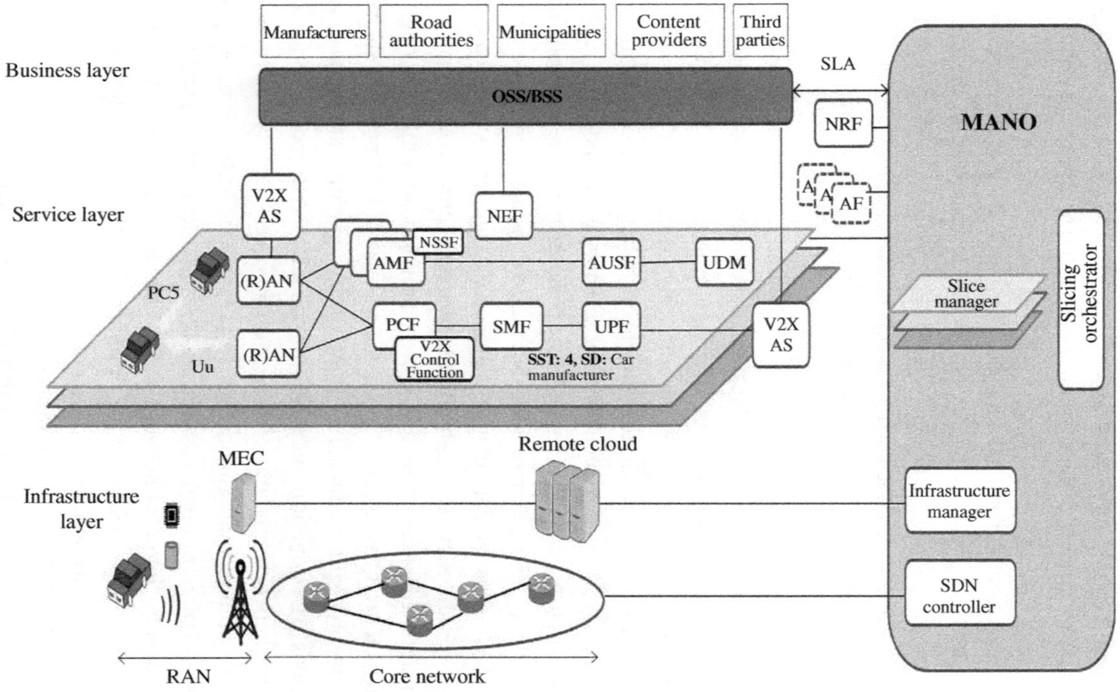

Figure 13.4 Network slicing architecture and slice instantiation for autonomous driving services. Source: Adapted from [20].

Figure 13.4 also graphically sketches the instantiation of a slice dedicated to autonomous driving. It relies on V2V as the prevalent RAT connection mode, and on additional RAN/CN functions, e.g. for network-controlled resource allocation over the PC5 interface (in the eNodeB), mobility (in the AMF), authentication, authorization, and subscription management (in the AUSF). Additionally, ultralow latency and highly reliable V2N connectivity with an edge-based V2X AS are required for video/data exchange that helps vehicles in high-definition map processing of the surrounding area and to extend the visual perception of each vehicle.

Departing from a mere theoretical analysis, the work in [16] focuses on resource allocation in the RAN segment for two types of slices to be served in the same cell, i.e. eMBB requested by conventional UEs and V2X slices by VUEs. A strategy based on offline Q-learning is proposed to ensure the adequate splitting of resources between the two slices, to meet their demands in terms of throughput (for the eMBB slice) and latency (for the V2X slice).

Different network slices need to be devised for different V2X services. Starting from this concept, in [15] the authors address the allocation of radio resources in the downlink direction for the infotainment and autonomous driving slices instantiated by VUEs. The former is supported through V2I links, whereas the latter can also rely on V2V communications, with some vehicles acting as relays for their neighbors to improve connectivity.

The work in [17] leverages the Euclidean norm theory for proposing a reliability and latency joint function to evaluate the impact on the performance. A network slicing scheme is designed to meet the URLLC demands for slices supporting autonomous vehicles. In [13] an air-ground integrated vehicular network is proposed to support location-based map and popularity-based content services. High-altitude platforms proactively broadcast popular contents to vehicles, while the ground RSUs provide services on demand through unicast.

In [14] an architectural framework is proposed to support V2X network slicing. There, the focus is on the procedures entailed to support the cooperative driving use case under roaming conditions. An evaluation study has been conducted as a proof-of-concept by leveraging the Mininet network emulator coupled with an overhauled Ryu SDN controller in charge of managing interoperator handovers. In [19] an architecture is developed to support traffic differentiation (through queuing and DSCP-packet marking) and flow isolation (through tunneling) for different network slices. The slices span from an MEC node to a remote end-server, hence emulating the traversal of a 5G CN and the main functions for slice identification, selection, and orchestration. The architecture has been realistically validated when considering different traffic generated by a vehicle and transmitted over different RATs, i.e. 4G and LoRAWAN. The work in [18] also advocates the need for end-to-end network slicing and proposes a framework leveraging SDN, fog, edge, and cloud computing to manage resources offered over heterogeneous RATs on top of which autonomous driving services are delivered. Resources for the basic functions required by autonomous driving (i.e. localization, perception, planning, and system management) are virtualized and scheduled using distinct service slices.

13.7 Lessons Learnt and Guidelines

Owing to the high dynamicity of the reference V2X environment and the variety of V2X services and applications, the design of V2X slices unveils much more complexity than other verticals and services. In the following, we discuss how V2X peculiarities and demands can be translated into network slicing design choices and guidelines, by also treasuring the main findings of the scanned related works and standard specifications.

13.7.1 Slice Mapping and Identification

V2X services do not easily allow straightforward mapping into ITU reference slices. Moreover, multiple slices may be required to be activated simultaneously by a single vehicle. For example, a driver starts her autonomous vehicle relying on a specifically designed autonomous driving slice, while passengers watch a movie offered as a vehicular infotainment slice. Besides the dedicated SST value already foreseen by 3GPP, a slice subtype may be required. The heterogeneity of the originating devices of a V2X slice asks for further *slice customization*. A V2X UE can be a smartphone in the case of a VRU, a transceiver unit embedded into the vehicle. The communication on the PC5 interface for example shall be optimized based on the type of UE: energy efficiency is critical for a smartphone UE engaged in V2P interactions for road safety purposes. Customization of slice on a per-device level should be enabled.

13.7.2 Multi-tenancy Management

New players such as road authorities, municipalities, and car OEMs will enter the V2X scene in addition to traditional operators. For instance, the road municipality may offer a V2V-based safety data exchange; a vehicle manufacturer can offer a vehicle diagnostics service; a content provider a video streaming, which can be mapped onto different slices and facilitated over the infrastructure owned by different network operators. A proper set of *flexible slice templates* should be defined, which allow the vertical to properly describe the services to be supported and the desired key performance indicators (KPIs). Then, the assurance of requirements is tracked by the OSS/BSS with respect to the template specifications agreed in the form of SLAs.

13.7.3 Massive Communications

The V2X slicing design shall enable *multiple AMF instances* to be flexibly deployed as VNFs and to be interconnected to meet the needs of the V2X slices with high density of moving vehicles. This prevents the AMF from being overloaded, with consequent increase in latency, while ensuring isolation from other (non-V2X) slices that leverage the same functionalities, but less aggressively (e.g. pedestrian/indoor UEs in case of an mMTC slice). The instantiation of AMFs could be boosted close to highways and urban areas to accommodate high vehicle density at peak hours, and it could be flexibly scaled down at off-peak hours to avoid over-provisioning. A lighter AMF deployment could be used, instead, in rural areas. Moreover, AMF functions can be co-located with the eNodeB to ensure low-latency signaling procedures, e.g. for autonomous driving use cases.

13.7.4 Transparent Mobility

The high vehicle speed requires frequent interactions with the network to manage mobility procedures affecting both UP and CP operations. Mobility poses serious issues to the run-time configuration of a slice. Quick path resource allocation algorithms in the backhaul/fronthaul segments should be designed, which allow to allocate/migrate resources in another area while reducing disruption periods. The effective and efficient instantiation of a sort of *moving slice* entails proper mobility prediction modules to trigger slice reconfiguration, while also pre-fetching UP functions close to vehicles. With the introduction of MEC platforms, live migration, and service redirection are necessary actions that increase the complexity of mobility management and handover procedures.

13.7.5 Isolation

The network operator should carefully prevent an abnormal situation in other applications from negatively affecting the QoS of V2X safety applications. 3GPP partially targets this issue by suggesting the allocation of a dedicated spectrum for V2V communications over the PC5 interface. The isolation of the V2X slices from other network slices must be also supported. Limiting the number of shared CP/UP functionalities, proper scheduling algorithms over the RAN and SDN-configured paths would contribute to achieving the required degree of isolation.

13.8 Conclusions

This chapter presented the main technology enablers and design principles of 5G network slicing for the challenging and promising automotive market. Particular attention has been paid to the role of edge cloud facilities to provide low-latency V2X use cases. Also, the emerging segment routing technology has been investigated as a further facilitator of network slicing for V2X communications. The most recent ETSI and 3GPP standard documents and the research and development results in the scientific literature have been analyzed and used to learn some guidelines for the future deployment of high-performing V2X-dedicated slices.

References

1 MacHardy, Z., Khan, A., Obana, K., and Iwashina, S. (2018). V2X access technologies: regulation, research, and remaining challenges. *IEEE Communications Surveys & Tutorials* 20 (3): 1858–1877.

2 Zhang, S. (2019). An overview of network slicing for 5G. *IEEE Wireless Communications* 26 (3): 111–117.

3 Campolo, C., Molinaro, A., Iera, A., and Menichella, F. (2017). 5G network slicing for vehicle-to-everything services. *IEEE Wireless Communications* 24 (6): 38–45.

4 Chowdhury, M., Rahman, M., Rayamajhi, A. et al. (2018). Lessons learned from the real-world deployment of a connected vehicle testbed. *Transportation Research Record* 2672 (22): 10–23.

5 Fallgren, M. et al. (2019). Multicast and broadcast enablers for high-performing cellular V2X systems. *IEEE Transactions on Broadcasting* 65 (2: 454–463).

6 Naik, G., Choudhury, B., and Park, J. (2019). IEEE 802.11bd & 5G NR V2X: evolution of radio access technologies for V2X communications. *IEEE Access* 7: 70169–70184.

7 Giust, F., Sciancalepore, V., Sabella, D. et al. (2018). Multi-access edge computing: the driver behind the wheel of 5G-connected cars. *IEEE Communications Standards Magazine* 2 (3): 66–73.

8 5G CARMEN (2019). Deliverable D2.1, 5G CARMEN Use Cases and Requirements.

9 Giust, F., Verin, G., Antevski, K. et al. (2018). MEC deployments in 4G and evolution towards 5G. *ETSI White paper* 24: 1–24.

10 Ventre, P.L., Salsano, S., Polverini, M. et al. (2019). Segment routing: a comprehensive survey of research activities, standardization efforts and implementation results. *arXiv preprint arXiv:1904.03471.*

11 Filsfils, C. (ed.) (2019). IPv6 segment routing header (SRH). Draft-ietf-6man-segment-routing-header-24, October 2019. https://datatracker.ietf.org/doc/draft-ietf-6man-segment-routing-header/ (accessed 02 June 2020).

12 Ali, Z. et al. (2019). Building blocks for slicing in segment routing network, Internet Engineering Task Force, Internet Draft draft-ali-spring-network-slicing-building-blocks, work in Progress. [Online]. Available: https://tools.ietf.org/html/draft-ali-spring-network-slicing-building-block.

13 Zhang, S., Quan, W., Li, J. et al. (2018). Air-ground integrated vehicular network slicing with content pushing and caching. *IEEE Journal on Selected Areas in Communications* 36 (9): 2114–2127.

14 Campolo, C., Fontes, R., Molinaro, A. et al. (2018). Slicing on the road: enabling the automotive vertical through 5G network softwarization. *Sensors* 18 (12): 4435.

15 Khan, H., Luoto, P., Samarakoon, S. et al. (2019). Network slicing for vehicular communication. *Transactions on Emerging Telecommunications Technologies*: 1–14

16 Albonda, H.D.R. and Pérez-Romero, J. (2019). An efficient RAN slicing strategy for a heterogeneous network with eMBB and V2X services. *IEEE Access* 7: 44771–44782.

17 Ge, X. (2019). Ultra-reliable low-latency communications in autonomous vehicular networks. *IEEE Transactions on Vehicular Technology* 68 (5): 5005–5016.

18 Chekired, D.A., Togou, M.A., Khoukhi, L., and Ksentini, A. (2019). 5G-slicing-enabled scalable SDN core network: toward an ultra-low latency of autonomous driving service. *IEEE Journal on Selected Areas in Communications* 37 (8): 1769–1782.

19 Sanchez-Iborra, R., Santa, J., Gallego-Madrid, J. et al. (2019). Empowering the internet of vehicles with multi-RAT 5G network slicing. *Sensors* 19 (14): 3107.

20 Campolo, C., Molinaro, A., Iera, A. et al. (2018). Towards 5G network slicing for the V2X ecosystem. *2018 4th IEEE Conference on Network Softwarization and Workshops (NetSoft)*, IEEE, pp. 400–405.

21 Soenen, T., Banerjee, R., Tavernier, W. et al. (2017). Demystifying network slicing: from theory to practice. *IFIP/IEEE Symposium on IM*.

14

Optimizing Resource Allocation in URLLC for Real-Time Wireless Control Systems

Bo Chang[1], *Liying Li*[2] *and Guodong Zhao*[3]

[1] *National Key Laboratory of Science and Technology on Communications, University of Electronic Science and Technology of China (UESTC), China*
[2] *Department of Mathematics, Physics and Electrical Engineering, Northumbria University, UK*
[3] *School of Engineering, University of Glasgow, UK*

14.1 Introduction

The imminent fifth generation (5G) cellular networks is expected to support three critical communication services with significantly different requirements – enhanced mobile broadband (eMBB), massive machine-type communications (mMTC), and ultra-reliable low-latency communication (URLLC). To achieve the above heterogeneous services, network slicing has been proposed in recent years, where these services are isolated from each other and the resources are accurately allocated to guarantee different system performance. In particular, network slicing can guarantee URLLC service by a combination of scheduling, which can ensure a certain amount of predictability in the available resources. Based on that, ultra-reliability and low latency can be maintained [1].

URLLC has shown the potential to provide service to real-time wireless control systems, which can provide significant advantage compared with traditional wired control systems. For instance, system deployment can be modified or upgraded flexibly while maintaining the commutation quality unaltered. In vehicle technologies, URLLC is one of the key enablers to in-vehicle wireless control, which can significantly reduce financial cost in manufacturing. This further improves fuel efficiency and reduces gas emission to provide timely, reliable, and accurate control and an open architecture for new applications [2] and [3]. Furthermore, URLLC is crucial for vehicle-to-anything (V2X) communication, which is essential for vehicle automatic driving via wireless communications [4] and [5].

There has been some research on wireless control systems, but most of it analyzes URLLC (e.g. Refs. [6–15]) and wireless networked control (e.g. Refs. [16–21]) separately. From the wireless communication perspective, the authors in [22] investigated the achievable channel coding rate for finite blocklength, which establishes one of the most important foundations of URLLC design in the physical layer. The authors in [23] further discussed the close-form expression of the achievable capacity in URLLC, which provides a guideline for URLLC design in different channel cases, e.g. additive white Gaussian noise (AWGN) channel and Rayleigh fading channel. Based on the above research, more works focus on

Radio Access Network Slicing and Virtualization for 5G Vertical Industries, First Edition.
Edited by Lei Zhang, Arman Farhang, Gang Feng, and Oluwakayode Onireti.

URLLC resource allocation design to maintain the extremely high quality-of-service (QoS) in URLLC [6–13]. For example, the authors in [6] studied the resource allocation for uplink communications and found that a huge amount of frequency and transmission power are needed to satisfy the extremely high QoS in URLLC. However, most of the research targets a latency or reliability constrained communication system without bringing specific control performance (e.g. control cost, stability, or state update rate) into consideration.

From wireless control perspective, communication latency and reliability are set as random variables caused by communication protocols [16–21]. For example, the authors in [16] studied the effect of communication packet loss on the control performance caused by transmission control protocol (TCP) or user datagram protocol (UDP). The authors in [18] further discussed the effect of both time delay and packet loss on the control performance caused by carrier sense multiple access/collision avoidance (CSMA/CA). In [17], the authors provided a survey on the wireless network design for the control systems, where latency and reliability were discussed in different communication protocols. The above research indicates that communication time delay and packet loss result in control performance loss since they enlarge the sample period of the control systems [17]. However, no close form expression is obtained to show exactly how communication latency and reliability affect the control performance.

The aforementioned work shows that design communication and control subsystems separately cannot guarantee optimal overall system performance. On the one hand, most of the research on URLLC intends to maintain extremely high QoS only in wireless communications. On the other hand, the effect of communication latency and reliability on control performance cannot be obtained exactly. In fact, co-design of URLLC and real-time wireless control is critical for both communication and control. First, communication resource is extremely limited, especially to meet the ultrahigh technical requirements for URLLC. A design tailored for a wireless control system can significantly reduce communication resource consumption. For instance, at some stages when the control performance is not constrained at an extremely high level, the allocated wireless resource can be reduced accordingly. Second, the control performance is determined by both the control sample period and the communication QoS, where the sample period should be adjusted accordingly to optimize the control performance when the communication QoS is different [17].

We notice that there is some research on communication-control co-design [24–30]. For instance, in [25], the authors obtained the optimal control sampling period and communication time delay and reliability by simulation analysis based on the existing communication protocols. In [26], we further found that different communication QoSs have different effects on control performance throughout the control process by simulation analysis, where the control process can be divided into two stages with different QoS service and the power consumption can be significantly reduced. Furthermore, in [27] and [28], the authors analyzed control stability with unreliability of communications in V2X communications. However, without close form expression indicating the relationship between communication and control by co-design, a high overall system performance cannot be obtained.

This chapter aims to optimize communication resource allocation by joint communication-control design. Taking uplink transmission as an example, our goal is to find the optimal communication bandwidth, communication power consumption, and control convergence rate[1] by maximizing the spectral efficiency (SE) while guaranteeing the control requirement. However, the idea can be extended into the downlink transmission scenario in a straightforward manner. Our problem formulation considers both URLLC QoS and control convergence rate constraints, while such a co-design is still an open problem to be explored. To achieve the goal, the key is to find a method to convert the constraints on the control subsystem into constraints that the communication subsystem can adopt to solve the optimization problem. The main contributions of this chapter are summarized as follows.

- We propose an optimal resource allocation scheme to maximize the SE by communication and control co-design, where both URLLC requirements and control convergence rate requirement are taken into account. The proposed scheme allows us to use optimal resource to support URLLC while guaranteeing the required control performance level.
- We analyze the effect of the relationship between the control convergence rate requirement and communication requirement on URLLC quality. We find that the lower bound of the communication reliability decreases linearly with the control convergence rate. Then, the requirement on control can be treated as a constraint on communication reliability, which allows us to convert the co-design problem into a regular communication optimization problem.
- We prove that the formulated optimization problem is concave based on the conversion. Subsequently, we develop an iteration method to find the optimal bandwidth and power allocation.

The rest of this chapter is organized as follows. In Section 14.2, the system model is presented. In Section 14.3, the communication-control co-design method is proposed, where the optimal resource allocation problem is formulated with control performance constraint. In Section 14.4, the relationship between the communication and control subsystems is first discussed, and then an iteration method is proposed to obtain the optimal resource allocation in terms of maximizing the communication SE. In Section 14.5, simulation results are provided to show the performance. Finally, Section 14.6 concludes the chapter.

14.2 System Model with Latency and Reliability Constraints

In this section, we propose the system model by taking communication latency and reliability into consideration. As shown in Figure 14.1, we consider a typical centralized real-time wireless communication-control system, where there are M static

1 The main control variables are state sampling period and control input gain, which jointly affect control convergence rate. Then, control convergence rate further affects the control performance, e.g. control cost or control stability. By constraint on control convergence rate, the state sampling period and control input gain are constrained. Thus, control convergence rate is considered as the control design in this chapter.

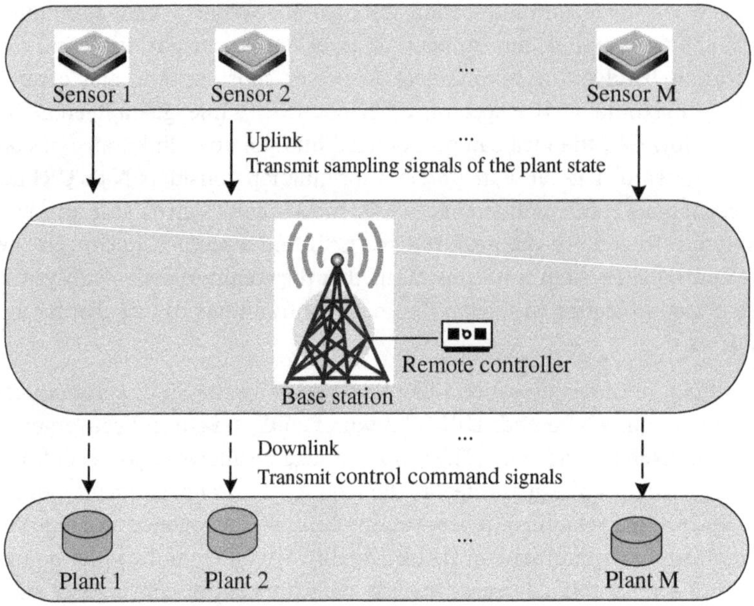

Figure 14.1 Real-time wireless control system model.

sensors transmitting sampling signals of the plant state to the base station (BS), the remote controller in the BS processing the sampling signals, and then the BS sending the calculated control command signals to the M corresponding static plants. With the control process continuing, the state of each plant will converge to a preset value. Note that all the notations to be used throughout the chapter for communication subsystem and control subsystem are summarized in Tables 14.1 and 14.2, respectively.

14.2.1 Wireless Control Model

In this section, we present the real-time wireless control model considering communication time delay and reliability. Except for the inherent control parameters, e.g. the mass and the speed of the plant, the main coefficients that contribute to control performance include sampling period at the sensor and communication QoS [17]. Since we focus on the control convergence rate requirement and communication QoS, a constant sampling period is adopted. Then, the continuous control function for the m-th plant is given by a linear differential equation as [25]

$$d\mathbf{x}_m(t) = \mathbf{A}_m\mathbf{x}_m(t)dt + \mathbf{B}_m u_m(t)dt + d\mathbf{n}_m(t), \tag{14.1}$$

where $\mathbf{x}_m(t)$ is the state of the m-th plant, $u_m(t)$ is the control input, and $\mathbf{n}_m(t)$ is the disturbance caused by AWGN with zero mean and variance $\mathbf{R}_m$. In addition, $\mathbf{A}_m$ and $\mathbf{B}_m$ represent the system parameter matrices. To illustrate the system parameters $\mathbf{A}_m$ and $\mathbf{B}_m$, we consider a controlled inverted pendulum system as shown in Figure 14.2.

Table 14.1 Summary of notations for communication subsystem

B_0	Bandwidth of each subcarrier	R	Coverage radius of the BS
$C_{m,n}$	Shannon capacity of the m-th sensor at time index n	T_{th}	Maximum transmission time delay
g_m	Path loss of the m-th sensor	$T_{m,n}$	Time resource of the uplink for the m-th sensor
$h_{m,n}$	Small scale fading for the m-th sensor	$V_{m,n}$	Channel dispersion for the m-th sensor
$N_{m,n}$	Total available number of subcarriers	$\alpha_{m,n}$	Indicator for packet loss
N_m	Number of subcarriers for the m-th sensor	$\gamma_{m,n}$	Received SNR of the m-th sensor
l_m	Distance between the m-th plant and the BS	$P_{\max}$	Maximum of transmission power
M	Total number of sensors in the coverage of the BS	$\varepsilon_{m,n}$	Packet error probability of the uplink for the m-th sensor
m	Index of the sensors ($1 \leq m \leq M$)	ε_{th}	Maximum packet error probability in communications
N_0	Variance of the AWGN on each subcarrier at the BS	η_m	Spectral efficiency of the uplink for the m-th sensor
p_m	Transmission power spectral density	λ_m	Payload information of the m-th sensor at each sample time

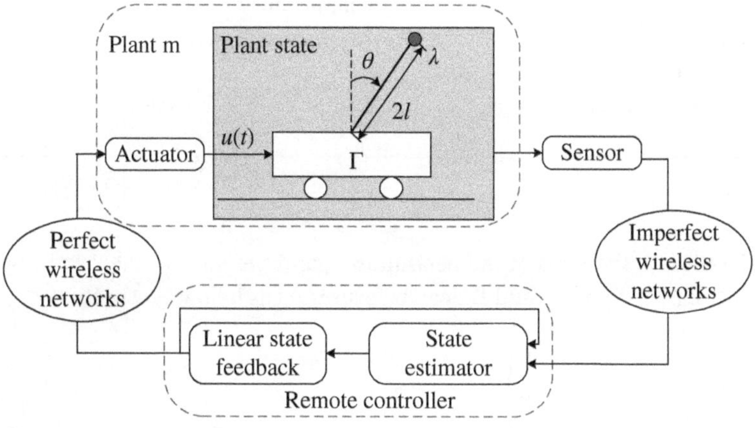

Figure 14.2 Wireless control system model for a single plant.

Example 14.1 The m-th controlled inverted pendulum system consists of an inverted pendulum and a motorized cart. We have the plant state $\mathbf{x}(t) = (c_t, \dot{c}_t, \theta_t, \dot{\theta}_t)$, where c_t represents the cart's position, $\dot{c}_t$ represents the cart's velocity, θ_t represents the pendulum's angle, and $\dot{\theta}_t$ represents the pendulum's angular velocity. The expression of $\mathbf{A}_m$ and $\mathbf{B}_m$ consists of the pendulum length $2l$, the inertia of the pendulum Ψ, the friction of the cart r, the

Table 14.2 Summary of notations for control subsystem

$\mathbf{A}_m$	System parameter on state in continuous control function	$u^a_{m,n}$	Control input calculated by the actuator
$\mathbf{B}_m$	System parameter on input in continuous control function	$u^c_{m,n}$	Feedback parameter calculated by the remote controller
c_m	Control performance constraint on wireless communications	$u_{m,n}$	Control parameter for simplification
J_{ave}	Average control cost	$\mathbf{x}_m(t)$	Plant state in continuous control function of the m-th plant
N	Total sampling time index in control process	$\mathbf{x}_{m,n}$	Plant state in discrete time control function
n	Sample time index	$\Delta_m(\cdot)$	Lyapunov-like function
$\mathbf{n}_m(t)$	Disturbance caused by AWGN in continuous control function	$\mathbf{\Phi}_0^{m,n}$	System parameter on input in discrete control function
$\mathbf{n}_{m,n}$	Disturbance in discrete time control function	$\mathbf{\Phi}_1^{m,n}$	System parameter on input caused by time delay
$\bar{\mathbf{n}}_{m,n}$	Generalized disturbance	$\mathbf{\Omega}_{e_0}$	General system parameter during packet losses
$\mathbf{Q}_m$	Given positive definite matrix	$\mathbf{\Omega}_{e_1}$	General system parameter when packet transmits successfully
$\mathbf{R}_m$	Variance of $\mathbf{n}_m(t)$	$\mathbf{\Omega}_{m,d}$	Generalized system parameter on state
$s_{m,n}$	Sample period of the m-th plant at time index n	$\mathbf{\Omega}_{m,n}$	System parameter on state in discrete control function
$\bar{s}_{m,n}$	Idle time before the sampling at time index n	$\xi_{m,n}$	Generalized plant state
$u_m(t)$	Control input in continuous control function	ρ_m	Control convergence rate

gravitational acceleration ϕ, the mass of the pendulum λ, and the mass of the cart Γ. By physical mathematics calculation, $\mathbf{A}_m$ and $\mathbf{B}_m$ can be expressed as follows, respectively:

$$\mathbf{A}_m = \begin{pmatrix} 0 & 1 & 0 & 0 \\ 0 & \frac{-(\Psi+\lambda l^2)r}{\Psi(\Gamma+\lambda)+\Gamma\lambda l^2} & \frac{\lambda^2 g l^2}{\Psi(\Gamma+\lambda)+\Gamma\lambda l^2} & 0 \\ 0 & 0 & 0 & 1 \\ 0 & \frac{-\lambda l r}{\Psi(\Gamma+\lambda)+\Gamma\lambda l^2} & \frac{\lambda\phi l(\Gamma+\lambda)}{\Psi(\Gamma+\lambda)+\Gamma\lambda l^2} & 0 \end{pmatrix}, \tag{14.2}$$

and

$$\mathbf{B}_m = \begin{pmatrix} 0 \\ \frac{(\Psi+\lambda l^2)r}{\Psi(\Gamma+\lambda)+\Gamma\lambda l^2} \\ 0 \\ \frac{\lambda l}{\Psi(\Gamma+\lambda)+\Gamma\lambda l^2} \end{pmatrix}. \tag{14.3}$$

To obtain the discrete time control model, we assume that $s_{m,n}$ represents the sample period of the m-th plant at time index n, which consists of wireless transmission time delay $T_{m,n}$ and an idle period $\bar{s}_{m,n}$. Their relationship can be expressed as

$$s_{m,n} = \bar{s}_{m,n} + T_{m,n}, \tag{14.4}$$

where $n = 1, 2, \dots, N$ represents the sampling time index in the control process and N is the maximum sampling time index. Then, the discrete time control model with time delay $T_{m,n}$ can be obtained as

$$\mathbf{x}_{m,n+1} = \mathbf{\Omega}_{m,n}\mathbf{x}_{m,n} + \mathbf{\Phi}_0^{m,n}u_{m,n} + \mathbf{\Phi}_1^{m,n}u_{m,n-1} + \mathbf{n}_{m,n}, \tag{14.5}$$

where $\mathbf{\Omega}_{m,n} = e^{\mathbf{A}s_{m,n}}$, $\mathbf{\Phi}_0^{m,n} = \left(\int_0^{\bar{s}_{m,n}} e^{\mathbf{A}_{m,n}t}dt\right) \cdot \mathbf{B}_{m,n}$, and $\mathbf{\Phi}_1^{m,n} = \left(\int_{\bar{s}_{m,n}}^{s_{m,n}} e^{\mathbf{A}_{m,n}t}dt\right) \cdot \mathbf{B}_{m,n}$. More details about the control model can be referenced in [17].

Assuming $\xi_{m,n} = (\mathbf{x}_{m,n}^T \;\; u_{m,n-1})^T$ is the generalized state, then the control function in (14.5) can be rewritten as

$$\xi_{m,n+1} = \mathbf{\Omega}_{m,d}\xi_{m,n} + \mathbf{\Phi}_{m,d}u_{m,n} + \bar{\mathbf{n}}_{m,n}, \tag{14.6}$$

where $\bar{\mathbf{n}}_{m,n} = (\mathbf{n}_{m,n}^T \;\; 0)^T$ and $\mathbf{\Phi}_{m,d} = \begin{pmatrix} \mathbf{\Phi}_0^{m,n} \\ \mathbf{I} \end{pmatrix}$. We assume $\mathbf{\Omega}_{m,n} = \mathbf{\Omega}_m$. Then, we have $\mathbf{\Omega}_{m,d} = \begin{pmatrix} \mathbf{\Omega}_m & \mathbf{\Phi}_1^{m,n} \\ 0 & 0 \end{pmatrix}$.

Considering the packet loss, we have $\Pr\{\alpha_{m,n} = 1\} = 1 - \varepsilon_{m,n} \geq 1 - \varepsilon_{th}$ and $\Pr\{\alpha_{m,n} = 0\} = \varepsilon_{m,n} < \varepsilon_{th}$, where "1" means that the packet is successfully transmitted and the control is under closed loop, and "0" means that the packet is lost and the control is under open loop. In addition, we assume that the state estimator is perfect, and then a linear feedback $u_{m,n} = \Theta_m\xi_{m,n}$ is used. Then, the closed-loop system in (14.6) can be rewritten as

$$\xi_{m,n+1} = \begin{cases} (\mathbf{\Omega}_{m,d} + \mathbf{\Phi}_{m,d}\Theta_m)\xi_{m,n} + \bar{\mathbf{n}}_{m,n}, & \text{if } \alpha_{m,n} = 1 \\ \mathbf{\Omega}_{m,d}\xi_{m,n} + \bar{\mathbf{n}}_{m,n}, & \text{if } \alpha_{m,n} = 0, \end{cases} \tag{14.7}$$

which can be rewritten in a general way as

$$\xi_{m,n+1} = \begin{cases} \mathbf{\Omega}_{e_1}\xi_{m,n} + \bar{\mathbf{n}}_{m,n}, & \text{if } \alpha_{m,n} = 1 \\ \mathbf{\Omega}_{e_0}\xi_{m,n} + \bar{\mathbf{n}}_{m,n}, & \text{if } \alpha_{m,n} = 0, \end{cases} \tag{14.8}$$

where $\mathbf{\Omega}_{e_1} = \mathbf{\Omega}_{m,d} + \mathbf{\Phi}_{m,d}\Theta_m$ is the parameter of the control system with transmission time delay included in $\mathbf{\Omega}_{m,d}$ when the data packet is successfully transmitted, and $\mathbf{\Omega}_{e_0} = \mathbf{\Omega}_{m,d}$ is the parameter of the control system with transmission time delay included in $\mathbf{\Omega}_{m,d}$ when the data packet fails. Furthermore, when there is no transmission time delay, i.e. $d_{m,n} = 0$, the expression (14.5) can be written as

$$\mathbf{x}_{m,n+1} = \mathbf{\Omega}_{m,n}\mathbf{x}_{m,n} + \mathbf{\Phi}_0^{m,n}u_{m,n} + \mathbf{n}_{m,n}. \tag{14.9}$$

Then, the expression (14.8) can be written as

$$\mathbf{x}_{m,n+1} = \begin{cases} \mathbf{\Omega}_{e_1}\mathbf{x}_{m,n} + \bar{\mathbf{n}}_{m,n}, & \text{if } \alpha_{m,n} = 1 \\ \mathbf{\Omega}_{e_0}\mathbf{x}_{m,n} + \bar{\mathbf{n}}_{m,n}, & \text{if } \alpha_{m,n} = 0, \end{cases} \tag{14.10}$$

where $\mathbf{\Omega}_{e_1} = \mathbf{\Omega}_{m,n} + \mathbf{\Phi}_0^{m,n}\Theta_m$ is the parameter of the control system without transmission time delay when the data packet is successfully transmitted, and $\mathbf{\Omega}_{e_0} = \mathbf{\Omega}_{m,n}$ is the parameter of the control system without transmission time delay when the data packet transmission fails. The above discussion indicates that the proposed presented model in this section can be used regardless of transmission time delay.

14.2.2 Wireless Communication Model

In the rest of this chapter, we focus on the uplink from the sensor to the BS, but the derivations can be extended into the downlink transmission in a straightforward manner. For the convenience of discussion, we assume that the uplink is imperfect, which experiences transmission time delay and packet loss, and the downlink is perfect. The scenario with both uplink and downlink experiencing imperfect channel will be considered as future work. Specifically, M plants are randomly distributed in the coverage of the BS with radius R. In addition, each plant is sampled by a corresponding sensor. To avoid interference, we consider orthogonal frequency division multiple access (OFDMA), where each sensor is allocated multiple subcarriers within a given continuous bandwidth, each subcarrier can be allocated to at most one sensor, and the given bandwidths for different sensors are not overlapping. Furthermore, we consider flat fading channel, where the channel gains over different subcarriers for one sensor are approximately identical and perfectly known for the sensor. We assume that the number of allocated bandwidth and transmission duration for the m-th sensor at time index n are $N_{m,n}$ and $T_{m,n}$, respectively. Then, for the m-th sensor at time index n, the received signal-to-noise-ratio (SNR) at the BS can be expressed as

$$\gamma_{m,n} = \frac{|h_{m,n}|^2 g_m N_{m,n} B_0 p_{m,n}}{N_0 N_{m,n} B_0} = \frac{|h_{m,n}|^2 g_{m,n} p_{m,n}}{N_0}, \tag{14.11}$$

where $h_{m,n}$ is the small scale fading for the m-th sensor at time index n, g_m is the path loss, $N_{m,n}$ is the number of allocated bandwidths, p_m is the allocated transmission power spectral density for the m-th sensor, B_0 is the separation among subcarriers, and N_0 is the single-sided noise spectral density. Then, the allocated bandwidth for the m-th sensor is $B_{m,n} = B_0 N_{m,n}$. Based on the received SNR, the Shannon capacity can be expressed as

$$C_{m,n} = \log(1 + \gamma_{m,n}). \tag{14.12}$$

In URLLC, with the constraints on ultra-reliability and low-latency, the blocklength is short and small packet sizes are adopted [8]. In addition, the size of payload data transmitted by sensors in control systems is usually small (e.g. about 100 bits [17]), which is suitable to be used in URLLC scenario. In such a scenario, the impact of the decoding error cannot be ignored. We assume that channel dispersion $V_{m,n}$ is adopted to represent the capacity loss caused by transmission error, which can be expressed as [23]

$$V_{m,n} = (\log e)^2 \left(1 - \frac{1}{(1 + \gamma_{m,n}^2)}\right). \tag{14.13}$$

When SNR is higher than 5 dB, we have $V_{m,n} \approx (\log e)^2$, which is very accurate [8]. When $V_{m,n} < (\log e)^2$, we can obtain a lower bound of the achievable rate with channel dispersion by substituting $V_{m,n} = (\log e)^2$. Thus, in the rest of this chapter, we adopt

$$V_{m,n} = (\log e)^2. \tag{14.14}$$

For the m-th plant, the available uplink rate with finite block length is the uplink capacity eliminating the error bits that are introduced by channel dispersion, which can be expressed as

$$R_{m,n} = C_{m,n} - \sqrt{\frac{V_{m,n}}{N_{m,n}T_{m,n}B_0}} f_Q^{-1}(\varepsilon_{m,n}) + \frac{\log(N_{m,n}T_{m,n}B_0)}{2N_{m,n}T_{m,n}B_0}, \tag{14.15}$$

where $\frac{\log(N_{m,n}T_{m,n}B_0)}{2N_{m,n}T_{m,n}B_0}$ is the approximation of the remainder terms of order $\log(N_{m,n}T_{m,n}B_0)/(N_{m,n}T_{m,n}B_0)$, and $f_Q^{-1}(\cdot)$ is the inverse of the Q-function. Then, the packet error probability can be expressed as

$$\begin{aligned} \varepsilon_{m,n} &= f_Q\left(\frac{N_{m,n}T_{m,n}B_0C_{m,n} - N_{m,n}T_{m,n}B_0R_{m,n} + (\log N_{m,n}T_{m,n}B_0)/2}{\sqrt{N_{m,n}T_{m,n}B_0V_{m,n}}}\right) \\ &= f_Q\left(\frac{N_{m,n}T_{m,n}B_0C_{m,n} - N_{m,n}T_{m,n}B_0R_{m,n} + (\log N_{m,n}T_{m,n}B_0)/2}{(\log e)\cdot\sqrt{N_{m,n}T_{m,n}B_0}}\right), \end{aligned} \tag{14.16}$$

where $f_Q(\cdot)$ is the Q-function.

The above channel model consists of path loss and small scale fading. According to [31], the path loss g_m can be expressed as

$$g_{m_{[dB]}} = -128.1 - 37.6\,\mathrm{lg}(l_m), \tag{14.17}$$

where l_m is the distance between the m-th plant and the BS with unit km and is larger than 0.035 km [31]. The small-scale fading $h_{m,n}$ follows Rayleigh distribution with mean zero and variance $\sigma_0^2 = 1$. In addition, small-scale fading is constant within coherence time, which is larger than the maximum *end-to-end* (E2E) time delay. Thus, we consider *quasi-static fading channel*, which is constant for each uplink subcarrier within a frame [32].

From (14.12), (14.14), (14.16), and (14.15), we can obtain SE of the uplink for the m-th plant at time index n, which can be expressed as

$$\eta_{m,n} = \frac{\lambda_m}{N_{m,n}B_0}(1 - \varepsilon_{m,n}), \tag{14.18}$$

where λ_m is the desired payload transmitted by the sensor, and SE means successful decoding bits at the BS per bit. In this chapter, we intend to obtain the uplink optimal wireless resource allocation by maximizing SE in (14.18).

14.3 Communication-Control Co-Design

Our goal is to maximize the communication SE and maintain good overall system performance. Thus, in this section, we first discuss the constraints from the perspectives of communication and control, respectively. Then, we formulate the co-design problem based on the constraints.

14.3.1 Communication Constraint

The main constraints from communication are the limited wireless resource and URLLC QoS. We assume that the total available transmission power and bandwidth without reuse among sensors for each sensor are $p^{\max}$ and $B^{\max}$ at each time index n, respectively, where the constraints can be expressed as

$$N_{m,n}B_0p_{m,n} \leq p^{\max}, \tag{14.19}$$

and

$$B_{m,n} = N_{m,n}B_0 \leq B^{\max}. \tag{14.20}$$

Based on the constraint of reliability in URLLC, successful transmission probability can be expressed as

$$\Pr\{\alpha_{m,n} = 1\} = 1 - \varepsilon_{m,n} \geq 1 - \varepsilon_{th}, \tag{14.21}$$

and the failed transmission probability can be expressed as

$$\Pr\{\alpha_{m,n} = 0\} = \varepsilon_{m,n} < \varepsilon_{th}, \tag{14.22}$$

where ε_{th} is the packet error probability bounded by the URLLC QoS requirement. Furthermore, communication time delay should also be bounded by the URLLC QoS requirement. Then, we have

$$T_{m,n} \leq T_{th}. \tag{14.23}$$

We assume that the sizes of payload λ_m for each sensor are identical, which can be rewritten as λ. Then, for a given λ, we have $N_{m,n}T_{m,n}B_0R_{m,n} = \lambda$. Then, (14.16) and (14.18) can be rewritten as

$$\varepsilon_{m,n} = f_Q\left(\frac{N_{m,n}T_{m,n}B_0C_{m,n} - \lambda + (\log N_{m,n}T_{m,n}B_0)/2}{(\log e)\cdot\sqrt{N_{m,n}T_{m,n}B_0}}\right), \tag{14.24}$$

and

$$\eta_{m,n} = \frac{\lambda}{N_{m,n}B_0}(1 - \varepsilon_{m,n}). \tag{14.25}$$

14.3.2 Control Constraint

In this chapter, we consider the control convergence rate ρ_m as the constraint from the control aspect, where the sampling period is fixed and the inherent parameters of the plant are physical reality and change only when the plants are different. Given a physical control system, the main control variables that can be optimized are state sampling period $s_{m,n}$ and the control input gain Θ_m, which jointly affect the control convergence rate ρ_m. Then, the control convergence rate further affects the control performance, e.g. control cost or stability. Through constraint on ρ_m, the state sampling period $s_{m,n}$ and the control input gain Θ_m are directly constrained. In other words, the relationship between control convergence rate and wireless resource allocation indicates the relationship between the main control variables and wireless resource allocation. Thus, the co-design of this chapter focuses on control convergence rate and wireless resource allocation.

To obtain the constraint on the control convergence rate, we consider a Lyapunov-like function for each plant, which can be expressed as [33]

$$\Delta_m(\xi_m) = \xi_m^T Q_m \xi_m, \tag{14.26}$$

where $\mathbf{Q}_m$ is a given positive definite matrix. The requirement for the Lyapunov-like function is that these functions should decrease at given rates $\rho_m < 1$ during the control process, which means that the control process guarantees the plant state decreasing to the preset point. However, affected by the control perturbation and stochastic communication coefficients, the Lyapunov-like function is random. Thus, for any possible value of the current plant states $\xi_{m,n}$, the Lyapunov-like function needs to satisfy [33]

$$\mathbb{E}[\Delta_m(\xi_{m,n+1})|\xi_{m,n}] \leq \rho_m \Delta_m(\xi_{m,n}) + \mathrm{Tr}(\mathbf{Q}_m \mathbf{R}'_m), \tag{14.27}$$

where $\mathbb{E}[\cdot]$ represents the expectation operator and $\mathbf{R}'_m = (\mathbf{R}_m \;\; 0)$. The control convergence is guaranteed with $\rho_m < 1$, which has been discussed in [33]. We conclude it by Lemma 14.1.

Lemma 14.1 ***(Control Convergence Lemma)*** *If (14.27) holds for each time step $n = 0, 1, 2, \ldots, N$, then by taking the expectation at both sides and by iterating backwards we can obtain*

$$\begin{aligned}\mathbb{E}[\Delta_m(\xi_{m,N})] &\leq \rho_m \mathbb{E}[\Delta_m(\xi_{m,N-1})] + \mathrm{Tr}(\mathbf{Q}_m \mathbf{R}'_m)\\ &\leq \cdots \\ &\leq \rho_m^N \mathbb{E}[\Delta_m(\xi_{m,0})] + \sum_{n=0}^{N-1} \rho_m^n \mathrm{Tr}(\mathbf{Q}_m \mathbf{R}'_m).\end{aligned} \tag{14.28}$$

Since the sum in (14.28) converges with $\rho_m < 1$, the second moments of the plant states decay exponentially, which is bounded by $Tr(\mathbf{Q}_m \mathbf{R}'_m)/(1 - \rho_m)$. □

14.3.3 Problem Formulation

According to the constraints from the communication and control aspects, we formulate the original communication-control co-design problem as

$$\max_{N_{m,n}, p_{m,n}, \rho_m} \quad \eta = \sum_{m=1}^{M} \sum_{n=1}^{N} \eta_{m,n} \tag{14.29a}$$

s.t.

$$\varepsilon_{m,n} \leq \varepsilon_{th}, \tag{14.29b}$$

$$N_m p_{m,n} \leq p^{\max}, \tag{14.29c}$$

$$B_{m,n} \leq B^{\max}, \tag{14.29d}$$

$$T_{m,n} \leq T_{th}, \tag{14.29e}$$

$$\mathbb{E}[\Delta_{m,n}(\xi_{m,n+1})|\xi_{m,n}] \leq \rho_m \Delta_m(\xi_{m,n}) + Tr(\mathbf{Q}_m \mathbf{R}'_m). \tag{14.29f}$$

In this problem, we intend to maximize the communication SE with optimal wireless resource allocation in URLLC and control convergence rate. More importantly, we jointly consider the control constraint on the control convergence rate ρ_m and communication constraint on URLLC QoS, which is difficult to deal with in the proposed communication-control co-design since the control constraint (14.29f) seems independent of the rest of the communication constraint terms in (14.29). However, the control constraint on the control convergence rate ρ_m can be affected and is determined by the communication QoS actually. In Section 14.4, we will deal with the problem in (14.29) in detail.

14.4 Optimal Resource Allocation for The Proposed Co-Design

In this section, we first explore the relationship between control and communication. Then, we discuss the solution for the problem in (14.29).

14.4.1 Relationship Between Control and Communication

From (14.8), we can understand that the plant state $\xi_{m,n}$ in (14.8) is determined by the control parameters and packet transmission. Then, the expression $\mathbb{E}[\Delta_m(\xi_{m,n+1})|\xi_{m,n}]$ depends on the packet transmission probability. From (14.8), we can find that the Lyapunov-like function can be expressed as

$$\begin{aligned}\mathbb{E}[\Delta_m(\xi_{m,n+1})|\xi_{m,n}] = {} & \Pr\{\alpha_{m,n} = 1\}\xi_{m,n}^T\boldsymbol{\Omega}_{e_1}^T\mathbf{Q}_m\boldsymbol{\Omega}_{e_1}\xi_{m,n} \\ & + \Pr\{\alpha_{m,n} = 0\}\xi_{m,n}^T\boldsymbol{\Omega}_{e_0}^T\mathbf{Q}_m\boldsymbol{\Omega}_{e_0}\xi_{m,n} \\ & + Tr(\mathbf{Q}_m\mathbf{R}'_m),\end{aligned} \tag{14.30}$$

which indicates that the communication reliability can affect the control Lyapunov-like function directly. Substituting (14.30) into (14.29f), we can obtain

$$\Pr\{\alpha_{m,n} = 1\} \geq \frac{\xi_{m,n}^T(\boldsymbol{\Omega}_{e_0}^T\mathbf{Q}_m\boldsymbol{\Omega}_{e_0} - \rho_m\mathbf{Q}_m)\xi_{m,n}}{\xi_{m,n}^T(\boldsymbol{\Omega}_{e_0}^T\mathbf{Q}_m\boldsymbol{\Omega}_{e_0} - \boldsymbol{\Omega}_{e_1}^T\mathbf{Q}_m\boldsymbol{\Omega}_{e_1})\xi_{m,n}}, \tag{14.31}$$

where $\xi_{m,n} \neq 0$.

From (14.31), we can obtain the relationship between the control requirement on the control convergence rate ρ_m and communication reliability requirement $\Pr\{\alpha_{m,n} = 1\}$. Here, the lower bound of the communication reliability decreases monotonically with ρ_m. This is reasonable since the plant state updates smoothly when ρ_m is small, which leads to good control performance [16]. In summary, small ρ_m means good control performance and needs high communication reliability to maintain the control performance. On the contrary, large ρ_m means loss in control performance and does not need high communication reliability to maintain the control performance. Furthermore, from (14.10) and (14.31), we can infer that the proposed relationship also works in the scenario where there is no transmission time delay, in a straightforward manner.

Let

$$c_m = \sup_{\mathbf{y}\in\mathbb{R}^n, \mathbf{y}\neq 0} \frac{\mathbf{y}^T(\boldsymbol{\Omega}_{e_0}^T\mathbf{Q}_m\boldsymbol{\Omega}_{e_0} - \rho_m\mathbf{Q}_m)\mathbf{y}}{\mathbf{y}^T(\boldsymbol{\Omega}_{e_0}^T\mathbf{Q}_m\boldsymbol{\Omega}_{e_0} - \boldsymbol{\Omega}_{e_1}^T\mathbf{Q}_m\boldsymbol{\Omega}_{e_1})\mathbf{y}} \tag{14.32}$$

represent the supremum of the right-hand term in (14.31). According to [34, 35], we can obtain the optimal c_m.

Based on the above discussion, we can obtain the following theorem about the relationship between control and communication.

Theorem 14.1 *In real-time wireless control systems, communication reliability is actually determined by the control performance constrained by the requirement of control convergence* *rate, rather than the suggested reliability in communications, i.e. being bounded by ε_{th}. Their relationship can be expressed as*

$$\Pr\{\alpha_{m,n} = 1\} \geq c_m^*. \tag{14.33}$$

□

Based on Theorem 14.1, we can obtain the following property.

Property 14.1 According to the relationship between the control performance constraint c_m^* and the reliability requirement $1 - \varepsilon_{th}$ in URLLC QoS, the actual resource consumption for URLLC in real-time wireless control system can be divided into the following three cases.

- *Case A*: When $c_m^* > 1 - \varepsilon_{th}$, the actual resource consumption is higher than the bound in the traditional URLLC resource allocation.
- *Case B*: When $c_m^* = 1 - \varepsilon_{th}$, the actual resource consumption is equal to the bound in the traditional URLLC resource allocation.
- *Case C*: When $c_m^* < 1 - \varepsilon_{th}$, the actual resource consumption is lower than the bound in the traditional URLLC resource allocation.

□

In the rest of this section, we solve the optimal problem in (14.29). First, we obtain the optimal communication resource allocation with constraint on control convergence rate. Then, the optimal control convergence rate can be obtained.

14.4.2 Optimal Resource Allocation

In this section, we first convert the optimal problem into a solvable problem. Then, we develop an algorithm to obtain the solution for optimal resource allocation.

14.4.2.1 Problem Conversion

Though the state update of the control is relevant in different time index n, the wireless resource allocation to guarantee the control requirement is independent. In addition, the available resource, i.e. the transmission power and bandwidth, is independent among different sensors. Thus, we can drop the time indices n and decompose Problem (14.29) into M subproblems. Furthermore, the constraint on communication reliability in (14.29b) should be replaced by the relationship in Theorem 14.1. Thus, (14.29) can be rewritten as

$$\max_{N_m, p_m} \quad \eta_m = \frac{\lambda}{N_m}(1 - \varepsilon_m) \tag{14.34a}$$

s.t.

$$\varepsilon_m \leq 1 - c_m^*, \tag{14.34b}$$

$$N_m p_m \leq p^{\max}, \tag{14.34c}$$

$$B_m \leq B^{\max}, \tag{14.34d}$$

$$T_m \leq T_{th}, \tag{14.34e}$$

where the control convergence rate ρ_m is omitted in the optimal variables since the expression (14.32) and the constraint in (14.34b) indicate that ρ_m can be obtained by $\varepsilon_m = 1 - c_m^*$. The optimal control convergence rate ρ_m will be discussed later in the Section 14.4.3.

Our goal is to maximize the wireless SE by optimal resource allocation, and meanwhile cause less resource consumption. To achieve this goal in solving (14.34), the time delay should be long enough, i.e.

$$T_m = T_{th}, \tag{14.35}$$

which is because large time domain resource can reduce other resource consumption. Then, we have

$$\begin{aligned} \varepsilon_m &= f_Q\left(\frac{N_m T_{th} B_0 C_m - \lambda + \log(N_m T_{th} B_0)/2}{(\log e)\sqrt{N_m T_{th} B_0}}\right) \\ &\leq 1 - c_m^*. \end{aligned} \tag{14.36}$$

Then, (14.34) can be rewritten as r

$$\max_{N_m, p_m} \frac{\lambda}{N_m}(1 - \varepsilon_m) \tag{14.37a}$$

s.t.

$$N_m p_m \leq p^{\max}, \tag{14.37b}$$

$$B_m \leq B^{\max}, \tag{14.37c}$$

$$\varepsilon_m \leq 1 - c_m^*. \tag{14.37d}$$

This is the final expression of the problem formulation. Next, we focus on the solution to this problem.

14.4.2.2 Problem Solution

In the sequel, we propose an iteration algorithm to find the global optimal solution of the problem (14.37). Before that, we need the following properties about η_m.

Property 14.2 Given N_m, η_m decreases monotonically with p_m. □

Proof: It is easy to infer that $f_Q(\cdot)$ in (14.36) decreases monotonically with p_m. Then, ε_m increases monotonically with p_m. Finally, η_m decreases monotonically with p_m. □

Property 14.3 Given p_m, (14.37) is a concave function with respect to N_m. □

Proof: See Ref. [34]. □

Based on **Property 14.2** and **14.3**, we propose an iterative algorithm to find the optimal solution for problem (14.37). Given p_m, the optimal values of N_m that maximizes (14.37) can be found via bisection method. By searching N_m, the optimal and minimum value of p_m can be obtained by (14.36). Thereby, we can obtain the optimal resource allocation to maximize the communication SE in (14.37). The details of the algorithm are provided in Algorithm 14.1.

Algorithm 14.1: The iterative algorithm for optimal resource allocation.

Input: $c_m^*, B_0, T_{th}, \lambda, N_m^l < 1, N_m^h = B^{\max}/B_0$, and the accuracy requirement of the bisection method v

1: Set $p_m = p^{\max}$

2: $N_m = \frac{N_m^h + N_m^l}{2}$

3: **while** $\varepsilon_m > 1 - c_m^*$ or $N_m p_m > p^{\max}$ **do**

4:    **while** $(|N_m^h - N_m^l| > v)$ **do**

5:       $N_m = \frac{N_m^h + N_m^l}{2}$

6:       $N_k^* = N_k$

7:       **if** $\frac{\partial(-\eta_m)}{\partial N_m} < 0$ **then**

8:          $N_m^l = N_m$

9:          $N_m^* = N_m$

10:      **else if** $\frac{\partial(-\eta_m)}{\partial N_m} > 0$ **then**

11:         $N_m^h = N_m$

12:      **else**

13:         $N_m^* = N_m$

14:         break

15:      **end if**

16:    **end while**

17:    Obtain p_m by solving
$$1 - c_m^* = f_Q\left(\frac{N_m T_{th} B_0 C_m - \lambda + \log(N_m T_{th} B_0)/2}{(\log e)\sqrt{N_m T_{th} B_0}}\right) \text{ to minimize transmission power}$$

18:    $\varepsilon_m = f_Q\left(\frac{N_m T_{th} B_0 C_m - \lambda + \log(N_m T_{th} B_0)/2}{(\log e)\sqrt{N_m T_{th} B_0}}\right)$

19: **end while**

20: $p_m^* = p_m$

Output: Optimal resource allocation N_m^* and p_m^*.

14.4.3 Optimal Control Convergence Rate

Once we obtain the maximum communication SE in (14.34), the communication reliability $(1 - \varepsilon_m)$ can be obtained accordingly. Then, we can obtain the optimal control convergence rate ρ_m^* to minimize control cost by solving the following equation:

$$\Pr\{\alpha_{m,n} = 1\} = 1 - \varepsilon_m = c_m^*$$
$$= \frac{\xi_{m,n}^T(\Omega_{e_0}^T Q_m \Omega_{e_0} - \rho_m Q_m)\xi_{m,n}}{\xi_{m,n}^T(\Omega_{e_0}^T Q_m \Omega_{e_0} - \Omega_{e_1}^T Q_m \Omega_{e_1})\xi_{m,n}}. \quad (14.38)$$

Thereby, we can obtain an optimal overall system performance.

14.5 Simulations Results

In this section, we provide simulation results to demonstrate the performance of the method that was explained earlier in Section 14.4, where the system models are the same as shown in Figures 14.1 and 14.2. In the rest of this section, we first illustrate the control performance and then discuss the optimal resource allocation to maximize the communication SE.

14.5.1 Control Performance

From the perspective of real-time wireless control, since M plants are independent in communication-control co-design, we assume that only two plants, i.e. $M = 2$ plants, are considered in the simulations. For simplicity, we assume that both plants have identical dynamics, where we assume that $\Omega_{e_0} = 1.5$ and $\Omega_{e_1} = 0.5$. We further assume that the given positive definite weight matrix is $\mathbf{Q}_m = 1$, the variance of the disturbance matrix $\mathbf{n}_m(t)$ is 1, i.e. $\mathbf{R}_m = 1$, and the sample period is $s_{m,n} = 100$ ms. In addition, we adopt the average control cost to evaluate the control performance [36], which can be expressed as

$$J_{\text{ave}} = \frac{1}{N}\Sigma_{n=1}^{N} x_{m,n}^2, \tag{14.39}$$

where $N = T/s_{m,n}$ is adopted, and T is the total time of the control process.

Figure 14.3 shows the average control cost J_{ave} with control time increasing, where we consider three different decreasing rates ρ_m, i.e. 0.01, 0.25, and 0.9. In the figure, all the curves increase at the initial time. This is because the control process is performed before the state returns to the preset point, where the plant state update leads to increasing J_{ave}. Furthermore, as the control time increases, when ρ_m is small, i.e. $\rho_m = 0.01$, the curve has a little drop and reaches a low approximative horizontal line smoothly. However, when ρ_m is large, i.e. $\rho_m = 0.9$, the curves have a little drop and reach a high approximative horizontal line roughly. This is reasonable since the plant state update is smaller when the plant turns more stable than that at the start phase, which leads to the average control cost J_{ave} having a little drop and reaching an approximative horizontal line. These phenomena indicate that small decreasing rate ρ_m leads to smooth state updating, where we can obtain small average

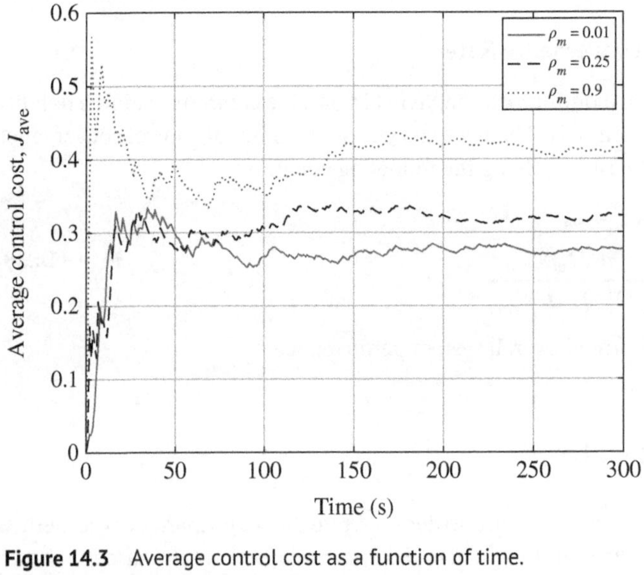

Figure 14.3 Average control cost as a function of time.

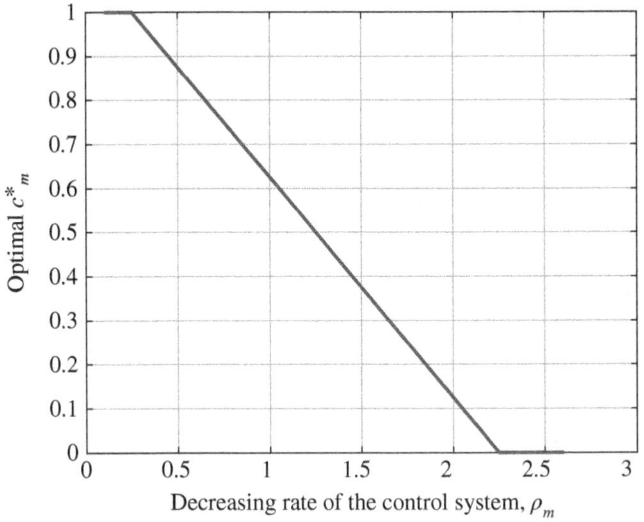

Figure 14.4 Optimal c^*_m with different control convergence rates ρ_m.

control cost J_{ave} updating. When ρ_m is large, the state updating is rough, which leads to large average control cost J_{ave} updating. In addition, when ρ_m is small, i.e. $\rho_m = 0.01$, the smooth state updating leads to low average control cost as the time is large enough, which means that small decreasing rate has better control performance than large decreasing rate.

Figure 14.4 shows the optimal c^*_m in (14.32) when decreasing rate ρ_m is different. In the figure, the curve decreases monotonically with ρ_m from $c^*_m = 1$ to $c^*_m = 0$, which means that small ρ_m results in high constraint on the communication reliability. This is reasonable since high reliability can guarantee smooth decrease of the plant state from control perspective, which leads to small decreasing rate ρ_m and small average control cost in Figure 14.3.

From the above control results, we can infer that Property 14.1 can be redescribed based on the different values of control decreasing rate ρ_m, where $c^*_m(\rho_{th}) = 1 - \varepsilon_{th}$. Then the three cases in Property 14.1 can be described as follows.

- *Case I*: $\rho_m < \rho_{th}$. In this case, $\rho_m < \rho_{th}$ means that $c^*_m > c^*_m(\rho_{th}) = 1 - \varepsilon_{th}$, where the allocated resource in traditional URLLC fails in guaranteeing the required control performance and more wireless resources are needed.
- *Case II*: $\rho_m = \rho_{th}$. In this case, $\rho_m = \rho_{th}$ means that $c^*_m = c^*_m(\rho_{th}) = 1 - \varepsilon_{th}$, where the allocated resource in traditional URLLC is just enough to guarantee the required control performance and more wireless resources are needed.
- *Case III*: $\rho_m > \rho_{th}$. In this case, $\rho_m > \rho_{th}$ means that $c^*_m < c^*_m(\rho_{th}) = 1 - \varepsilon_{th}$, where the allocated resource in traditional URLLC is more than that needed in guaranteeing the required control performance and wireless resource waste is presented.

In the rest of this section, we discuss the communication performance of the method that was explained earlier in Section 14.4. Furthermore, we use $c^*_m = 1 - 10^{-9}$, $c^*_m = 1 - 10^{-5}$, and $c^*_m = 1 - 10^{-3}$ to discuss the communication performance in the above three cases, respectively.

14.5.2 Communication Performance

From the perspective of wireless communication, we assume that the bandwidth of each subcarrier is 1 kHz, the single-sided noise spectral density is −174 dBm/Hz, and the distance between the BS and the plants is 100 m. For the URLLC, the maximum packet transmission error probability is $\varepsilon_{th} = 10^{-5}$, the maximum transmission time delay for the uplink is $T_{th} = 0.5$ ms, and the maximum number of allocation subcarriers is 60. In addition, we consider the traditional method only considering the communication aspect in [7] for comparison, where the authors intended to minimize transmission power.

Figure 14.5 demonstrates the wireless resource allocation when payload information λ_m is different. Figure 14.5a shows the total power allocation. In the figure, the total resource allocation (i.e. multiplying transmission power by frequency bandwidth) increases with the communication reliability increasing from Case III to Case I. This is reasonable since high reliability needs more communication resource to support. Figure 14.5b,c show the allocated transmission power on each subcarrier and the allocated number of subcarriers for the m-th sensor. From Figures 14.5a–c, both the allocated subcarriers and allocated transmission power of the method that was explained earlier in Section 14.4 increase

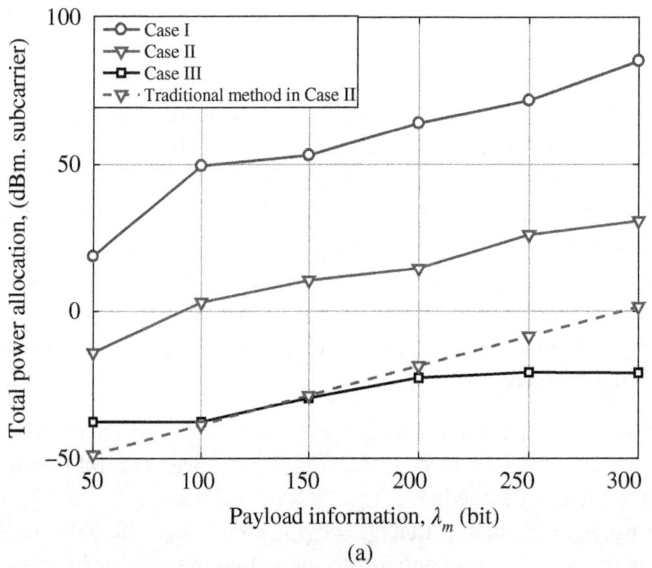

Figure 14.5 Optimal resource allocation with different payload information λ_m. (a) Total resource allocation with different payload information λ_m. (b) Optimal power allocation with different payload information λ_m. (c) Optimal subcarrier allocation with different payload information λ_m.

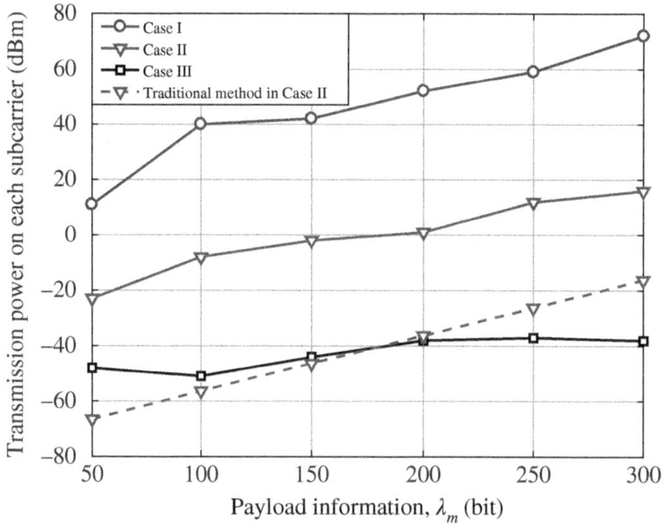

(b)

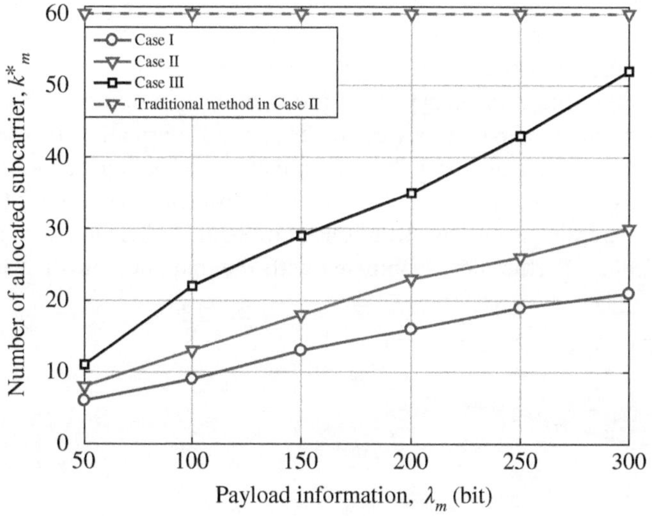

(c)

Figure 14.5 (*Continued*)

monotonically with λ_m, which means that more payload information needs more resources for transmission. From Figure 14.5b, the allocated transmission power on each subcarrier of the method that was explained earlier in Section 14.4 increases with communication reliability increasing from Case III to Case I. However, from Figure 14.5c, the allocated subcarriers of the method that was explained earlier in Section 14.4 decrease with communication reliability increasing from Case III to Case I. This is reasonable since the method that was explained earlier in Section 14.4 in Algorithm 14.1 leads to an increasing power allocation from Case III to Case I, which further leads to the subcarrier allocation reducing from Case III to Case I while maintaining total resource allocation increasing from Case III to Case I as shown in Figure 14.5a. In summary, higher control performance needs more communication resources. In addition, compared with the traditional method in [7], the number of allocated subcarriers of the method that was explained earlier in Section 14.4 is significantly reduced. For instance, the number of the allocated subcarrier is reduced from 60 to 8 when the payload information is 50 bits in Figure 14.5c, while the allocated transmission power on each subcarrier of the method that was explained earlier in Section 14.4 in Figure 14.5b is increased from −65 to −22 dBm, and the total allocated transmission power of the method that was explained earlier in Section 14.4 in Figure 14.5a is increased from −50 to −20 dBm. This is reasonable since the traditional method aims to minimize the transmission power with maximum allocated bandwidth, which leads to lower total power allocation for the traditional method as shown in Figure 14.5a.

Figure 14.6 shows the spectral efficiency when payload information λ_m is different. From the figure, it is easy to infer that all the curves increase with the payload information increasing. In addition, spectral efficiency increases with the communication reliability increasing from Case III to Case I. This is reasonable since the allocated subcarriers decrease with communication reliability increasing from Case III to Case I in Figure 14.5c, which leads to increasing spectral efficiency. Furthermore, compared with the traditional method in

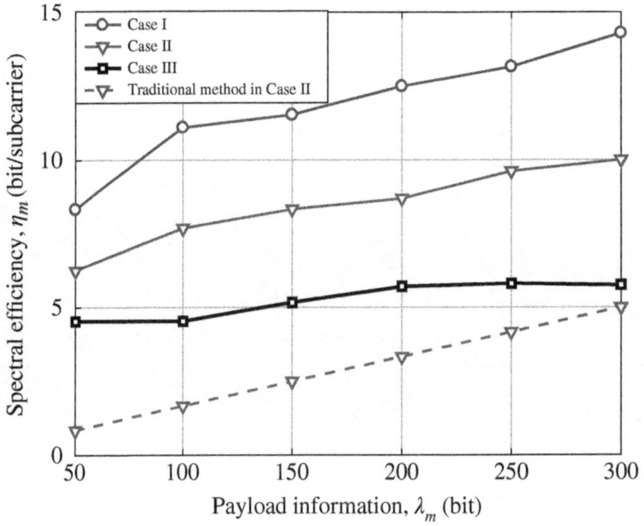

Figure 14.6 Spectral efficiency with different payload information λ_m.

Case II, the method that was explained earlier in Section 14.4 has larger spectral efficiency, i.e. at most 700% spectral efficiency performance increases when payload information is 50 bits. This is reasonable since the method that was explained earlier in Section 14.4 is optimal in maximizing SE.

14.6 Conclusions

This chapter proposed an optimal resource allocation scheme to maximize the communication uplink spectral efficiency in URLLC for real-time wireless control systems. In this scheme, we considered that the URLLC service should satisfy the requirement on control convergence rate, which is formulated as a communication-control co-design problem. To solve the hybrid problem, the control requirement was converted into a constraint on the wireless communication reliability. Then, the hybrid optimization problem can be replaced by a regular wireless resource allocation problem. We proved that the converted problem is concave and an iteration algorithm is developed to obtain the optimal solution. Based on that, the optimal control convergence rate can be obtained. The simulation results showed that the method that was explained earlier in Section 14.4 achieves the maximum spectral efficiency while maintaining the control performance. The proposed co-design approach established a theoretic foundation for the URLLC serviced real-time wireless control system performance analysis and algorithm design.

References

1 Popovski, P., Trillingsgaard, K., Simeone, O., and Durisi, G. (2018). 5G wireless network slicing for eMBB, URLLC, and mMTC: a communication-theoretic view. *IEEE Access* 6: 55765–55779.

2 Li, X., Yu, Y., Sun, G., and Chen, K. (2018). Connected vehicles' security from the perspective of the in-vehicle network. *IEEE Network* 32 (3): 58–63.

3 Sadi, Y. and Ergen, S.C. (2013). Optimal power control, rate adaptation, and scheduling for UWB-based intravehicular wireless sensor networks. *IEEE Transactions on Vehicular Technology* 62 (1): 219–234.

4 Wang, P., Di, B., Zhang, H. et al. (2019). Energy efficient V2X-enabled communications in cellular networks. *IEEE Transactions Vehicular Technology* 68 (1): 554–564.

5 Li, S., Xu, S., Huang, X. et al. (2015). Eco-departure of connected vehicles with V2X communication at signalized intersections. *IEEE Transactions on Vehicular Technology* 64 (12): 5439–5449.

6 She, C., Yang, C., and Quek, T.Q.S. (2016). Uplink transmission design with massive machine type devices in tactile internet. *IEEE Globecom Workshops (GC Wkshps)*, Washington, DC, United States, December 2016, pp. 1–6.

7 She, C., Yang, C., and Quek, T. (2018). Cross-layer optimization for ultra-reliable and low-latency radio access networks. *IEEE Transactions on Wireless Communications* 17 (1): 127–141.

8 Sun, C., She, C., Yang, C. et al. (2019). Optimizing resource allocation in the short blocklength regime for ultra-reliable and low-latency communications. *IEEE Transactions on Wireless Communications* 18 (1): 402–415.

9 Swamy, V.N. Suri, S.; Rigge, P. et al. (2017). Real-time cooperative communication for automation over wireless. *IEEE Transactions on Wireless Communications* 16 (11): 7168–7183.

10 Yu, Y., Chen, H., Li, Y. et al. (2017). On the performance of non-orthogonal multiple access in short-packet communications. *IEEE Communications Letters* 22 (3): 590–593.

11 Mendis, H.V. and Li, F. (2017). Achieving ultra reliable communication in 5G networks: a dependability perspective availability analysis in the space domain. *IEEE Communications Letters* 21 (9): 2057–2060.

12 Dosti, E., Wijewardhana, U., Alves, H., and Latva-aho, M. (2017). Ultra reliable communication via optimum power allocation for type-I ARQ in finite block-length. *2017 IEEE International Conference on Communications (ICC)*, Paris, France, May 2017, pp. 1–6.

13 Trillingsgaard, K. and Popovski, P. (2017). Downlink transmission of short packets: framing and control information revisited. *IEEE Transactions on Communications* 65 (5): 2048–2061.

14 Chen, Y., Cheng, L., and Wang, L. (2017). Prioritized resource reservation for reducing random access delay in 5G URLLC. *2017 IEEE 28th Annual International Symposium on Personal, Indoor, and Mobile Radio Communications (PIMRC)*, Montreal, QC, Canada, October 2017, pp. 1–5.

15 Kotaba, R., Manchón, C., Balercia, T., and Popovski, P. (2018). Uplink transmissions in URLLC systems with shared diversity resources. *IEEE Wireless Communications Letters.* 7 (4): 1–4.

16 Schenato, L., Sinopoli, B., Franceschetti, M. et al. (2007). Foundations of control and estimation over lossy networks. *Proceedings of the IEEE* 95 (1): 163–187.

17 Park, P., Ergen, S., Fischione, C. et al. (2017). Wireless network design for control systems: a survey. *IEEE Communications Surveys & Tutorials* 20 (2): 978–1013.

18 Cetinkaya, A., Ishii, H., and Hayakawa, T. (2017). Networked control under random and malicious packet losses. *IEEE Transactions on Automatic Control* 62 (5): 2434–2449.

19 Feng, Y., Chen, X., and Gu, G. (2017). Output feedback stabilization for discrete-time systems under limited communication. *IEEE Transactions on Automatic Control* 62 (4): 1927–1932.

20 Su, L. and Chesi, G. (2015). On the robust stability of uncertain discrete-time networked control systems over fading channels. *American Control Conference (ACC)*, Chicago, Illinois, United States, July 2015, pp. 6010–6015.

21 Shi, L., Yuan, Y., and Chen, J. (2013). Finite horizon LQR control with limited controller-system communication. *IEEE Transactions on Automatic Control* 62 (7): 1835–1841.

22 Polyanskiy, Y., Poor, H.V., and Verdu, S. (2010). Channel coding rate in the finite blocklength regime. *IEEE Transactions on Information Theory* 56 (5): 2307–2359.

23 Durisi, G., Koch, T., and Popovski, P. (2016). Toward massive, ultrareliable, and low-latency wireless communication with short packets. *Proceedings of the IEEE* 104 (9): 1711–1726.

24 Zhao, G., Imran, M.A., Pang, Z. et al. (2019). Toward real-time control in future wireless networks: communication-control co-design. *IEEE Communications Magazine* 57 (2): 138–144.

25 Park, P., Araújo, J., and Johansson, K.H. (2011). Wireless networked control system co-design. *2011 International Conference on Networking, Sensing and Control (ICNSC)*, Delft, the Netherlands, pp. 486–491.

26 Chang, B., Zhao, G., Imran, M. et al. (2018). Dynamic wireless QoS analysis for real-time control in URLLC. *IEEE Globecom Workshops (GC Wkshps)*, December 2018, pp. 1–5.

27 Zeng, T., Semiari, O., Saad, W., and Bennis, M. (2018). Integrated communications and control co-design for wireless vehicular platoon systems. *2018 IEEE International Conference on Communications (ICC)*, Kansas City, Kansas, United States, May 2018, pp. 1–6.

28 Firooznia, A., Ploeg, J., Wouw, N., and Zwar, H. (2017). Co-design of controller and communication topology for vehicular platooning. *IEEE Transactions on Intelligent Transportation Systems* 18 (10): 2728–2739.

29 Chang, B., Zhao, G., Chen, Z. et al. (2019). D2D transmission scheme in URLLC enabled real-time wireless control systems for tactile internet. *2019 IEEE Global Communications Conference (GLOBECOM)*, December 2019, Puako, Hawaii, United States, pp. 1–6.

30 Chang, B., Zhao, G., Chen, Z. et al. (2019). Packet-drop design in URLLC for real-time wireless control systems. *IEEE ACCESS* 7 (1): 1–10.

31 3GPP (2016). Study on Scenarios and Requirements for Next Generation Access Technologies. *Technical Report 38.913*. Technical Specification Group Radio Access Network, Release 14.

32 Yang, W., Durisi, G., Koch, T., and Polyanskiy, Y. (2014). Quasi-static multipleantenna fading channels at finite blocklength. *IEEE Transactions on Information Theory* 60 (7): 4232–4264.

33 Gatsis, K., Pajic, M., Ribeiro, A., and Pappas, G.J. (2015). Opportunistic control over shared wireless channels. *IEEE Transactions on Automatic Control* 60 (12): 3140–3155.

34 Chang, B., Zhang, L., Li, L. et al. (2019). Optimizing resource allocation in URLLC for real-time wireless control systems. *IEEE Transactions on Vehicular Technology* 68 (9): 8916–8927.

35 Chi, C., Li, W., and Lin, C. (2017). *Convex Optimization for Signal Processing and Communications: From Fundamentals to Applications*. Boca Raton, FL: CRC Press.

36 Sarangapani, J. and Xu, H. (2015). *Optimal Networked Control Systems with MATLAB (Automation and Control Engineering)*. Boca Raton, FL: CRC Press.

Index

Radio Access Network Slicing and Virtualization for 5G Vertical Industries, First Edition.
Edited by Lei Zhang, Arman Farhang, Gang Feng, and Oluwakayode Onireti.